C0-BRL-499

Handbook of Ion Exchange Resins: Their Application to Inorganic Analytical Chemistry

Volume I

Author

Johann Korkisch, Ph.D.
Professor of Analytical Chemistry
Institute of Analytical Chemistry
University of Vienna
Vienna, Austria

CRC Press, Inc.
Boca Raton, Florida

Library of Congress Cataloging-in-Publication Data

Korkisch, Johann.
Handbook of ion exchange resins.

Includes bibliographies and indexes.
1. Ion exchange chromatography. 2. Ion exchange resins. I. Title.
QD79.C453K67 1989 543′.0893 87-27829
ISBN 0-8493-3191-9 (v. 1)
ISBN 0-8493-3192-7 (v. 2)
ISBN 0-8493-3193-5 (v. 3)
ISBN 0-8493-3194-3 (v. 4)
ISBN 0-8493-3195-1 (v. 5)
ISBN 0-8493-3196-X (v. 6)

Direct all inquiries to CRC Press, Inc., 2000 Corporate Blvd., N.W., Boca Raton, Florida, 33431.

International Standard Book Number 0-8493-3191-9 (Volume I)
International Standard Book Number 0-8493-3192-7 (Volume II)
International Standard Book Number 0-8493-3193-5 (Volume III)
International Standard Book Number 0-8493-3194-3 (Volume IV)
International Standard Book Number 0-8493-3195-1 (Volume V)
International Standard Book Number 0-8493-3196-X (Volume VI)

Library of Congress Card Number 87-27829
Printed in the United States

PREFACE

In writing this handbook, the author's primary objective was to provide the reader with essentially all available information on the application of ion exchange resins to inorganic analytical chemistry. This information was extracted from more than 6000 original publications in many languages which included Japanese, Chinese, and Russian so that additional translation work was required in numerous cases. These publications comprised the entire relevant literature published until 1986, and the material as described in more than 1000 tables is presented in the six volumes of the book. In most of these tables comprehensive analytical methods involving the assay of virtually all elements of the Periodic Table are described in a concise manner and in a condensed and digested form. The ion exchange characteristics of the elements as well as other important information required by analysts using ion exchange resins are presented in separate tables which appear in the appropriate sections of the *Handbook*.

In the presentation of the analytical procedures special emphasis and attention has been paid to those methods which allow the multielement analysis of complex matrices such as geological, biological, and industrial materials. This is a treatment of the subject which is generally not adopted in books dealing with ion exchange resins. Therefore, the practical part of the *Handbook* is expected to find broad application in analytical laboratories of all types performing analyses which require the utilization of ion exchange resins (both for routine and research work).

The practical part describes analytical methods involving the rare earth elements and actinides, and then a systematic treatment is presented of the other elements which have been subdivided into transition metals and elements of the main groups of the Periodic Table.

The practical part in Volume I is preceded by a general description of the theoretical, instrumental, and other principles underlying the various applications of ion exchange resins in inorganic analytical chemistry. This general portion of the book was the basis of a very successful course on ion exchange separations in analytical chemistry held by the author at the Catholic University of Rio de Janeiro, Brazil, in 1986. Therefore, it is expected that this and other large portions of the *Handbook* can be used as a text at universities and colleges, and for reference and study by all who are familiar with inorganic analytical chemistry. These include research workers, graduate and undergraduate students of all branches of chemistry and of the other natural sciences, e.g., geology, environmental sciences, and medicine, as well as investigators engaged in analyses involving the quantitation of the elements in industrial products.

Since this book is devoted exclusively to ion exchange resins and their application to inorganic analytical chemistry, ion exchanging materials other than synthetic polymers containing ionogenic groups are not considered, except in a very few cases. Also not treated in this handbook are preparative procedures such as ion exchange methods used for the production of deionized water and the purification of inorganic and/or organic reagents. Included, however, are techniques for the isolation of radioisotopes which are required for radioanalytical purposes.

ACKNOWLEDGMENTS

While working on this handbook valuable help in many respects was provided by the following persons (listed in alphabetical order): J. J. Alberts (Director, Marine Institute, University of Georgia, Sapelo Island), P. Sanchez Batanero (Ph.D., Professor of Chemistry, Department of Analytical Chemistry, Faculty of Science, University of Valladolid, Spain), T. F. Cummings (Ph.D., Professor of Analytical Chemistry, Department of Chemistry, Bradley University, Peoria, Ill.), O. Egert (Typist, Vienna, Austria), R. Frache (Ph.D., Professor of Analytical Chemistry, Institute of General and Inorganic Chemistry, University of Genoa, Italy), J. M. Fresco (Ph.D., Professor of Chemistry, Department of Chemistry, McGill University, Montreal, Quebec, Canada), O. W. Lau (Ph.D., Professor of Chemistry, Department of Chemistry, University Science Centre, The Chinese University of Hong Kong, Shatin NT, Hong Kong), G. E. Janauer (Ph.D., Professor of Analytical Chemistry, Department of Chemistry, State University of New York at Binghamton, New York), R. Keil (Ph.D., Federal Institute for Reactor Research, Würenlingen, Switzerland), G. Korkisch (Typist, Vienna, Austria), G. Kurat (Ph. D., Professor of Geology, Museum of Natural History, Vienna, Austria), Chuen-Ying Liu (Ph.D., Professor of Chemistry, Department of Chemistry, National Taiwan University, Taipei, Taiwan), J. D. Navratil (Ph.D., Manager, Chemical Research, Rockwell International Rocky Flats Plants, Golden, Colo.), Kunio Ohzeki (Ph.D., Professor of Analytical Chemistry, Department of Chemistry, Faculty of Science, Hokkaido University, Sapporo, Japan), K. Orlandini (Ecological Sciences Section, Radiological and Environmental Research Division, Argonne National Laboratory, Argonne, Ill.), Isao Sanemasa (Ph.D., Professor of Analytical Chemistry, Department of Chemistry, Faculty of Science, Kurokami, Kumamoto, Japan), F. W. E. Strelow (Ph.D., National Chemical Research Laboratory, Pretoria, South Africa), Kyoji Tôei (Ph.D., Professor of Analytical Chemistry, Department of Chemistry, Faculty of Science, Okayama University, Tsushima, Okayama, Japan), N. E. Vanderborgh (Ph.D., Los Alamos Scientific Laboratory, Los Alamos, N.M.), Hirohiko Waki (Ph.D., Professor of Analytical Chemistry, Department of Chemistry, Faculty of Science, Kyushu University, Hakozaki, Higashiku, Fukuoka City, Japan), and H. Weiss (Ph.D., Department of the Navy, Naval Ocean Systems Center, San Diego, Calif.). The contributions of these 21 scientists and typists made it possible for me to finish the six volumes of the book in a 4-year period of most intensive and sometimes agonizing work. This assistance and additional help, with respect to typing and copying, received from some employees of the Institute of Analytical Chemistry, is most gratefully acknowledged.

INTRODUCTION

Ion exchange phenomena have been observed since about the middle of the 19th century, but the practical significance of ion exchange was not recognized immediately, and it was not until the first part of the 20th century that natural or synthetic ion exchangers were widely known. By then a number of relatively pure minerals such as zeolites and clays were found or synthesized that exhibited exchange characteristics. Although these exchangers were commonly used in water treatment and are still employed today, they are unstable in acid and alkaline solutions and can only be used satisfactorily under nearly neutral conditions. Consequently these aluminosilicates are of only little significance for the analytical separation of metal ions.

A more significant development took place in 1934, however, when it was discovered that some synthetic high-molecular weight organic polymers containing a large number of ionogenic groups, as an integral part, could be employed as ion exchange resins which can be considered as gel-like dispersed systems. The dispersed medium is usually water and the dispersed portion is the three-dimensional polymer of the ion exchange resin, which is of organic origin. Cross-linked bonds (e.g., divinylbenzene bridges) between the polymer chains form a three-dimensional matrix (network; skeleton), which hinders the motion of the polymer chains and formation of a solution in contact with the solvent (water). Only swelling of the matrix in contact with the solvent occurs. This swelling is governed especially by the character, number, and length of the cross-linked bonds.

An important feature differentiating the ion exchange resins from other types of gels is the presence of functional groups (also called ionogenic or exchangeable groups). The groups (e.g., $-SO_3H$, $-\overset{+}{N}R_3$, etc.) are attached to the matrix. The ion exchange process between the ions in the solution takes place on these functional groups.

The exchange of ions between the ion exchange resin and the solution is governed by the following two principles:

1. The process is reversible (only rare exceptions are known).
2. The exchange reactions take place on the basis of equivalency in accordance with the principle of electroneutrality. The number of millimoles of an ion sorbed by an exchanger should correspond to the number of millimoles of an equally charged ion that has been released from the ion exchanger.

Ion exchange resins, on account of their property to exchange ions in solutions, can be applied in various fields of chemistry. Concerning analytical chemistry, which is the topic of the present book, ion exchangers can be successfully applied not only in the quantitative separation of complex ionic mixtures (which is the main application of ion exchange resins) but also be used for other analytical purposes, e.g., for the microchemical detection of elements.

Ion exchange is an important and modern tool in every contemporary analytical laboratory.

THE AUTHOR

Johann Korkisch, Ph.D., is a Professor in the Institute of Analytical Chemistry at the University of Vienna, Austria.

Dr. Korkisch received his higher education at the University of Vienna. His undergraduate work was completed in Chemistry with a minor in Physics. His thesis research concerned the use of ion exchange resins in the microanalysis of gallium and uranium. After having obtained a Ph.D. in 1957 he was an Assistant Professor at the University of Vienna (Insitute of Analytical Chemistry) from 1957 to 1961.

In 1961—1962 Dr. Korkisch was a Visiting Professor at the Scripps Institution of Oceanography, University of California, La Jolla, where he performed research on the chemical analysis of marine sediments and seawater.

After his appointment in 1966 as an Associate Professor at the Institute of Analytical Chemistry (University of Vienna) Dr. Korkisch joined (from 1967 to 1968) the Argonne National Laboratory of the U.S. Atomic Energy Commission, Argonne, Ill. There, as a Visiting Scientist, he performed research work connected with the application of ion exchange resins to separations of uranium, plutonium, and trans-plutonium elements in mixed aqueous-organic solvent systems.

This work, as well as previous and later research on similar lines, was sponsored by the U.S. Atomic Energy Commission mainly acting through the International Atomic Energy Agency (IAEA) in Vienna. Research support was also provided by other sources such as the Fund for Peaceful Atomic Development USA, the American Petroleum Research Fund, as well as by the Austrian Fund for the Promotion of Scientific Research.

In 1970 Dr. Korkisch worked as a Visiting Scientist at the State University of New York at Binghamton, and in the following year was appointed by the IAEA to the Mexican Atomic Energy Commission to act as a Technical Assistance Expert in Mexico City. There, he introduced the use of ion exchange resins for separations involving the analysis of nuclear raw materials.

In 1973 he joined the University of Kumamoto, Japan as a Visiting Scientist where he performed research work connected with the application of ion exchange resins to the analysis of natural waters polluted with heavy metals.

In the same year Dr. Korkisch was promoted to the rank of Full Professor of Analytical Chemistry at the University of Vienna.

Since 1957 Dr. Korkisch has published over 200 scientific papers mainly relating to analytical applications of ion exchange resins. His numerous collaborators come from Austria, the U.S., Japan, Egypt, India, Brazil, Turkey, Bulgaria, Israel, Mexico, Romania, Italy, and Greece.

Dr. Korkisch is also the author and co-author, respectively, of the books entitled *Modern Methods for the Separation of Rarer Metal Ions,* Pergamon Press and *Handbook of the Analytical Chemistry of Uranium,* Springer-Verlag. Other publications are, in addition to numerous reports and two U.S. patents, three chapters in books published by CRC Press, Wiley-Interscience, and Pergamon Press.

At present his research in inorganic analytical chemistry is mainly concerned with practical applications of ion exchange resins to the analysis of natural materials (geological and biological samples) and also industrial products (uranium and thorium). Concerning natural materials Dr. Korkisch's special attention is directed towards the analysis of manganese nodules using ion exchange resins and ICP-emission spectroscopy.

At the University of Vienna, Dr. Korkisch is teaching courses in quantitative analysis and in the application of ion exchange resins to analytical chemistry. The latter course which is based on this 6-volume *Handbook* is also the topic of 3- to 4-week courses which he likes to present at foreign universities and institutions.

TABLE OF CONTENTS

Volume I

Volume II

Volume III

Volume IV

Volume V

Volume VI

General Discussion

ION EXCHANGE RESINS AND FUNDAMENTAL CONCEPTS OF ION EXCHANGE PROCESSES

ION EXCHANGE RESINS

Preparation of Ion Exchange Resins

The matrix of the vast majority of ion exchange resins employed in analytical chemistry is a cross-linked copolymer consisting of styrene and divinylbenzene (DVB) which is prepared by a perlpolymerization. The particle size and particle size distribution of the spherical beads thus obtained depend on the extent of mechanical agitation and reaction conditions during the polymerization.

The sulfonation of this copolymer can be performed using various sulfonating agents, e.g., fuming sulfuric acid, chlorosulfonic acid, S trioxide in nitrobenzene, or concentrated sulfuric acid, and, on the average, about one sulfonic acid group per benzene ring will be present in commercial products of this type. This corresponds to an exchange capacity of approximately 5 meq/g dry resin in the H form. On sulfonation, structures of the type shown in Figure 1 and Table 1 are obtained.

Anion exchangers of the quarternary ammonium type are usually prepared by first chloromethylating the cross-linked polymer and then treating the product with a tertiary amine, such as trimethylamine (see reactions in Figure 1 and formulas in Table 1). If the chloromethylated polymer is treated with a secondary amine instead of with a tertiary amine, the exchanger formed is a weakly basic tertiary amine. Treatment with ammonia or primary amines results in the formation of polyfunctional anion exchange resins. Chloromethyl methyl ether in the presence of a catalyst such as Al chloride is usually employed for the chloromethylation. The exchange capacity of the resin obtained depends on the extent of this chloromethylation which is, however, not so easily controlled as the sulfonation reaction, and a secondary reaction may take place between a chloromethyl group attached to one benzene ring and another aromatic nucleus. As a result, methylene bridges are formed between neighboring aromatic nuclei, and if these are in different polymer chains, additional cross-linkages will be introduced into the polymer network.

A summary of the synthetic processes that are utilized for the preparation of the analytically most important ion exchange resins is presented in Figure 1.

The exchange capacity of strongly basic ion exchange resins, based on cross-linked polystyrene, is usually about 3.4 to 4 meq/g dry resin.

In Tables 1 and 2 the types of ion exchange resins most frequently used for analytical purposes are presented.

Marking of Ion Exchange Resins

At present, the majority of ion exchanger manufacturers indicate their own products by the company name (e.g., Dowex, Wofatit, etc.). This name is supplemented by symbols and numbers describing the type of the exchanger. Additional symbols or abbreviations are used in order to describe the products more exactly. These symbols or abbreviations reflect the purity of the starting raw materials, the methods of their treatment, the grain size of the resulting products, or other special properties of the materials.

General Symbols Most Often Used

X =	weight percentage of DVB in the copolymer (e.g., Dowex® 50-X8)
W =	the product is of white or yellowish color (e.g., Dowex® 50W-X8)
MP =	macroporous (macroreticular) matrix
AG, AR, pa =	ion exchange resin of "analytical grade" purity
CG =	ion exchangers for chromatographic purposes

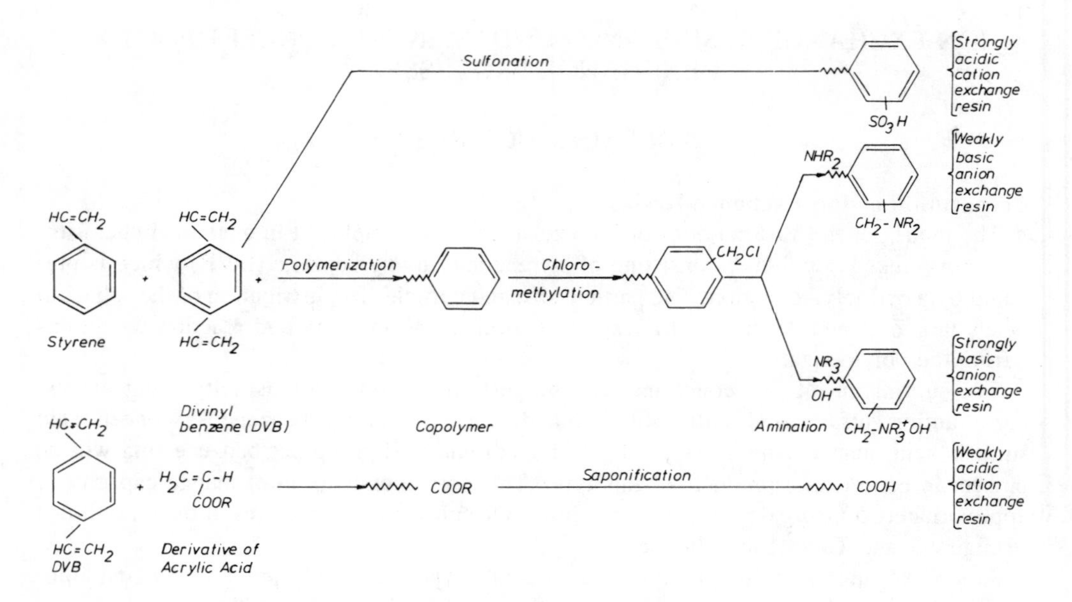

FIGURE 1. Preparation of ion exchange resins.

Basic Markings of Some Resins

Markings of "Dowex" Exchangers

The name Dowex® is followed by a symbol composed of one or two numerals; the following letter indicates the special properties of the exchanger. For a more precise description of the cross-linking of the exchanger skeleton, the symbol X followed by a numeral, which indicates the DVB content in the exchanger matrix and which completes the name of the exchanger, is often used.

Numeral:

1 — strongly basic anion exchanger of type 1
2 — strongly basic anion exchanger of type 2
3 — weakly basic anion exchanger
21 — strongly basic anion exchanger of type 1, lower cross-linking
44 — weakly basic anion exchanger
50 — strongly acidic cation exchanger

The symbol following the numeral indicates:

W — colorless or slightly yellowish type of resin

Markings of "Amberlite" Exchangers

The indication begins with the name Amberlite® followed by a group of letters and numbers. Macroporous exchangers suitable for work in nonaqueous media are indicated by the name Amberlyst® followed by the letter A (for anion exchangers) and a group of numbers.

Group of letters:

IR — strongly acidic cation exchangers; exceptions are two types of weakly basic anion exchangers also marked by the letters IR, e.g., IR-4B
IRC — weakly acidic cation exchanger

Table 1
STRUCTURAL FORMULAS OF COMMERCIAL ION EXCHANGE RESINS

Trade name (®)	Structural formula	Classification (resin type)
Bio-Rad AG 50W, Dowex 50, Amberlite IRA 120, Zeo-Karb 225, Cationite KU-2, Diaion SK1, Wofatit KPS, Lewatit S-100, Ostion KS	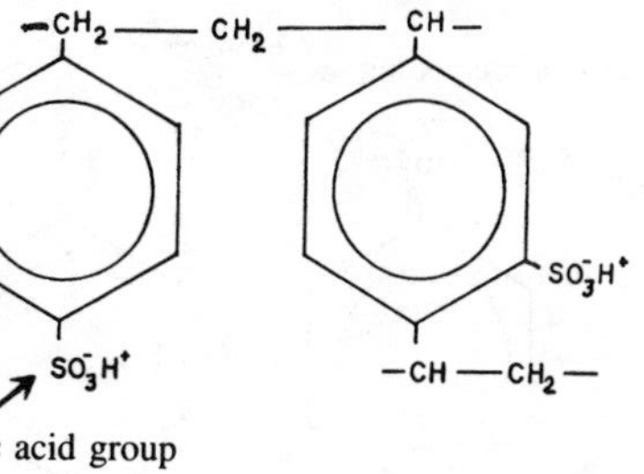 $-CH_2-CH_2-CH-$; $SO_3^-H^+$; $SO_3^-H^+$; $-CH-CH_2-$; Sulfonic acid group	Monofunctional strongly acidic cation exchange resin (H^+ form)
Cationite KU-1, Duolite C3, Diaion-BK, Zeo-Karb 315, Bio Rex 40	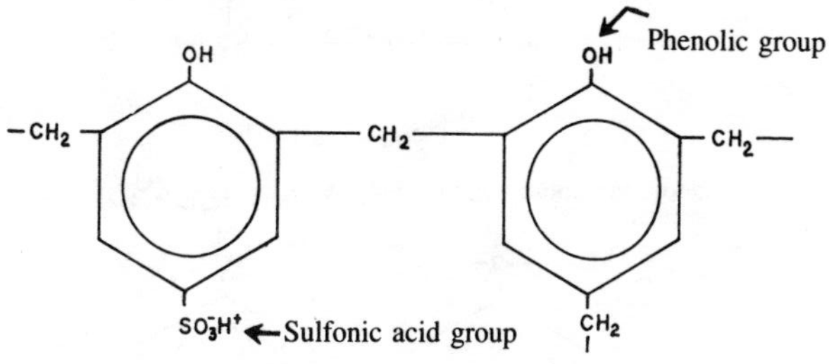 OH; OH; Phenolic group; $-CH_2$; CH_2; CH_2-; $SO_3^-H^+$ ← Sulfonic acid group; CH_2	Bifunctional strongly acidic cation exchange resin (H^+ form)
Bio-Rad AG1, Dowex 1, Amberlite IRA-400, Deacidite FF, Anionite AV-17, Diaion SA1, Wofatit SBW, Lewatit M 500, Ostion AT, Resin 717	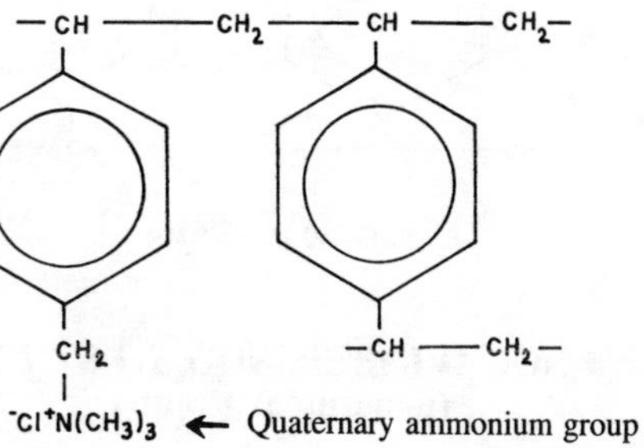 $-CH-CH_2-CH-CH_2-$; CH_2; $-CH-CH_2-$; $^-Cl^+N(CH_3)_3$ ← Quaternary ammonium group	Monofunctional strongly basic anion exchange resin (Cl^- form) (type 1)

Table 1 (continued)
STRUCTURAL FORMULAS OF COMMERCIAL ION EXCHANGE RESINS

Trade name (®)	Structural formula	Classification (resin type)
Bio-Rad AG 2, Dowex 2, Amberlite IRA-410, Diaion SA-2, Wofatit SBK, Lewatit M 600, Ostion AD	$-CH-CH_2-CH-CH_2-$; CH_2 ; $-CH-CH_2-$; $^{-}Cl\,^{+}N(CH_3)_2$ ← Quaternary ammonium group ; CH_2-CH_2-OH	Monofunctional strongly basic anion exchange resin (Cl^- form) (type 2)
Chelex 100, Dowex A-1, Wofatit CM-50, ANKB 50	$-CH-CH_2-CH-CH_2-$; CH_2 ; $-CH-CH_2-$; $N(CH_2-COOH)(CH_2-COOH)$ ← Iminodiacetic acid group	Chelating resin (H^+ form)

Table 2
SELECTION OF VARIOUS TYPES OF ION EXCHANGE RESINS FROM DIFFERENT MANUFACTURERS[a]

Manufacturer	Trade name (®)	Strongly acidic resins		Weakly acidic resins	Strongly basic resins	Medium and weakly basic resins
		Polystyrene	Phenolic			
Dow Chemical Co., Midland, Mich.	Dowex	50 50W	30 —	— —	1 2	3 —
Rohm and Haas Co., Philadelphia, Pa.	Amberlite	IR-120 IR-200	IR-100 IR-105	IRC-50 IRC-84	IRA-400 IRA 401 IRA 410 IRA-411 IRA-910	IR-4B IR-45 IRA-93
Diamond Alkali Co., Western Div., Calif.	Duolite	C-20 C-26	C-3 C-10	CS-101 CS-100	A-42, A-101, A-40, A-162	A-2, A-4, A-5, A-14, A-30, A-43, A-47, A-368
Permutit Co., New York	Permutit (Ionac)	Q RS	C-200 P	H-70, CN CNO, C-265	S-1, ESB A-550	E, W, A-320
United Water Softeners, London	Zerolit (Zeo-karb)	225	315	226 216 236	FF	B, E, G, F
Farbenfabrik Bayer, Leverkusen, West Germany	Lewatit	S-100 SP-120	KS KSB KSN, PN	CNO CNP	MN M 600 MP 600	MP-62
Veb Chemie, Kombinat, Bitterfeld, East Germany	Wofatit	KPS-200	F, P, KS	CP-300, H, CA-20 CN	SBW SKB	N, MD L-165 AD-41, L-150, L-160
U.S.S.R. product	U.S.S.R.	KU-2	KU-1	KM, KB4 KB-2	AV-17 AV-18	AN-2F EDE-10P

[a] Other manufacturers producing ion exchange resins which are also frequently used in analytical chemistry include: Bio-Rad Laboratories, Richmond, Calif. (Bio-Rad® resins equivalent to Dowex® resins) and Mitsubishi Chemical Industries Ltd., Tokyo (Diaion® resins). An essentially complete list of ion exchange resins hitherto prepared by various manufacturers is presented in the appendix to Volumn VI of this *Handbook*.

IRA — anion exchanger
AR — analytical grade; exchangers for analytical purposes

Markings of "Diaion" Exchangers

The marking is composed of the name Diaion® followed by a group of letters and numbers:

SK-1A — a fundamental type of strongly acidic cation exchangers; gel-like type
SK-1B — the same exchanger as the preceding type; better physical properties
SK-100 — strongly acidic cation exchanger of the gel-like type; the last two digits (in the range 02 to 16) give the DVB content (in %) in the exchanger matrix
SA-10A — a fundamental type of strongly basic anion exchanger (type 1); gel-like matrix
SA-10B — the same as the previous type; better physical properties
SA-20A — a fundamental type of strongly basic anion exchanger (type 2); gel-like type
SA-20B — the same as the previous type, better physical properties
PK 200 — strongly acidic cation exchanger with a macroporous matrix; the last two digits (in the range 04 to 28) give the DVB content (in %) in the exchanger matrix
PA 300 — strongly basic anion exchanger with a macroporous matrix (type 1); the last two digits (in the range 02 to 20) give the DVB content (in %) in the exchanger matix
PA400 — strongly basic anion exchanger with a macroporous skeleton (type 2); the last two digits (in the range 04 to 20) give the DVB content (in %) in the exchange matrix.

Exchangers Produced in the Soviet Union

Ion exchange resins produced in the Soviet Union are not marked with a company name. The marking used during the research and development work is used.

The following symbols are used for the exchangers (anionites and cationites) produced at the Institute for the Research and Development of Plastics:

AN — weakly basic anion exchange resin (anionite)
AV — strongly basic anion exchanger (anionite)
KB — weakly acidic cation exchanger (cationite)
KU — strongly acidic cation exchanger containing the $-SO_3H$ group (cationite)

The Resin Matrix

From Table 1 it is seen that the ion exchange resin is composed of a matrix and functional (ionogenic, ion exchangeable) groups. These groups are attached to the matrix (resin network: skeleton).

Basically, the properties of this elastic, three-dimensional network are determined by the ratio of the amounts of the individual monomers used during its synthesis. Resins with skeletons containing a low proportion of DVB swell strongly in aqueous solutions. Thus, large ions can easily diffuse through the exchanger and the exchange kinetics is fast. Mechanical strength of the matrix decreases with decreasing percentage of DVB.

Ion exchange resins with skeletons containing a higher proportion of cross-linking agent (>15% DVB) swell in aqueous solutions to a substantially lower degree. Their mechanical strength is higher. Considering the high cross-linking, the number of functional groups which can be attached to the matrix is lower and decreases with increasing percentage of DVB. At the same time, the rate of diffusion of the exchangeable ions through the resin matrix decreases. A compromise is to use 5 to 8% of DVB for the synthesis of the matrix.

Numerous properties of the matrix are also affected by the purity of the cross-linking agent. Technical DVB is a mixture of the *m*- and *p*-isomers. In addition to these isomers, a considerable amount of diethylbenzene and other impurities are present. The cross-linking values stated for commercial products are not the real values, but values of the so-called nominal cross-linking. Considering that the incorporation of DVB into the growing particles of the copolymer is faster than that of styrene, the resulting matrix has a so-called "island" structure. Furthermore, the copolymeric particles formed at the beginning of the reaction are cross-linked to a higher degree as compared with particles formed later.

The matrix heterogeneity mentioned above varies mostly in the range of $\pm 0.5\%$ of the cross-linking stated for the commercial products.

When synthesizing the styrene-DVB matrix in the absence of other substances (especially of common solvents), a gel-like matrix is formed during the copolymerization of both components. This type of matrix can be composed of mutually penetrating networks formed by the individual chains. The sieve size is considerably heterogenous. Generally speaking, the structure of the matrix formed is composed of low-porosity islands in a more porous medium. The pore size (given by the distance between the individual polymeric chains) is very low and porosity appears only after swelling the resin matrix in an appropriate solvent. This type of molecular porosity can be considered as latent (microporous or microreticular).

Ion exchange resins having a matrix with the properties mentioned are called gel-like (or microreticular or microporous) resins.

A matrix of sufficiently large pores which are statistically distributed throughout the whole matrix volume can be obtained by additional cross-linking and by the modification of the porous structure (chloromethylation and additional cross-linking by means of these groups) in the preformed styrene-DVB mixture. This type of matrix is called isoporous.

Another matrix type of modern organic ion exchangers is the so-called macroporous (macroreticular) structure. Skeletons of this type are formed by the addition of an appropriate solvent (which dissolves the monomer easily) to the polymerizing system during the polymerization. The liquid phase is successively separated from the copolymer formed during the polymerization. The gels formed exhibit a characteristic porous (spongy) structure composed of aggregates of spheres of normal gel-like porosity which are penetrated by a continuous nongel porous structure. However, these macropores are not a part of the gel structure of the polymer. The size of the pores formed can be controlled during the matrix preparation. Pore sizes of several hundred of 10^{-10} m in diameter can be obtained. Macroreticular skeletons exhibit a large inner surface. Surfaces up to 100 m^2/g and more can be obtained. To prevent collapse of the structures, a larger proportion of cross-linking agent needs to be used. Due to their structure, the exchangers of this matrix type are much more resistant to osmotic shock. They also exhibit a smaller swelling difference in polar and nonpolar solvents, a smaller loss of volume during drying, and a higher oxidation resistance. Furthermore, reactions connected with introducing the ionogenic groups into the skeleton occur more easily and with a higher yield as compared with the gel-like types.

Higher porosity facilitates diffusion of ions within resin particles to provide more accessible exchange sites and faster attainment of equilibrium. Therefore, one of the major advantages of using microporous resins of low cross-linkages such as Bio-Rad® AG1-X2 and -X4 and of macroporous resins, e.g., Bio-Rad® AGMP-1 (~20 to 25% cross-linkage), over the usual AG1-X8 and -X10 resins is their much better exchange kinetics.

Significant differences in distribution as a function of resin cross-linkage occur and are most striking in hydroiodic acid because of the large radii of anionic iodide complexes. However, the distribution coefficients of some metals in hydrochloric acid (HCl) media were also found to vary considerably with the cross-linkages of the resins. For example, on the Bio-Rad® resins AG1-2, -X4, -X8, -X10, and AG MP-1 the distribution coefficients of some metals in HCl media were found to vary considerably with the cross-linkages of the resins. Thus, on the resins AG1-X2, -X4, -X8, -X10, and AGMP-1 the distribution coefficients of U(VI) in 9 *M* HCl are 312, 470, 540, 812, and 1220, respectively. A similar effect was observed in the case of adsorption of Fe(III), Zn, Cd and other elements. Generally speaking, the distribution coefficients of metals increase with increase in cross-linkage, while exchange kinetics show a marked improvement with decrease in cross-linkage.

One of the major disadvantages of the resins with higher cross-linkages is the fact that serious tailing often occurs when an ion has to be eluted at a relatively high distribution coefficient (K_d between 15 and 40). By using a resin of lower cross-linkage, the exchange kinetics improves considerably and, in addition, the distribution coefficient becomes significantly smaller (see above). This tends to lead to better elution. For this reason an ion which cannot be eluted satisfactorily from an 8% cross-linked resin can often be eluted quantitatively by using a resin of lower cross-linkage.

For the determination of the degree of cross-linking of styrene-DVB resins, mostly infrared spectroscopy employing various experimental techniques is used. A method utilizing pyrolitic gas chromatography has also been developed. Structural analysis of the exchanger material can also be carried out by this method.

Among various new instrumental techniques, nuclear magnetic resonance is a useful method for the investigation of ion exchange resins suspended in water. This technique is suitable for the characterization of anion or, preferably, of cation exchangers with $-SO_3H$ groups based on styrene-DVB copolymers; properties such as homogeneity, cross-linking, rate of exchange, water of hydration, and concentration of counter ions in the resin phase can be evaluated.

Electron microscopy is a suitable method for studying and determining the porosity of ion exchange resins.

IR spectroscopy is suitable for the investigation of the molecular structure of the exchangers, including solvation at the exchange sites. Species retained by the exchanger sites can also be examined in some cases.

Functional Groups of Ion Exchange Resins

According to the character of ionogenic groups, the ion exchangers can be classified in the following way:

- Cation exchangers — contain acid ionogenic (functional) groups such as $-SO_3H$ (see Table 1) and $-COOH$
- Anion exchangers — contain basic ionogenic groups such as $-\overset{+}{N}R_3\ OH^-$ (see Table 1), $-NH_2$, $=NH$, $\equiv N$, etc.)
- Amphoteric exchangers — contain acid and basic ionogenic groups
- Chelating (selective) ion exchange resins — contain ionogenic groups reacting only with a relatively small group of ions (see Table 1)

Ion exchangers containing only one type of inogenic group are called monofunctional (or homoionic) exchangers. Resins containing more than one type of exchangeable (ionogenic) group (e.g., $-SO_3H$ and $-OH$ as in cationite KU-1; see Table 1) are called polyfunctional exchangers. Monofunctional ion exchange resins are always preferred in analytical chemistry, especially in separation procedures.

Investigating the ionogenic group of an ion exchange resin more thoroughly, we observe that the group can be divided into parts. The first part of the group is fixed firmly to the resin matrix by a covalent bond forming, together with the matrix, a certain type of macroion. This part of the ionogenic group is called the "fixed ion". Ions of opposite charge are bound to the macroion by electrostatic forces. They are called "counter ions". These counter ions are also a part of the ion exchanger and can be exchanged for an equivalent amount of various ions (of the same type and charge) from the solution.

Together with the counter ions, oppositely charged ions (as compared with counter ions) diffuse from the solution into the interior of the resin during the ion exchange process. These ions (called "co-ions") are not considered as part of the ion exchange resin.

This nomenclature can be illustrated by the following example: denoting the resin matrix as R_s and the attached sulfonic acid, ionogenic groups as $-SO_3H$, the ion exchange resin can be written symbolically as R_s-SO_3H. Contacting the resin with, for example, an aqueous Na chloride solution, a macromolecular fixed ion $R_sSO_3^-$ is formed on the one hand, and the initial counter ion H^+ is exchanged for the counter ion Na^+ on the other. At the same time, Cl^- ions form co-ions which diffuse from the solution into the resin.

Various exchangeable ions (counter ions) may be present in the ionogenic group. In order to denote which type of ion is attached to the resin, the name of the appropriate ion, followed by the term "form" (or sometimes also "cycle") is used. The symbol R_s-SO_3H, mentioned above, denotes that the ion exchanger is in the H^+ form. Exchanging the H^+ ion for Na^+, the ion exchanger is in the Na form.

Anion exchange resins of the type shown in Table 1 can be in the hydroxide, chloride, nitrate, etc. forms.

According to the dissociation ability of various ionogenic groups of ion exchangers in the form of free acids or bases, the ion exchange resins can be classified as strongly, medium, or weakly acidic (or basic).

Acidic ionogenic groups in the H^+ form dissociate with the release of the H^+ ion:

Strongly acidic cation exchange resin: $R_sSO_3H \rightleftharpoons R_sSO_3^- + H^+$ [apparent dissociation constant pK ($= -\log K$): 1]

Weakly acidic cation exchange resin: $R_sCOOH \rightleftharpoons R_sCOO^- + H^+$ (pK 4 to 5)

Similarly, the basic ionogenic groups in the OH^- form (free-base form) release OH^- ions:

Strongly basic anion exchange resin: $R_sNR_3OH \rightleftharpoons R_sNR_3^+ + OH^-$ (pK 1)

Weakly basic anion exchange resin: $R_sNH_2 + H_2O \rightleftharpoons R_sNH_3^+ + OH^-$ (pK 6 to 9)

The type of the exchanging groups of an ion exchange resin can be estimated from the neutralization (titration) curve as can the exchange capacity and some other parameters such as the dissociation constant.

The experimental potentiometric determination of the titration curve is carried out with the H^+-form cation exchangers and the OH^--form anion exchangers, respectively.

Titration curves of monofunctional strongly acidic (or basic) exchangers resemble curves obtained by potentiometric neutralization titration of strong mineral acids or Na hydroxide.

The titration curve on an ion exchanger can be obtained in two ways: the method of direct titration can be used for exchangers with easily dissociated functional groups and with fast exchange kinetics. Exchangers containing weakly dissociated functional groups and of slow exchange equilibration are treated by the method of individually weighed samples. However, this method can be used for any type of exchanger.

Exchange Capacities

Every ion exchange resin contains a certain amount of exchangeable (ionogenic) groups. Their number is quantitatively expressed by the term "exchange capacity". The following definitions are used.

Theoretical weight exchange capacity (Q_0) — This capacity is expressed by the total number of millimoles (milliequivalents = meq) of ionogenic groups related to 1 g of dry resin in the H^+ or Cl^- forms.

Theoretical volume exchange capacity (Q_v) — Its significance is similar to Q_0; it is related to 1 mℓ of swollen resin (wet resin).

Analytical weight exchange capacity (Q_A) — It is expressed as the total amount of ions (in millimoles) exchangeable by 1 g of dry resin under given specific conditions (which should always be stated).

Breakthrough capacity (Q_B) — (see "Breakthrough Capacity of the Ion Exchange Column")

The exchange capacity of the resin is closely related to the composition of its matrix which affects a number of processes taking place during the ion exchange process. Typical capacities in millimoles of H^+ per gram of dry resin (millimoles of H^+ per milliliter of wet resin) for ion exchangers are

Strongly acidic resins:	5 (1 to 2.5) (polystyrene); 4 to 4.5 (1) (phenolic)
Weakly acidic resins:	10 (3.5 to 4.0)
Strongly basic resins:	3 to 4 (1.5)
Weakly basic resins:	5 to 10 (2.5)

Usually, the exchange capacity ion exchange resins is determined via an ion exchange reaction corresponding to the functional group of the resin in question.[1-10] The most widely used methods for capacity determination are titrimetric methods that make use of the back-titration of the excess of NaOH or HCl in the effluent in the case of cation and anion exchange resins, respectively (see below). Other methods that can be used to determine resin capacities are based on simple weighings and on elemental analysis of the exchanger.

To determine the total weight exchange capacity of a cation exchange resin the air-dried exchanger (1 to 2 g; H^+ form) is allowed to swell in deionized water. It is also possible to use wet exchanger (1 to 3 g) which has been previously sucked on a Büchner funnel. At the same time, 0.3 to 1 g of the exchanger is weighed to determine the dry substance. The swollen resin is transferred to an ion exchange column (6 to 10 × 1 cm). A known amount (200 to 300 mℓ) of 0.1 *M* NaOH is passed through the resin bed at a flow rate of 2 to 3 mℓ/min. After the total amount of NaOH solution has passed through the column, the residual liquid present in the column is ejected with compressed air. In an aliquot portion of the NaOH passed through, the loss of OH^- ions (due to the reaction $R_sSO_3^-H^+ + NaOH \rightarrow R_sSO_3^-Na^+ + H_2O$) is determined by titration with 0.1 *M* HCl, and the total weight exchange capacity, expressed in millimoles of H^+ per gram of dry exchanger, is calculated.

The capacity of anion exchange resins can be determined by using the following procedure: the resin (1 to 2 g) in the Cl^- form (either air dried or fully swollen with any excess liquid removed by suction) is transferred to an ion exchange column (if a dry exchanger has been used, it should be fully swollen in water before placing it in the column). Then, 1 *M* NaOH (200 mℓ) heated to 40°C is passed through the column to convert the resin to the OH^- form (flow rate 3 mℓ/min). Subsequently, the resin bed is washed with deionized water and 0.1 *M* HCl (200 mℓ) is passed through at the same rate of 3 mℓ/min. The exchanger is then washed with water (30 mℓ), and in an aliquot of the combined effluent and washings the HCl concentration is determined by titration with 0.1 *M* NaOH using methyl orange as indicator. The total weight exchange capacity of the exchanger is evaluated from the known starting and final concentrations of HCl. The dry residue of the exchanger is determined in a separate sample.

As mentioned above, capacities can also be determined by weighing the same amount of dried exchange resin in various ionic forms. For this purpose the resin sample (2 to 3 g) is loaded to saturation with an ionic species A, then washed and air dried. A weighed portion ($\simeq$1 g) of the product is fully loaded with ionic species B, washed, and dried in a vacuum oven at 110°C for 3 days; a further (weighed) portion ($\simeq$1 g) of A-saturated material is also dried and the two weights are measured. The ion exchange capacity of the resin is calculated from these weights, i.e., the weight of the two air-dried portions and the equivalent weights of species A and B. For a cation exchange resin, A and B can be, e.g., H^+ and K^+; for an anion exchange resin, Cl^- and NO_3^- are suitable.

The determination of ion exchange capacities using elemental analysis is based on the quantitative determination of the elements forming the fundamental component of the ion exchanger functional group. However, this method is limited mainly to monofunctional ion exchangers. S and N are the most frequently determined elements. The value of the exchange capacity of ion exchange resins obtained by elemental analysis may differ from those obtained by other methods. The reason lies in the fact that not all groups containing the element determined behave as ion exchanging groups. Thus, slightly higher capacity values are obtained by this method.

Unless classical procedures of combusion of organic materials are used, the mineralization of ion exchange resins is carried out by wet techniques.

Conversion of ion exchange resins to a standard state and to other ionic forms can be effected by means of the following procedures:

- Strongly acidic cation exchangers into the H^+ form: 5 to 8 volumes (vol) (related to the exchanger treated) of 7% HCl. The resin is then washed with deionized water until an acid reaction (methyl red) of the effluent is no longer observed.
- Weakly acid cation exchangers into the H^+ form: 6 to 8 vol of 2% HCl. The exchanger is washed with deionized water until an acid-free (methyl red) effluent is obtained.
- Strongly basic anion exchangers into the Cl^- form: 10 vol of 7% HCl. The resin is washed with deionized water until the washings give a negative reaction to Cl^- ions.
- Weakly basic anion exchangers into the free-base form: 10 vol of 4% NaOH. The resin is washed with deionized water until the washings give a negative reaction to phenolphthalein.

For acidic or basic solutions the flow rate is 0.1 to 0.15 vol of the acid or base per 1 vol of the resin per minute.

If it is necessary to convert the resin into another ionic form, the procedure usually starts from the standard state. However, the Na^+ form (cation exchangers) and OH^- form (anion exchangers) are often more advantageous. In order to completely convert the ion exchanger into the ionic form required, a 2.5 to 3-fold surplus of the appropriate agent (as compared with its theoretical need) is used; pH adjustment of the solution used is necessary in some cases.

Hydrolysis of some ionic forms of weakly acidic or basic ion exchangers occurs during their washing with water. A minimum amount of water or washing with methanol or ethanol is used in such a case.

Contacting the exchange resin with water (or some other appropriate solvent) causes solvation of ionogenic groups and a ''solution'' of functional groups in the resin is formed. The resin exhibits a tendency to go into solution. However, this is prevented by the elastic three-dimensional structure of the resin matrix. The resulting effect of these two opposing forces is the uptake of a certain amount of the solvent by the resin, manifesting itself as swelling.

The swelling can be considered as a result of the osmotic pressure difference between the interior of the exchange resin and the external, more dilute solution. The internal ion concentration decreases by swelling.

If an ''air-dried'' ion exchange resin is immersed in water, the quantity of regained water is primarily influenced by the matrix structure, i.e., the degree of cross-linking. The lower the cross-linking the more water is taken up by the resin, so that low cross-linked resins swell much more than resins containing higher percentages of the cross-linking component (e.g., DVB).

If an electrolyte is present in water, the swelling depends also on the electrolyte concentration. If the electrolyte concentration is increased, the solvent uptake will decrease, since the osmotic pressure difference between the interal and external solutions is lower.

The swelling is also affected by the nature of the electrolyte, i.e., swelling is also a function of the ionic form. Thus, a resin in the H^+ form swells much more than when present in the metal form, e.g., Na form.

The swelling of weakly acidic or basic resins in H^+ – (OH^-) forms is low due to the low hydration of their weakly dissociated ionogenic groups.

With nonaqueous solutions, the swelling of strongly acidic resins generally decreases with decreasing hydrophility of the solvent used. Weakly acidic cation exchangers and anion exchangers swell strongly in ethanol; anion exchangers also swell in less polar solvents.

If the resin swells in a mixture of two solvents, one of them penetrates preferentially into the resin phase. This leads to a different composition of solvents inside and outside the resin. Distribution of the solvents between the resin and the solution depends on their solvation ability as well as on the type and ionic form of the resin. (For example, in a mixture of water and a less polar solvent such as acetone, the solution inside the resin can be enriched in water depending on the type and ionic form of the resin and on the amount of cross-linking agent.)

It can be concluded that the swelling of ion exchange resins depends on the following factors:

1. The surrounding medium (nature of solvent, electrolyte concentration)
2. The nature of resin matrix (type of matrix, cross-linking)
3. The nature and concentration of ionogenic groups
4. The type of counter ion

To determine resin swelling the fully swollen resin is transferred to a measuring cylinder which is then pounded on a hard rubber washer until the volume of the resin does not change. The volume of the stationary layer is read. The resin is then dried to constant weight. The volume swelling is expressed as milliliters per gram of the dry exchanger.

Resin swelling can also be determined by the centrifugation and electrolyte (dye dilution) methods as well as by means of microscopy. Using a good microscope with appropriate practical techniques, the microscopic method gives results with a fairly high degree of accuracy.

A series of methods exists for the determination of water in ion exchange resins, differing in practical technique as well as (to a certain degree) in the results obtained.[11-23]

The most commonly used method consists of drying the resin at a certain temperature to constant weight. The resins are dried either at 105°C (cation exchangers) or at 60 to 80°C (anion exchangers). It is not recommended to dry strongly basic anion exchangers in the OH^- form owing to their low thermal stability. It is more convenient to dry the resins under reduced temperature and pressure in a desiccator over P pentoxide.

The Karl Fischer method, as used for determinations of the water content in other systems, gave good results for the water content of styrene-DVB resins. Results obtained by this method exhibited lower deviation ($\pm$0.75%) than those obtained by drying the resin at 100°C ($\pm$2%). The Karl Fischer reagent is a solution of I, S dioxide, and pyridine in methanol, with the strength of any preparation being dependent on the I concentration. As the Fischer reagent reacts with water in a particular sample, the red-brown color of I is consumed until all water has been reacted. The reagent thus serves as its own indicator during visual titration.

The water content of an ion exchanger can also be determined by azeotropic distillation (using toluene, xylene, etc.), nuclear magnetic resonance, calorimetry, and by the use of isopiestic methods.

Depending on the type of functional groups present the various ion exchange resins show the following fundamental properties.

Strongly Acidic Cation Exchangers

1. Two types of exchangers with different skeletons are commercially available: styrene-DVB resins containing the $-SO_3H$ groups and phenolic polycondensates containing $-SO_3H$ and phenolic OH groups (see Table 1).
2. Styrene-DVB resins are monofunctional.
3. The $-SO_3H$ ionogenic group is bound either directly to the benzene core or through a methylene bridge.
4. Exchangers in the H^+ form split neutral salts to corresponding acids and the resin is converted to the corresponding salt form.
5. The degree of ionization of the ionogenic groups corresponds to that of strong mineral acids. The sulfonic acid group (bound directly to the benzene core) dissociates more easily as compared with the $-CH_2SO_3H$ grouping.
6. Chemical reactions of the resin are similar to that of sulfuric acid.
7. The resin exchange capacity does not depend appreciably on the pH value of the solution. The resin can be used in acidic, neutral, or alkaline solutions.
8. Selectivity of the sulfonic acid group increases with atomic number, valency, and degree of ionization of the exchanged element. Selectivity decreases with increasing ionic radius of the hydrated ion. Generally, selectivity decreases in the series (Dowex® 50-X8):

   $$Ag^+ > Cs^+ > Rb^+ > K^+ > NH_4^+ > Na^+ > H^+ > Li^+;$$

   $$Ba^{2+} > Sr^{2+} > Ca^{2+} > Mg^{2+} > Be^{2+};$$

   $$Ba^{2+} > Pb^{2+} > Sr^{2+} > Ca^{2+} > Ni^{2+} = Cu^{2+} > Cd^{2+} > Co^{2+} > Zn^{2+}$$

   $$= Mg^{2+} > Mn^{2+} > Be^{2+} > UO_2^{2+} > Hg^{2+}; \quad La^{3+} > Ce^{3+} > Cr^{3+};$$

   $$Th^{4+} > Fe^{3+} > Al^{3+} > Ba^{2+} > Tl^+ = Pb^{2+} > Sr^{2+} > Ca^{2+} > Co^{2+} >$$

   $$Ni^{2+} = Cu^{2+} > Zn^{2+} = Mg^{2+} > UO_2^{2+} = Mn^{2+} > Ag^+ > Cs^+ > Be^{2+}$$

   $$= Rb^+ > Cd^{2+} > NH_4^+ = K^+ > Na^+ > H^+ > Li^+$$

   The selectivity series mentioned above are valid for dilute solutions (~ 0.1 *M*). The adsorption of monovalent ions in concentrated solutions is higher as compared with polyvalent ions.
9. A large excess of a strong acid of relatively high concentration is necessary to convert the resin completely into the H^+ form.
10. Styrene-DVB resins exhibit sufficient resistance towards oxidizing agents and solutions of acids and hydroxides.

Medium Acidic Cation Exchangers

1. Ion exchange resins containing the $-PO(OH)_2$ or $-OPO(OH)_2$ groups attached to various type skeletons are commercially available.
2. The chemical properties of these resins are similar to those of phosphoric acids.
3. The dissociation ability of the H^+ form is between that of strongly and weakly acidic cation exchangers.
4. The exchange capacity depends on the pH of the external solution. The effective pH value is in the region of pH > 5 for most of the mono- and divalent cations.

5. Selectivity of the ionogenic groups depends strongly on the type of the adsorbed ion and on the pH of the solution:

   $-PO(OH)_2$: $Pb^{2+} > Cu^{2+} > Zn^{2+} > Cd^{2+} > Mn^{2+} > Co^{2+} > Ni^{2+}$;

   $Th^{4+} > U^{4+} > UO_2^{2+} > Fe^{3+} >$ rare earths $> H^+ > Cu^{2+} > Zn^{2+} >$

   $Cd^{2+} > Mn^{2+} > Co^{2+} > Ni^{2+} > Ca^{2+} > Mg^{2+} > Sr^{2+} > Ba^{2+} > Na^+$

6. A small excess of a strong mineral acid as compared with the stoichiometry is necessary to convert the resin completely into the H^+ form from ionic forms positioned behind the H^+ ion in the selectivity series. A considerably larger excess of an acid is needed in the case of other ions.

Weakly Acidic Cation Exchangers

1. The commercially available resins are either monofunctional (–COOH groups; carboxylic acid grouping) prepared by copolymerization of acrylic or metacrylic acid with DVB or phenolic polycondensates of resorcylic acid (–COOH and –OH groups).
2. Resins in the H^+ form do not split neutral salts.
3. The degree of ionization corresponds to that of acetic acid.
4. The exchange capacity depends strongly on the pH value of the external solution. The effective pH is in the range of 6 to 14.
5. High selectivity to H^+ ions is a characteristic property. The selectivity series of alkali metal ions was found to be reversed as compared with that for strongly acidic cation exchangers. The resin also show a relatively high affinity for the alkaline earth elements.

   Selectivity series at pH = 7: Mg < Ca < Ni < Co < Cu
   General series: $H^+ >$ Ca > Mg > Na

6. A small stoichiometric excess of a strong mineral acid is sufficient to convert the resin into the H^+ form.
7. On changing the resin H^+ form for other ionic forms, great volume changes occur.

Strongly Basic Anion Exchangers

1. Commercially available resins are based mostly on styrene-DVB copolymers (see Table 1). The functional groups $-\overset{+}{N}(CH_3)_3Cl^-$ (type 1), $-\overset{+}{N}(CH_3)_2C_2H_4OH{\cdot}Cl^-$ (type 2), or pyridinium groups can be present.
2. Resins in the hydroxide (hydroxyl; free-base) form separate neutral salts with the formation of the corresponding bases. The resin is converted to the corresponding salt form.
3. The degree of ionization of ionogenic groups corresponds to that of the alkali metal hydroxides. Resins in the OH^- form adsorb even very weakly dissociated acids (e.g., boric acid; see in chapter on Boron, Volume VI). The basicity of resin type 1 is higher than that of type 2.
4. The exchange capacity does not depend appreciably on the pH of the solution.
5. Selectivity of the resin towards anions increases with increasing valency of the ion and decreases with increasing hydrated ionic radius. Selectivity decreases in the following series:

Dowex® 1-X8 (type 1): $I^- > HSO_4^- > NO_3^- > Br^- > CN^- > HSO_3^- > NO_2^- >$

$Cl^- > HCO_3^- > H_2PO_4^- > HCOO^- > CH_3COO^- > OH^- > F^-$; $SO_4^{2-} > CrO_4^{2-} >$

citrate > tartrate > $NO_3^- > AsO_4^{3-} > PO_4^{3-} > MoO_4^{2-} >$ acetate;

$I^- > Br^- > Cl^- > F^-$

Dowex® 2-X8 (type 2): $ClO_4^- > I^- > HSO_4^- > SCN^- > CCl_3COO^- >$

$CF_3COO^- > = NO_3^- = Br^- > NO_2^- = CN^- > Cl^- \geqslant BrO_3^- > OH^- >$

$HCO_3^- > H_2PO_4^- > IO_3^- > CH_3COO^- > F^-$

6. A large excess of NaOH solution (concentration at least 2% solution of NaOH) is needed to convert the resin completely into the OH^- form, especially when the resin is in the Cl^- form. Type 2 is converted more readily than type 1.
7. Strongly basic resins operate at all pH values up to 13 and can be used in highly concentrated acids (e.g., 12 *M* HCl). They are sufficiently resistant towards solutions of acids and hydroxides as well as towards oxidizing agents. Type 1 is more resistant as compared to type 2, especially in the OH^- form. It is not recommended to work with the resins in the OH^- form above 60 (type 1) or 50°C (type 2). The OH^--form resin absorbs CO_2 readily from air.

Medium Basic Anion Exchangers

1. The chemical properties of these resins lie between those of the strongly and weakly basic ion exchangers. The resins contain strongly as well as weakly basic ionogenic groups (predominantly tertiary amine groups).
2. If the ionogenic groups have been regenerated with NaOH solution, the resin will separate neutral salts and adsorb weak acids proportionally to the content of the strongly basic group. If the resins have been regenerated with solutions of Na carbonate or ammonium hydroxide they behave as weakly basic resins.

Weakly Basic Anion Exchangers

1. Styrene-DVB, polamine-epichlorhydrine resins, or phenolformaldehyde carrier skeletons are used in commercial products.
2. The resins contain the following ionogenic groups: $-NH_2$ or substituted amines $-NHR$, $-NR_1R_2$ (primary or secondary amines).
3. The resins do not split neutral salts with the formation of the corresponding acids.
4. The degree of ionization of the ionogenic groups is analogous to that of ammonia.
5. Resins, in the free-base form, adsorb strong mineral acids with the formation of the appropriate salts. Acidic forms of the resins liberate acids when in contact with water due to hydrolysis.
6. The exchange capacity of the resins depends strongly on the pH of the solution and on the valency of the adsorbed ion.
7. A series of this type of resin is able to bind some substances by van der Waals' forces. The amine groups of these resins are able to form strong complexes with various ions (Cu^{2+}, Ag^+, etc.).
8. Regeneration of the resin to a free base can be carried out with only a small stoichiometric excess of Na carbonate, NaOH, or ammonia (or aromatic amines).
9. The affinity of various anions towards the resin decreases in the following series: OH^-

$> SO_4^{2-} > CrO_4^{2-} >$ citrate $>$ tartrate $> NO_3^- > AsO_4^{3-} > PO_4^{3-} > MoO_4^{2-} > CH_3COO^- > I^- = Br^- > Cl^- > F^-$.

10. The resins are sufficiently stable in dilute solutions of acids and hydroxides.

Chelating Ion Exchange Resins

Under the term "chelating ion exchangers" should be understood resins carrying functional groups which are able to form complex (or inner-complex, i.e., chelate) compounds with selected ions. These exchangers which are also called complexing or selective resins combine two analytical processes, i.e., ion exchange and complexing reactions.

A chelating resin is distinguished from the ordinary type of ion exchange resin by three main properties:

1. The affinity of a particular metal ion to a functional group of the chelating resin depends mainly on the nature of this chelating group. The size of the ion, its charge, and physical properties are of secondary importance.
2. The strength of the binding forces in chelating resins are much higher than in ordinary exchangers.
3. The exchange process in a chelating resin is often slower than in, e.g., strongly acidic or basic resins and is controlled either by a particle diffusion mechanism or a second-order chemical reaction.

Details concerning structures, selectivities, and practical uses of chelating resins of a large variety of types, of which some are commercially available (e.g., Chelex® 100 and Srafion® NMRR), are presented throughout this *Handbook* (for typical examples see chapters on Actinides [Volume II], Platinum Metals [Volume III], Copper [Volume III], and Mercury [Volume IV]).

A new type of selective ion exchange resin has been prepared by Blasius et al.[24-34] Their functional groups are formed by crown compounds or cryptands. Polycondensates of formaldehyde with resorcinol or styrene-DVB copolymers were used as the resin matrix. An example is shown below:

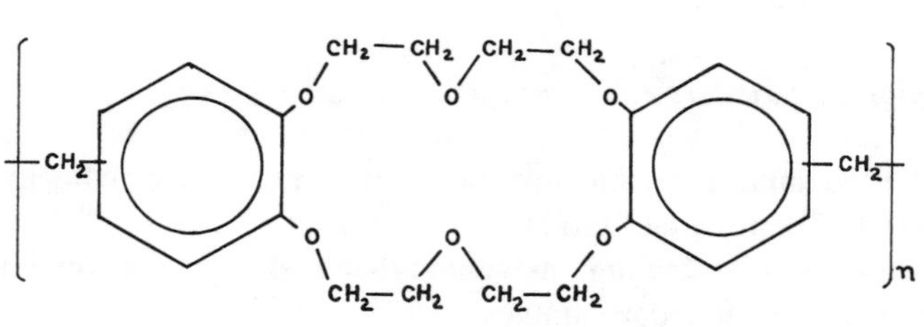

dibenzo-18-crown-6 (2 benzene rings; 18-membered ring, 6 of them are O atoms).

These resins differ from the common type of ion exchanger in that anions are adsorbed together with the exchanged cations, due to the preservation of electroneutrality in the exchange system. Thus, alkali and alkaline earth cations, together with the accompanying anions, can be exchanged by these resins. The adsorbed salts can then be separated chromatographically, often with only water as the eluent.

The complexes which the polyethers form with salts are held together by interactions between the cations and the C–O dipoles of the polyethers; that means that the polyethers are neutral ligands. Variable ring sizes as well as the type, the number, and positions of the donor atoms in the ether ring permit the selective adaption to a certain cation.

Redox Polymers

These polymers (electron exchangers; redoxites) are high molecular weight resins which are able to take part in oxidation-reduction processes. However, their analogy to ion exchange resins is only formal, especially in practical work. They contain oxidation-reduction groups built into the macromolecular resin matrix. Types formed by polycondensation as well as by the polymerization of suitable monomers (pyrogallol, hydroquinone, oxianthraquinone, alizarine, methylene blue, vinylhydroquinone, etc.) exist.[35-40] Working with these resins requires a special technique (their contact with the O in the air must be prevented). The resin is ready for use when treated with a suitable reducing agent (e.g., 10% solution of dithionite in 1 *M* ammonia, 10% solution of Ti[III] chloride in 0.5 *M* sulfuric acid, or a solution of Na sulfite in sulfuric acid). All the water used should be freed from dissolved O by boiling.

The Ion Exchange Resin Bead

According to the outer shape, ion exchange resins can be divided into grained and ungrained types. Spherical or granular particles (granules of irregular shape) belong to the first type. The second group of materials is formed by ion exchanging materials in the form of membranes, paper, plates, etc.

In addition to the gel-like ion exchange resins of various porosity and macroreticular-type ion exchangers (macroporous resins), pellicular resins have been developed for ion chromatographic separations (see next chapter, Special Analytical Techniques Using Ion Exchange Resins).

The bead (particle; grain) size is in most cases given as the standard mesh size through which it will pass. This mesh size (in the case of U.S. standard screen) is related to the diameter of the bead by the approximate equation:

$$\frac{16}{\text{mesh}} \cong \text{diameter of the resin bead in mm}$$

For analytical separations, resins with particle sizes of 50-100 (0.14 to 0.29 mm), 100-200 (0.074 to 0.14 mm), and 200-400 mesh (0.04 to 0.074 mm) are most frequently used and are commercially available.

Sometimes the mesh size of resins consisting of particles of very small diameter is given in "minus" mesh. Thus, Dowex® 50W-X12, -400 mesh means that this commercial product contains resin beads with sizes in the range of ~10 to 75 μm with ~40- to 45-μm beads on the average. Resins of this grain size are used in ion exchange chromatography, because elution curves with sharper peaks are obtained with the finer and more homogeneous particles. On the other hand, coarser fractions allow faster rates of elution.

The determination of particle (bead) size and particle distribution is performed by mesh analysis. For this purpose the air-dried ion exchange resin (cation and anion exchangers in the Na^+ and Cl^- forms, respectively) is sieved through a series of standard sieves and portions retained by the individual sieves are weighed. Using the values of the weight of the individual fractions, a distribution curve can be plotted.

To determine the sphericity of an exchanger, the resin is poured (at a constant rate) onto a metallic sheet. The sheet is at 10° to the horizontal so that the spheres move freely down the sheet without mutual hindrance. The weight of the grains which roll off the metallic sheet is determined.

The amount of defective spheres (beads) is determined by a microscope or better by a microphotograph. Thus, the number of spheres which are undamaged and those showing surface cracks is determined. Simultaneously, the number of particles formed by disintegration of the spheres is obtained.

The specific weight of ion exchange resins is determined by use of the pycnometric method. This determination is carried out on the dry as well as on the swollen resin.

To determine the bulk density and specific bulk volume, the fully swollen resin in a suitable ionic form is sucked onto a Büchner funnel and its weight is determined. Then the resin is transferred to a measuring cylinder filled with water and pounded on a hard rubber filler until its volume does not change. The bulk density is expressed by the ratio of the weight of the resin to its stationary volume (g/mℓ). The specific bulk volume is expressed in units of mℓ/g.

Ion exchange resins are produced mostly in the form of spheres of various grain size. The resin beads are graded in a range of sizes by the manufacturers. Resins of small particle size (varying over a narrow range) are necessary in some cases, especially for chromato graphic separation of various mixtures. If resins of an appropriate particle size are not available, they can be ground in some cases in order to adjust their particle size. This mechanical treatment should be avoided as far as possible in the case of ion exchange resins in the form of spheres. The spherical shape of the undamaged particles insures, among other things, a lower hydraulic resistance of the ion exchanger column (also in the case of very low grain size) as compared with the grains of an irregular shape. Furthermore, a large amount of very fine particles is always formed during grinding and these particles should be considered as waste.

If the resin is sufficiently fragile it is very easily crushed. Small amounts of the resin can be readily crushed to particles of a required size in an agate grinding mortar. A manual mechanical mill proved successful in the grinding of a small amount of resin. A large amount of resin can be ground in a ball mill, which can be used for the grinding of dry as well as of wet resin. It is necessary to note that in metallic mills, only resins in the Na form (cation exchangers) or chloride form (anion exchangers) can be ground. The grinding of cation exchangers in the H form leads to an increased wear of the metallic milling area as well as to a contamination of the resin with metal.

A large amount of a very fine and useless waste is often formed during grinding. It is therefore necessary (on repeated grinding of the resin) to preliminarily separate particles of a required size by appropriate sieves.

Resin particles can be graded according to their size by dry (or wet) sieving.

Dry sieving — Air-dried resin is sieved through a series of standard sieves either manually or by a mechanical vibrator. The resin fraction of the particle size required is suspended and decanted with water several times before further use. The adhering powder particles are removed from the resin in this way. Dry sieving is effective to about 400 mesh (38 μm).

Wet sieving — The resin, swollen in water, is sieved through standard sieves by a stream of distilled water. Another method consists of immersing the sieve in a sufficiently large vessel filled with water. The sieve is moved slowly up and down so that smaller resin particles go through the sieve mesh. It is not recommended to use this method for sieving cation exchangers in the H^+ form. The resin may react with the sieve material and may be contaminated with the metallic material. Similarly, anion exchangers containing polyamine groups strongly attack bronze sieves. Wet sieving has been used successfully to give a fraction of 500-625 mesh (20 to 25 μm) which yielded excellent chromatograms. Consistent production of material smaller than this is difficult to achieve with sieving, however.

Another method which can be used is elutriation (hydrodynamic sedimentation). This consists of allowing the resin particles to drift downward in a tall column which is tapered at the bottom and filled with a liquid (usually water). A constant upflow of liquid in the column allows the large particles to settle to the bottom, but carries the smaller particles out via overflow at the top. With careful attention to flow rates, quite good size separations can be achieved, especially if the nonsized particles are spherical or at least uniform in shape. However, even nonuniformly shaped particles can deliver good results.

Purification of the Ion Exchange Resins[41,42]

Commercial-grade resins are seldom pure enough to be directly used in analytical work.

They contain metallic impurities (e.g., Ni, Fe, Cu, Zn, and Pb) and, even more important, organic impurities, including incompletely polymerized materials that leak slowly in use. This is especially true of quaternary amine anion exchange resins. These must be washed repeatedly with acid, alkali, and methanol before use.

The best solution for the analyst is to purchase prepurified resins from supply houses such as the Bio-Rad Corp., Richmond, Calif. or Mallinkrodt Chemical Works, St. Louis. These resins need only brief washing to be ready for use.

If for some reason an exchanger has to be purified in the analytical laboratory, a suspension of the swollen resin in water is transferred to a glass column of large diameter and which is equipped at the bottom with a porous glass disk and a outlet tap. After sedimentation of the resin in the column, the water layer is adjusted so that it is 5 to 10 cm above the resin layer. Then the exchanger is slowly washed with distilled water until the outlet liquid is clear and colorless. It is advantageous, at this stage of purification, to wash the exchanger column from below (upflow mode).

The resin is then converted to the H^+ form (Cl^- form) with about 3 *M* HCl (downflow mode). An approximately threefold amount of the acid is used as compared with the exchanger volume. The flow rate should be chosen so that the liquid amount mentioned flows through the column within 30 to 40 min. The resin is then washed with water (15 to 20 exchanger volumes) and converted into the $Na^+(OH^-)$ form with 1.5 *M* NaOH.

The next operation consists of a thorough washing with water (up to the removal of an alkaline reaction of the washing water; in the case of weakly acidic or basic exchangers up to a pH of the washing water of 8). The whole purification procedure is repeated two to three times. In the case of low cross-linked resins, a large volume change may occur due to strong swelling of the exchanger in hydroxide solution. Rupture of the glass column can occur in some cases. It is therefore convenient to transfer the resin into a beaker after finishing the washing procedure in the first stage of cycling and then to treat it with the hydroxide solution. Having reached maximum swelling, the exchanger is transferred again into the column and washed with the hydroxide solution. Usually 2 to 3 vol of the solution of the concentration mentioned are sufficient to convert the exchanger.

The final operation of one purification cycle consists of the conversion of the $Na^+(OH^-)$ form of the resin into the $H^+(Cl^-)$ form using 3 *M* HCl followed by water (see above).

This purification procedure is quite sufficient for most purposes.

A high degree of purification can be obtained by washing the resin with hot water for about 30 min after the cycling described above (H^+-form cation exchanger: 90 to 95°C, Cl^--form anion exchanger: 60 to 80°C). The exchanger is then washed with a small amount of cool, deionized water and with methanol, ethanol, or acetone until the effluent is clear and colorless. After repeated washing with water, the ion exchanger is converted into the desired ionic form. This procedure can be used provided that the ion exchange resin is stable in the solvents used and at the temperature given. The majority of the styrene-DVB-type ion exchangers is stable under the conditions mentioned.

Various authors recommend in some cases to carry out final purification of the ion exchange resins with solutions containing chelating agents such as EDTA or citric acid.

Chemical, Thermal, and Radiation Stability of Ion Exchange Resins[43-54]

The chemical, thermal, and radiation stability of ion exchange resins depends on the type of resin matrix, its degree of cross-linking, the type of the ionogenic group, and their counter ion.

In addition to these fundamental factors, the stability of ion exchangers is affected by their synthesis; important are the purity of the original raw material and the method of preparation.

The change in resin stability becomes especially evident in the change of the volume exchange capacity, the loss of ionogenic groups, swelling changes, in the formation, eventually, of new groups (especially weakly acidic), or in the destruction of the resin matrix to various degrees.

Chemical Stability

According to the nature of degradation processes which occur during the chemical corrosion of ion exchangers, the following reaction types can occur:

1. Degradation of the macromolecular chain between the spatial bridges
2. Chemical changes in groups belonging indirectly to the macromolecular chain (the ionogenic groups remain unaffected)
3. Substitution or degradation of the functional groups
4. Formation of new (especially weakly acidic) functional groups

From the point of view of chemical stability of ion exchangers, the most important characteristic is their resistance towards oxidizing agents.

If the resin is degraded in its polymer matrix because of insufficient oxidation resistance, its cross-linking is decreased. The resin swelling increases and, in a limited case, the resin is dissolved. The phenol-formaldehyde skeletons undergo this degradation more easily than the styrene-DVB matrices. The degree of degradation of the styrene-DVB resins depends on the amount of DVB present in the polymer as well as on the purity of monomers used in their synthesis. The effect of the DVB isomer used (or mutual ratio of the individual isomers in the technical DVB used) has also been observed.

Resins of lower DVB content exhibit a higher oxidation degradation. The degradation process is accelerated by the presence of elements which are able to act as catalysts (especially Fe and Cu).

Phenol-formaldehyde, strongly acidic cation exchanger KU-1, on heating to 70°C in 3% H_2O_2 and in the presence of Fe(III), loses 97% of its weight during 24 hr. The following weight losses were found under the same conditions for styrene-DVB-SO_3H: 1% DVB - 62% weight loss, 2% DVB - 46%, 4% DVB - 26%, 8% DVB - 11.6%.

Strongly basic anion exchange resins are especially sensitive to H peroxide. Resins containing the pyridinium groups are more resistant than other strongly basic exchangers.

If the resins are contacted with acidic solutions of chromates, permanganates, and vanadates, these anions are reduced with the oxidation of some components of the resin.

In a medium of nonoxidizing acids, phenol-formaldehyde-type resins are less stable (especially on heating) than the styrene-DBV types which exhibit good stability as cation and anion exchangers. In a medium of oxidizing acids (especially on heating) the phenol-formaldehyde resins undergo strong degradation, especially at high acid concentrations. The styrene-DVB anion exchangers Dowex® 1 and Permutit® SK showed satisfactory stability in flowing 7 *M* nitric acid up to 60°C, provided that nitrite ions were not present. Batch experiments in 7 *M* nitric acid showed severe damage in 2 weeks, due apparently to autocatalytic build-up of nitrite.

Polycondensation resins slowly dissolve in solutions of alkalies, especially at higher hydroxide ion concentrations. Styrene-DVB resins are more resistant. However, strongly basic anion exchangers, when in contact with alkaline solutions of high concentration (>1.0 *M*), split off their strongly basic groups. It is not advisable to allow such a resin in the OH^- form to stand in a strong alkaline solution for a prolonged period of time.

Macroreticular, strongly basic anion exchange resins exhibit a higher stability to alkaline solutions than do gel-like resins.

Ion exchangers are not appreciably damaged by the usual organic solvents (at room temperature). However, a loss of a small amount of organic material occurs which may (in some cases) interfere with further analytical procedures.

Thermal Stability

Air-Heating Stability

Sulfonated cation exchangers of the phenol-formaldehyde type (in the H^+ form) show first symptoms of desulfonation at 100°C. No change in the exchange capacity was found when this type of resin had been heated to 70°C in sealed ampules for 60 days. The opposite behavior was found for a styrene-DVB strongly acidic cation exchanger.

The exchange capacity of the latter resin decreases with increasing temperature due to increased desulfonation:

$$R_sSO_3H \rightarrow R_sH + H_2SO_4$$

Increasing the temperature accelerates the desulfonation process, and, at temperatures >150°C, a reaction between the sulfuric acid formed and oxidizable resin components begins. S dioxide is formed.

$$R_{red} + H_2SO_4 \rightarrow R_{ox} + H_2O + SO_2$$

Desulfonation, together with other reactions, as a consequence of the air heating of the resin, was also observed for styrene-DVB cation exchangers. For this type of exchange resin in the H^+ form, differential thermal analysis revealed three endothermic changes: dehydration at 100 to 120°C, desulfonation at 270 to 310°C, and oxidative degradation of the matrix at temperatures above 430°C. In addition to these three endothermic effects, an exothermic effect at 370°C was observed corresponding to the depolymerization of the matrix.

The evidence suggests that at the higher temperatures, the $-SO_3H$ groups react with the formation of sulfones and additional cross-links. Further rise in temperature causes the breakdown in the sulfone to SO_2 as well as causing an extensive degradation of the polymer matrix.

The course of desulfonation of strongly acidic cation exchangers is affected by the nature of their matrices, the location of the sulfo group, and on the presence of other substituents on the aromatic core (in the case where the matrix is formed by aromatic compounds). Sulfonated phenol-formaldehyde ion exchangers containing both $-SO_3H$ and $-OH$ groups are less stable on air heating as compared with strongly acidic cation exchangers of the styrene-DVB type.

Converting the H^+ form of exchangers into their salt forms increases the thermal stability.

Relatively little attention has been paid to the reactions which occur during the air heating of weakly acidic cation exchangers. Resins based on methacrylic acid and DVB are sufficiently stable on prolonged heating at 120°C (H^+ form). A further increase in temperature leads to the loss of hydration water. At approximately 200°C, condensation reactions between the ionogenic groups occur; and further heating at this temperature leads to total decarboxylation of the resin, followed by depolymerization.

The thermal stability of weakly acidic resins is also increased when they are converted into a salt form.

Comparing the thermal analysis data obtained during the air heating of various strongly and weakly acidic cation exchangers, the following conclusions can be made: all the data show dehydration over a broad temperature range, desulfonation (or loss of CO_2 for weakly acidic cation exchangers), and oxidative degradation of the matrix. The temperature range corresponding to this process will vary depending on the type and cross-linking of the resins. The destructive reactions always occur first in the ionogenic groups, since bonds in these groups are usually weaker than those in the polymer chains.

Air drying at room temperature of type 1, strongly basic anion exchange resins in the OH^- form, leads to a decrease in the volume exchange capacity of their functional groups. The loss of capacity accelerates with increasing temperature.

This decrease in the exchange capacity can be characterized as a consequence of deamination and degradation of these groups:

1. $R_s{-}CH_2\overset{+}{N}(CH_3)_3OH^- + H_2O \rightarrow R_s{-}CH_2OH + \overset{+}{N}H(CH_3)_3OH^-$
2. $R_s{-}CH_2\overset{+}{N}(CH_3)_3OH^- + H_2O \rightarrow R_sCH_2\overset{+}{N}H(CH_3)_2OH^- + CH_3OH$

However, it is probable that the main reaction is as follows:

$$R_s{-}CH_2\overset{+}{N}(CH_3)_3OH^- + H_2O \rightarrow R_sCH_2\overset{+}{N}H(CH_3)_2OH^- \searrow \overset{+}{N}H(CH_3)_3OH^- + R_sCH{=}O$$

The following reaction occurs when a strongly basic anion exchanger of type 2 in the OH^- form (the $-CH_2CH_2OH$ group is bound to the N atom less firmly than the $-CH_3$ group) is air heated:

$$R_s{-}CH_2\overset{+}{N}(CH_3)_2(CH_2CH_2OH)OH^- + H_2O \rightarrow R_sCH_2\overset{+}{N}H(CH_3)_2OH^- + HOCH_2CH_2OH$$

Salt forms of strongly basic anion exchange resins are more stable on air heating than the OH^- form of exchangers. Their stability decreases in the series $R_sCl > (R_s)_2SO_4 > (R_s)_3BO_3$.

On prolonged heating of type 1 anion exchangers in the Cl^- form, a degradation of a part of the quaternary ammonium groups occurs at 130°C:

$$R_sCH_2\overset{+}{N}(CH_3)_3Cl^- \rightarrow R_sCH_2Cl + N(CH_3)_3$$

A further increase in temperature to 180°C leads to an appreciable decrease of the amount of strongly basic groups as a consequence of the reaction:

$$R_sCH_2N^+(CH_3)_3Cl^- \rightarrow R_sCH_2N(CH_3)_2 + CH_3Cl$$

If the salt form of the resin is derived from an oxidizing acid (nitric acid, etc.) the stability of this nitrate form depends strongly on temperature: up to 140°C it is higher than that of the OH^- form; pyrolytic decomposition and self-ignition of the resin occurs at 160 to 200°C.

The following stability series of various types of anion exchange resins in the nitrate form has been published: Dowex® 2 > Permutit® SK > Amberlite® IRA-411. It is not recommended to dry the exchangers in the nitrate form at temperatures higher than 60°C.

Anion exchangers containing phosphonium or sulfonium ionogenic groups are much less stable on air heating than the similar ammonium analogs.

The thermal stability of weakly basic anion exchange resins is better than that of strongly basic resins.

The following practical recommendations can be given, taking into account the information presented above:

1. When determining the dry residue of the H^+ form of cation exchangers, it is useful to dry the resins at low temperatures (80°C, vacuum, P_2O_5). Temperatures above 100°C are to be avoided.
2. It is not recommended to dry or even store strongly basic resins in the OH^- form. Destruction of the quaternary ammonium groups occurs even at room temperature.

3. It is very difficult to prepare a water-free ion exchange resin by drying at increased temperature without damaging it. For example, the styrene-DVB strongly acidic cation exchange resin KU-2, dried at a constant weight at 100°C, still contains 9% water in the form of the monohydrate of the sulfo group.

Thermal Stability in the Presence of Solvents

All authors of papers dealing with the thermal stability of ion exchangers state that the polymeric types of strongly acidic cation exchangers, especially those with a styrene-DVB matrix, are fairly resistant when heated in water. Thus, Dowex® 50 is stable when heated in water at 105 to 115°C. Heating Amberlite® IR-120 in the H^+ form in water at 150°C, a total loss in the exchange capacity occurred after 12 days. At 180°C, the total loss occurred after 24 hr.

It has also been established that macroreticular ion exchange resins heated in water are more stable than the microreticular resins. For some types of macroreticular ion exchangers, a change in the pore volume has been observed after thermal treatment. The volume exchange capacity did not change.

The thermal stability of ion exchange resins is strongly influenced by the nature of the counter ion.

The mechanism of the thermal decomposition of the salt form of a strongly acidic cation exchange resin can be illustrated by the following reactions:

$$R_sSO_3Me + H_2O \rightarrow R_sH + MHSO_4$$

$$R_sSO_3Me + MHSO_4 \rightarrow R_sSO_3H + M_2SO_4$$

The liberated H^+ ions enter the exchange reaction with the counter ion in the resin and thus, converting the resin into the H^+ form, indirectly reduces its thermal stability (H^+ form is thermally less stable).

Many different counter ions can act as catalysts in the hydrolysis and in other degradation steps. It has been found that the rate of desulfonation of the phenol-formaldehyde cation exchanger $-SO_3H$ groups in alkali-metal ionic forms increases in the series $Cs < Rb < K < Na < Li$, i.e., with increasing effective radius of the hydrated counter ion.

Cation exchangers containing phosphonic groups on the styrene-DVB matrix are more thermally stable than the sulforesins. Differences between the stability of the H^+ and salt forms are lower and the effect of the DVB content is smaller.

In the presence of concentrated solutions of strong acids or bases and at increased temperature, hydrolysis of ionogenic groups also takes place. The hydrolysis rates are faster, in some cases, than those observed in water. In addition to hydrolysis of the functional groups, an oxidative destruction of the resin matrix can occur in the presence of oxidizing acids.

A decrease in exchange capacity is also found on heating cation exchangers in nonaqueous solvents. The solvents can react with the functional groups, which may also catalyze reactions between solvent molecules to yield higher molecular weight products. These products block the ionogenic groups or can be retained in the porous structure of the resin. Heating a strongly acidic cation exchange resin of the styrene-DVB type in acetone can be taken as an example. At 150°C, the loss in its exchange capacity is as follows: 2.8% after 24 hr, 8% after 48 hr, 11.2% after 120 hr. At the same time a condensation of acetone molecules catalyzed by the resin occurs even at temperatures below 100°C. The condensation products are fixed in the porous structure of the resin.

If the resin is heated in the presence of aliphatic alcohols, its thermal stability decreases with an increasing number of C atoms in the alcohol used.

The thermal stability of anion exchangers in water or aqueous solutions depends especially on the nature of the polymeric matrix, on the basicity of the functional groups, and on their ionic form.

Strongly basic anion exchangers in the OH^- form undergo the classical Hofmann degradation, giving a tertiary amine and alcohol as degradation products.

Heating a type 1 anion exchanger, degradation of the exchanger can proceed in the following way:

$$R_sCH_2N^+(CH_3)_3OH^- \nearrow R_sCH_2N(CH_3)_2 + CH_3OH \quad \text{(degradation)} \quad (1)$$

$$R_sCH_2N^+(CH_3)_3OH^- \searrow R_sCH_2OH + N(CH_3)_3 \quad \text{(deamination)} \quad (2)$$

According to Reaction 1, the resin acquires weakly basic properties. Reaction 2 results in a high molecular weight alcohol (without exchanging properties) and trimethylamine.

In the case of type 2 resins in the OH^- form, the Hofmann degradation will probably proceed according to the empirical law of β-elimination — mainly in the following way:

$$R_sCH_2\underset{\displaystyle CH_2CH_2OH}{\underset{|}{N^+}}(CH_3)_2OH^- \rightarrow R_sCH_2N(CH_3)_2 + CH_2{=}CHOH + H_2O$$

The resin assumes weakly basic properties. The other product will rearrange to acetaldehyde ($CH_3C{\overset{\displaystyle O}{\diagup\!\!\diagup}}\!\!\diagdown_{H}$).

Strongly basic anion exchange resins are more stable in a salt form than in the OH^- form. Their stability decreases with increasing DVB content in the resin matrix.

Anion exchangers in the OH^- form undergo deamination and degradation in aqueous-organic solvents. The loss in the exchange capacity increases with increasing concentration of the organic solvent in the system (under the same conditions, a greater loss in the exchange capacity was found in absolute ethanol or acetone than in water).

Radiation Stability

The interaction of radioactive radiation with ion exchange resins leads to their radiation damage. In the presence of a solvent (water), this primary degradation is accompanied by secondary damage as a consequence of the reaction of the solvent radiolytic products with the resin.

The consequences of radiation damage of the resin can be summarized as the following effects:

1. Loss in the exchange capacity due to the degradation of functional groups
2. Formation of new functional groups
3. Degradation of the resin matrix (Solubility of the resin increases due to the decrease in the molecular weight. On the other hand, the reverse effect occurs when the cross-linking of the matrix increases [especially in resins of low cross-linking].)
4. All the effects ([1] to [3]) leading to a change in the resin swelling
5. Gaseous products also formed by radiolytic degradation of the resins

The swollen resin also undergoes an indirect radiolytic damage (radiolysis of water, presence of hydrogen peroxide, etc.). The gross result of the resin irradiation is manifested in changes in mechanical and physicochemical properties, selectivity, exchange kinetics, etc.

The stability of various types of ion exchangers towards radioactive radiation decreases in the following order: inorganic ion exchangers > sulfonated coal > cation exchange resins > anion exchange resins.

According to their resistance towards ionizing radiation, cation exchange resins can be divided into the following groups:

a) Sulfonated, monofunctional, strongly acidic cation exchange resins of the polymer type
b) Sulfonated, polyfunctional, strongly acidic cation exchange resins of the polycondensation type
c) Phosphorylated, mono- and polyfunctional, medium acidic cation exchange resins of the polymer and polycondensation types
d) Weakly acidic, mono- and polyfunctional, cation exchange resins of the polymer and polycondensation types

Resins Sub a)

Considerable changes appear at doses $>10^6$ Gy. Only 5% of the exchange capacity remained when Dowex® 50W-X8 had been irradiated by 3.9×10^7 Gy. The total extent of damage is given by a series of resin properties and its structure. The effect of the cross-linking agent is that resins cross-linked with DVB are less stable than those cross-linked with butadiene. If DVB is used for the resin preparation, the radiation stability of the exchanger will also depend on the ratio of the DVB isomers present. The effect of the amount of the cross-linking agent is that resins with a higher content of cross-linking agent are relatively more stable. The effect of the ionic form is that resins in the H^+ form are the most sensitive. Their stability can be increased by converting them to a suitable ionic form.

If the resin is irradiated in air, new functional groups (predominantly carboxylic) are formed. Radiolysis of dry resins leads to the formation of gases (SO_2, SO_3, and especially H_2).

On irradiating the swollen resin, new functional groups are also formed. However, the formation of hydroxyl groups predominates.

It can be concluded from the numerous literature data that substantial changes occur for this type of resin at integral doses $> 10^6$ Gy. Doses higher than 10^7 Gy usually cause such a great damage that the resin cannot be used.

The stability of the macroporous-type resins towards ionizing radiation does not differ appreciably from that of gel-like resins.

Resins Sub b)

These types of resin are more resistant as compared with the preceding type. Appreciable damage occurs only at doses greater than 5×10^6 Gy. There is no difference in the radiation stability of H^+ and salt-form resins in contradistinction of the previous type. New types of functional groups, especially carboxylic groups, are formed during irradiation.

Resins Sub c)

These types exhibit a high radiation stability. The exchange capacity is practically unchanged up to an integral dose of 10^7 Gy. The swelling changes in water depend on the type of matrix. However, this change is appreciably lower than that of strongly acidic cation exchange resins. Some of these resins are able to work even at doses higher than 10^7 Gy.

Resins Sub d)

Weakly acidic cation exchangers are more sensitive towards ionizing radiation as compared with other types of cation exchange resins. The methacrylic acid-DVB resins exhibit appreciable changes of their exchange properties even at an integral dose of 10^5 Gy. Other types are more stable. However, they lose up to 90 to 95% of their exchange capacity at doses around 10^6 Gy. Concerning various ionic forms, the highest stability has been found for the H^+ form with the stability of the salt forms being appreciably lower (especially for swollen resins).

As with cation exchange resins, the radiation stability of anion exchange resins depends on the matrix type and on the nature of the functional groups. The stability of resins containing aromatic cores in their matrix is higher than that of anion exchange resins based purely on aliphatic hydrocarbons. A special position is occupied by anion exchangers containing pyridine cores. They are appreciably resistant towards ionizing radiation. The stability of the functional group of anion exchangers increases with decreasing basicity of the group.

The exchange capacity of strongly basic anion exchange resins (e.g., types 1 and 2) is appreciably reduced at doses of 10^5 to 10^6 Gy. They cannot be used at doses around 10^7 Gy.

The swelling ability decreases with increasing absorbed dose. Monofunctional, strongly basic resins are changed to polyfunctional types with functional groups of different basicity. At the same time, ammonia and various low molecular weight amines are split off. New functional groups of the –OH and –COOH type are often formed. Thus, the resin is changed to a bipolar (amphoteric) ion exchanger.

The radiation stability of anion exchange resins of the polycondensation type is similar to that of the styrene-DVB type. In contradistinction to the polymeric type of resins, degradation of this type of resin leads to an increase in swelling ability as a consequence of opening the overtight structure of the unirradiated resin.

A special position is occupied by anion exchangers containing pyridine or its derivatives. They show a high radiation stability. Considerable changes occur only at doses above 5×10^6 Gy. Monofunctional types are more resistant than polyfunctional anion exchangers containing amines other than the pyridinium bases.

An inner irradiation by α particles (e.g., after the adsorption of Pu, dose of 2.27×10^6 Gy) of the anion exchanger AV-23M (methylated copolymer of 2,5-methylvinylpyridine and DVB) caused no change in basicity. However, cross-linking of the macromolecular matrix was damaged and the swelling of the resin was increased. Pu was gradually desorbed from the exchanger due to the gradual degradation of the resin and reduction of Pu(IV) to Pu(III).

Weakly basic anion exchange resins are very sensitive towards ionizing radiation. A remarkable degradation occurs usually at a dose of 10^5 Gy. However, several anion exchangers are substantially damaged even at doses of 10^2 Gy.

Dissolution (Ashing) of Ion Exchange Resins

The mineralization of ion exchange resins can be achieved by wet or dry ashing.[55-57] These techniques are occasionally used to recover adsorbed metals which are difficult to elute from the resins and employed in connection with the determination of impurity elements in ion exchangers. Sometimes dissolution of the resin is also desirable, for example, for remote removal of contaminated resin from nuclear processing facilities and for preparing counting samples of sorbed α and β emitters.

Wet ashing of cation and anion exchange resins can be effected by means of the following general procedure: the air-dried resin is heated with concentrated sulfuric acid (2 to 3 mℓ/0.5 g resin) in a Kjeldahl flask of suitable size until white fumes are evolved. Then 70% perchloric acid is carefully and slowly added in small drops. Every portion of perchloric acid added should be completely reacted before the addition of the next portion. If necessary, the mixture is heated. In order to prevent the danger of explosion, the perchloric acid should be added very carefully. Nitric acid can be used instead of perchloric acid. The decomposition can, in some cases, be carried out at room temperatures. If it is necessary to heat the mixture, it should be carried out carefully (the content of the flask can boil over). Nitric acid is added at a rate of 0.5 mℓ/min.

A wet ashing procedure for anion exchange resins, e.g., Bio-Rad® AGl-X8 to recover adsorbed elements, is based on the use of nitric and perchloric acids. The resin (1 g) is mixed with concentrated nitric acid (20 mℓ) and concentrated perchloric acid (4 mℓ) and digested. Once the water has evaporated, the resin begins to dissolve. More concentrated nitric acid and more perchloric acid (2 mℓ) are then added to the mixture until the resin dissolves completely. Subsequently, the solution can be diluted with water to, for example, 50 mℓ.

Complete dissolution of resins of the sulfonic acid type can also be achieved by refluxing with 30 to 90% nitric acid, by treatment at 60°C with 50% H_2O_2 containing Fe, and by digestion with 6% K permanganate. In the first two methods clear solutions are obtained; in the last method, a slurry of Mn dioxide is formed. Also, mixtures of 50% H_2O_2 and concentrated nitric acid as well as a 2:1 mixture of nitric and perchloric acids have been used for the dissolution.

Ammonium- or Na-form cation exchange resins of the strongly acidic type can be completely dissolved in about 1 hr at 80°C with 6% H_2O_2 and 0.001 *M* Fe(II) or Fe(III) in 0.1 *M* nitric acid (Fe-catalyzed dissolution).[56]

Wet-ashing procedures should also be employed whenever small particles of the resin from the column or its degradation products are present in the eluate containing the elements to be determined, especially when present in trace amounts. Otherwise the organic material may cause interferences in the final determination of the elements. This interference may be due to a partial readsorption of the metals on the resin particles from the solution in which the elements are determined, e.g., by spectrophotometric, fluorimetric, titrimetric, or other means. Another source of interference is the resin degradation products which may contain compounds with chelating abilities exceeding that of the organic color reagents frequently used in connection with spectrophotometric, complexometric, and other procedures. In Table 51 in the chapter on Actinides (Volume II) an example is presented of the application of a wet-ashing technique following elution of a metal (Th) from an anion exchange resin column. The resin degradation products are destroyed with K permanganate in HCl medium. Traces of strongly acid cation exchange resins passing into eluates can be destroyed with fuming nitric acid and 30% H_2O_2 solution.

Dry ashing of ion exchange resins (for examples see chapter on Gold, Volume III) is much less popular, because combustion in air is often incomplete and volatilization of a number of elements occurs, which include Hg, As, Zn, Cd, and Pb. Therefore, this ashing technique, even when performed in pure O (to ensure complete combustion), is not of general applicability for the analytical determination of elements that have been adsorbed on ion exchange resins.

ION EXCHANGE PROCESSES

Choice of Appropriate Ion Exchange Resin

The following three basic criteria are usually considered when selecting the most suitable type of resin for a given analytical purpose:

1. The type of ionogenic groups and their ionic form
2. The matrix, its cross-linking, and suitable porosity
3. The resin grain size and uniformity of particles

Additional criteria specifying more precisely the applicability of the resin to a given purpose should be taken into account. Thus, emphasis is laid upon the monofunctionality of the resin, its purity, mechanical, chemical, thermal, or radiation stability, etc.

The choice of ion exchange resins in inorganic analysis is usually limited to strongly acid and basic ion exchange resins and to chelating resins of the types illustrated in Table 1.

Procedures based on using either cation or anion exchange resins can be chosen in a series of applications. A typical example is the separation of a mixture of metal cations which can be carried out successfully using a cation exchanger. However, the use of anion exchangers may often be more advantageous when the cations can be converted to suitable anionic complexes. Separation of U(VI) from Na ions can be shown as an example: a mixture of these two cations which has been adsorbed on a strongly acidic cation exchanger can be separated by successive elution with HCl of different concentration. Using a strongly basic anion exchanger it is possible to adsorb only the U(VI) ion by choosing a suitable HCl concentration. The Na ions are not adsorbed at all. Thus, the separation of this mixture can be carried out in only one operation. Similarly, separating orthophosphate from, e.g., Na ions, only Na is adsorbed on a cation exchanger. The phosphate ions remain in the solution. Using an anion exchanger, the behavior of the ions is reversed.

The resin degree of cross-linking especially affects the degree of swelling, the selectivity change, and the rate of exchange (equilibration). A low degree of resin cross-linking leads to a high swelling in contact with the solvent. The resin volume is considerably changed when changing the properties of the external solution. This volume change is an unwelcome phenomenon, especially when column work is considered. A low degree of cross-linking in the resin also diminishes the affinity of various ions to the ion exchanger. It is also one of the reasons for the diffuse zone distribution during the chromatographic separation of ions.

On the other hand, a high degree of resin cross-linking leads to a reduction in the swelling in a solvent as well as to an increase in selectivity. Due to small volume changes (even under great changes of the external solution properties) these exchangers are suitable for column work. The disadvantages of the high cross-linked resins are their lower weight exchange capacity and restriction of ionic diffusion inside the resin which results in slow exchange kinetics. Furthermore, large ions cannot penetrate into the exchanger due to the dense network of cross-bonds.

A compromise between these two extremes has been chosen with the styrene-DVB resins. Cross-linkages of 8% DVB (X-8) (strongly acidic cation exchangers) and 4 to 8% DVB (X-4 to X-8) (strongly basic anion exchangers) are most frequently used.

Another important factor affecting the choice of a suitable resin is the particle size. Resin grain size of 50 to 400 mesh is used for most analytical operations. Low grain size (in extreme cases even several microns), and if possible with a narrow range of particle sizes, is required for chromatographic applications.

The selectivity of ion exchanging processes can be increased by the use of selective ion exchangers (chelating resins). Thus, for example, resins containing iminodiacetic acid groups (see Table 1) retain selectively even traces of heavy metals in the presence of large amounts of alkali metal or alkaline earth metal ions.

Selectivity of Ion Exchange Resins

If the ion exchange resin is added to an aqueous solution (free from complex-forming substances) of an electrolyte, an exchange reaction between the ion exchanger and the ions in the solution takes place.

The exchange reaction of ions of the same charge can be written as

$$R_sA + B^+ \rightleftharpoons R_sB + A^+ \qquad (R_s = \text{resin matrix})$$

If this reaction is carried out in a closed system, the reaction goes to equilibrium due to the fact that the ion exchange is a reversible process. Equilibrium concentrations of ions taking part in the exchange are not identical under equilibrium conditions. They are dependent on the value of relative affinity of the reacting ions to the ion exchanger as well as on their initial concentration.

It has been shown experimentally that the affinity (or ion exchange potential) of various ions to the same resin (in dilute solutions, <0.1 *M*) increases with the ionic charge of the ion investigated. Polyvalent ions are, therefore, attached to the exchange resin more firmly (under the same conditions) as compared with monovalent ions. For ions of the same charge, it has been found that affinities are inversely proportional to the radius of the hydrated ions. However, this statement is only of general validity; exceptions exist.

Investigations have shown that the affinity of cations to a series of cation exchange resins is similar or identical to the so-called lyotropic series.

The affinity of anions is governed by similar rules. Moreover, it increases with increasing polarizability of the anions.

The following general affinity series have been found for various ions:

Strongly acidic cation exchange resins:

$$Li^+ < Na^+ < NH_4^+ < K^+ < Rb^+ < Cs^+ < Tl^+ < Ag^+;$$

$$Mg^{2+} < Ca^{2+} < Sr^{2+} < Ba^{2+} < Ra^{2+};$$

$$Fe^{2+} < Co^{2+} < Cu^{2+} < Zn^{2+};$$

$$Al^{3+} < Sc^{3+} < Lu^{3+} < Yb^{3+} < Tm^{3+} < Er^{3+} < Ho^{3+}$$

$$< Y^{3+} < Dy^{3+} < Tb^{3+} < Gd^{3+} < Eu^{3+} < Sm^{3+} < Pm^{3+}$$

$$< Nd^{3+} < Pr^{3+} < Ce^{3+} < La^{3+};$$

$$Th^{4+} > La^{3+} > Ca^{3+} > Na^+$$

Strongly basic anion exchange resins:

$$F^- < Cl^- < Br^- < I^-; \quad SO_4^{2-} > AsO_4^{3-} > MoO_4^{2-} > CrO_4^{2-};$$

$$I^- > NO_3^- > Br^-$$

The order of affinities in the individual series varies slightly depending on the individual nature of the exchange resin and the conditions.

The difference in affinity between the individual ions is called selectivity!

The selectivity depends on the type and concentration of the reacting ions as well as on the quality of the solvent and the nature of the exchange resin.

In principle, the selectivity of an ion exchange resin depends on factors mentioned below (*selectivity rules*):

1. Selectivity of the resin increases with increasing content of the cross-linking substance, e.g., DVB.
2. Ions with a smaller effective hydrated ionic radius are preferentially adsorbed.
3. If the ionogenic group of the resin forms an ionic pair with the reacting ion, the selectivity of the resin to this type of ion is increased.
4. If substances in the solution form slightly dissociated compounds, then ions forming a more dissociated compound are preferentially adsorbed.
5. Selectivity of the resin decreases with increasing temperature. This effect can be explained as a consequence of the decrease in the ionic hydration shells leading to a diminishing of the difference between the effective hydrated ionic radii of the reacting ions.

From the point of view of analytical practice, the control of resin selectivity is of great importance. The most convenient way is to add a complex-forming agent to the ion exchange system. By choosing a suitable agent it is possible to separate easily a mixture of ions (either by selective sorption or by selective elution) due to the change in the selectivity of the ionogenic groups of the resin.

For the theoretical and practical description of ion exchange equilibria, and hence of selectivities, the so-called selectivity coefficients (quotients), distribution coefficients, and separation factors can be used. While the selectivity coefficients are mainly employed for theoretical purposes, the distribution coefficients are of enormous practical significance. The latter, therefore, are listed in this *Handbook* for many elements under varying experimental conditions.

Selectivity Coefficient

If the exchange reaction of two ions A and B is described by the equation

$$bR_sA + aB^{b+} \rightleftharpoons aR_sB + bA^{a+}$$

then the selectivity coefficient, $k_{B,A,}$ is expressed by the relation

$$k_{B,A} = \frac{(R_sB)^a(A)^b}{(R_sA)^b(B)^a}$$

where ( ) = the analytical concentration (molarity, molality) of the ion A or B, and a,b = the absolute charges of ions A and B, respectively.

Experimentally, the same batch method is used for the determination of the selectivity coefficients as is employed in connection with the determination of distribution coefficients (see below). Thus, a known weight of the ion exchange resin in a required ionic form is mixed with a known volume of the solution containing a known amount of the ion to be investigated, and the mixture is shaken until equilibrium is established. The equilibrium concentration of the ion investigated is then determined in an aliquot portion of the solution. The value $k_{A,B}$ is evaluated from the data obtained.

It is recommended to carry out preliminary kinetic experiments to ascertain the time necessary to attain equilibrium. Usually 1 to 2 hr is sufficient for strongly acid or basic ion exchangers. However, up to several days may be necessary in the case of weakly acidic (basic) and selective exchangers.

If strongly as well as weakly acidic (basic) groups are present in the exchanger (e.g., phenolic polymers such as cationite KU-1), and if it is used in the $H^+(OH^-)$ form, the distribution of ions between the exchanger and the solution depends on the pH value of the system.

The aliquot portion of solution to be analyzed should be carefully separated from the ion exchanger particles. If the ion exchanger is analyzed, some difficulties connected with the separation of liquid phase occluded by the resin arise. This liquid phase cannot be removed by washing the exchanger with water without affecting the equilibrium established. This circumstance can be neglected only in the case of very dilute solutions.

If sufficiently low concentrations of the ions investigated are used, it is possible to determine the equilibrium concentration of only one of the ions. The concentration of the second ion can be calculated. However, when using high concentrations, both solution components should be determined. The errors in the determinations appreciably affect the calculated value of the selectivity coefficient, especially in the case of the exchange of ions of unequal charges.

Equilibrium Distribution Coefficient

The equilibrium state in the system, electrolyte solution - ion exchange resin, can be expressed in various ways. Ion exchange equilibria are very often expressed in terms of the "distribution coefficient". This quantity is given by the ratio of the equilibrium concentrations of the same ion in the exchanger phase and in the solution.

The equilibrium distribution coefficient D is defined by the following equation:

$$D = \frac{\text{amount of component in exchanger phase at equilibrium}}{\text{amount of component in liquid phase at equilibrium}}$$

If the distribution coefficient is related to a unit weight (1 g) of the exchanger and to a unit volume (1 mℓ) of the solution, it is denoted by the symbol D_g (in this *Handbook* the symbol K_d or K_D is most frequently used). If this quantity is related to a unit volume (1 mℓ) of the exchanger, the symbol D_v is used. However, no uniform symbol exists for this quantity as yet and, therefore, a wide variety of symbols can be found in the literature.

The relation between the "weight" and "volume" distribution coefficients can be expressed by the equation

$$D_g = D_V \cdot \rho$$

where ρ is the specific weight of the ion exchanger (the weight of the dry residue of the exchanger in 1 mℓ of the column filling).

The distribution coefficients are determined either by the static (batch) method or by the dynamic (column) method.

In the batch method a known amount of the ion exchange resin in a known ionic form is added to a solution of known volume and qualitative and quantitative composition. After equilibration (mechanical shaking machine), the change in concentration of the cation in question is determined (usually by analyzing the filtrate). For more precise measurements the amount of the ion sorbed by the resin should also be determined. The occluded solution is removed from the exchanger by centrifuging and the adsorbed ion is eluted with an appropriate solution. At the same time, the dry residue of the exchanger is determined. Resin quantities up to 1 g and solution volumes up to 250 mℓ are most frequently used for laboratory, static measurements. The concentration of the ions studied varies between trace carrier-free amounts (for radioactive elements) and 10^{-3} to 10^{-2} *M*.

The results are expressed as the weight distribution coefficient, defined according to the following equation:

$$D_g(K_d) = \frac{\text{amount of component/gram of dry exchanger}}{\text{amount of component/milliliter of liquid phase}}$$

This method is especially well suited for the determination of distribution coefficients in the order >10 to 10^6.

At distribution coefficients below 10 to 20 the column method can be successfully used. For this purpose a submilligram amount (e.g., 500 μg) of the element in question is dissolved in a small volume of solvent (e.g., 1 or 2 mℓ) and this solution is passed through a small column (e.g., 1 g) of the resin using a suitable flow rate. This is followed by washing with a solution of the same composition (but not containing this element). By determining the metal ion content in successive small fractions of the effluent, the elution curve is constructed whereby the volume of eluent V_{max} is obtained which has passed through the column to elute the maximum of the elution peak (peak of the band). (In ion exchange chromatography this V_{max} is called retention volume.) From this value the distribution coefficient is calculated using the approximate relationship:

$$V_{max} = D_g(K_d) \times m \quad (m = \text{mass of dry resin in the column})$$

For more exact determinations of the distribution coefficient by this column method the following equation can be used:

$$D_g(K_d) = \frac{V_{max} - V_0}{m}$$

where V_0 = void volume of the column (the volume occupied by the mobile phase in the packed section of the column; this volume is also called dead, free, or interstitial volume) and m = mass of the ion exchange resin in the column (in grams).

In a procedure which is suitable for the determination of very low D_g values the eluting solution containing a small amount of the ion investigated is passed through the exchanger column until the compositions of the influent and effluent solutions are equal. The exchanger is then washed with water or another suitable liquid and the amount of adsorbed ion (after elution) is determined.

In another procedure which is suitable for the determination of medium D_g values (10 to 10^2) the volume V_{max} is determined in the following way: eluting solution of known composition containing a small amount of the ion investigated is slowly passed through the ion exchange column. When the concentration of the ion attains one half of the starting (initial) value, the effluent volume is determined. This value is equal to V_{max} in an ideal case.

The interstitial volume V_0 can be determined by direct measurement, titration, or calculation. For example, a solution containing a suitable ion which is not adsorbed under the conditions given is passed slowly through the resin column. The volume of effluent is measured and the moment when the ion appears in the effluent is followed. This very simple procedure is suitable for cation as well as anion exchangers.

Separation Factors

The separation factor α is given by the ratio of the distribution coefficients or retention volumes of two different elements which were determined under identical experimental conditions.

$$\text{Separation factor} = \alpha = D_g(K_d)(a)/D_g(K_d)(b) = V_{max}(a)/V_{max}(b)$$

where a and b may, for instance, be Na and K or any other pair of elements. This ratio determines the separability of two elements by ion exchange. Separations can only be achieved if this ratio shows a value which is different from unity.

Kinetics of Ion Exchange

The ion exchange process may be quite rapid and equilibrium may under certain conditions be reached almost instantaneously. On the other hand, ion exchange resins having the functional groups accessible only through very small pores exhibit a low speed of exchange. Thus, for tightly cross-linked resins the rate of reaction is slower than for resins with a low degree of cross-linking.

Study of the rates at which ion exchange reactions reach equilibrium have suggested that the process of equilibration consists of five steps:

1. Diffusion of the ion to the resin surface
2. Diffusion of the ion through the resin to the exchange site
3. Exchange of ions at the site
4. Diffusion of the exchanged ion through the resin to the surface
5. Desorption and diffusion of the exchanged ion into the external solution

In most cases, the exchange step (3) is rapid so that the reaction is controlled by the rate of diffusion of the ions, primarily through the resin phase.

Other facts which are of importance from an analytical point of view are that the rates of exchange increase considerably with decreasing particle size and increasing temperature. The rate of ion exchange reaction is inversely proportional to the square of the radius of the resin particles. Therefore, finely divided ion exchange resin particles can collect ions from solutions much more rapidly than beads of larger size.

The rates of ion exchange have a great influence upon the efficiency of ion exchange separations by column operation. Thus, the rate determines the sharpness of elution peaks and the degree of overlap in chromatographic separations.

Ion Exchange Separations

In order to initiate the desired ion exchange process, the ion exchange resin should be in contact with a solution containing ions capable of being retained by it. Two methods exist for bringing an ion exchanger into contact with ions in a solution: the batch method and the column method. Batch ion exchange is never complete, but in certain cases it can be accepted as practically quantitative.

Ion exchange resins are nearly always used in columns. As a solution passes down an exchanger column it continually meets fresh unreacted exchanger, so that the equilibrium is displaced in the direction one wants. By this multistage effect, even an unfavorable equilibrium can give exchange which is complete within the limits of analytical detection.

The Batch Method

The ion exchange resin is equilibrated with a solution in a suitable vessel with stirring or shaking (i.e., forming a slurry). The ion exchange reaction proceeds in a closed system until the equilibrium between the ion exchanger and the ions in solution is established. After equilibrium is reached, the exchanger is separated from the solution phase (by filtration, centrifugation, or settling). A quantitative exchange of ions present in the solution can be obtained either by using a large excess of the ion exchange resin (single-batch process) or by adding smaller resin portions successively to the solution. However, both phases should be separated after each equilibration. This method is called the multistage (cascade) batch process. This modification of the batch process is laborious, time consuming, and may lead to numerous experimental errors. Its use is therefore limited; the single-stage operation is preferred. The multistage process is of certain importance in the analysis of systems liberating gases in contact with the ion exchange resin (e.g., decomposition of alkali carbonate solutions with H^+-form cation exchange resins; liberation of CO_2). It can also be used in special cases when the exchange equilibrium between the resin and ions in the solution is shifted in the forward direction by the ion exchange reaction. Neutralization reactions, formation of stable complexes, and formation of insoluble products can be mentioned as examples. The reverse of the last process can be imagined, i.e., conversion of insoluble substances to a soluble form. Ba, for example, can be brought into solution from insoluble Ba sulfate:

$$2R_sSO_3Na + BaSO_4 \rightarrow (R_sSO_3)_2Ba + Na_2SO_4$$

The resin is washed with water and the Ba ion can then be eluted with 3 to 4 *M* HCl. However, the operation should be carried out with a large excess of the resin and at increased temperature. A similar process can also be used in the analysis of Pb, Sr, or Ca sulfates, Pb chloride, insoluble phosphates of divalent metals, etc. Several examples of the batch adsorption of both cations and anions on ion exchange resins directly from aqueous suspensions of finely dispersed, solid samples, e.g., soils and minerals, are illustrated in the chapter on Copper, Volume III (see Table 40), Phosphorus, Volume VI (see Table 6), and Sulfur, Volume VI (see Table 10).

The batch process is often used in the determination of various physicochemical parameters. It is especially suitable in systems where the course of the exchange reaction need not be quantitative (e.g., determination of distribution coefficients, determination of the structure, stability of complexes, etc.).

Due to its operating simplicity, the batch process is widely used in the collecting of equilibrium data for column operations.

The Column Method

The dynamic (column) arrangement of separation is used in the vast majority of procedures of ion exchange separation of inorganic mixtures. Solutions to be analyzed are fed into an ion exchange column (containing a suitable amount of the ion exchange resin) using either gravity flow (keeping a constant liquid level in the reservoir) or at a flow rate controlled by a low-pressure pump. The effluent flowing out of the column is then divided, if necessary, into a series of fractions. These fractions or the entire effluent are then analyzed in the usual way.

The ion exchange column is a tube of appropriate design and filled with the ion exchange resin. In the simplest case, the ion exchange column may be represented by a glass tube of suitable size containing the swollen resin.

Glass or quartz are the most suitable materials for the manufacture of laboratory ion exchange columns. In the case of corrosion danger (e.g., presence of hydrofluoric acid etc.) the column is made of other appropriate materials (Teflon® or other plastics, stainless steel, etc.). In any case, the material should be resistant towards the solutions flowing through the column.

The sample solution is usually led to the upper part of the column, above the resin surface (downflow mode). The solution is also sometimes passed through the column from below (upflow mode).

Column Terminology

The *column volume* V_L is the volume of the part of the column that contains the ion exchange resin (packing, i.e., the resin bed [resin column]). (Occasionally the volumes of solutions that are passed through the column are given in column volumes [an example is shown in Table 25 in the chapter on Actinides, Volume II]).

Interstitial volume V_0 is the volume occupied by the solution (mobile phase) in the resin bed, i.e., in the packed section of a column. Sometimes this volume is also called *void , free, or dead volume.* $V_0 = V_L$ minus the volume of the swollen resin.

Interstitial fraction ϵ_i (relative interstitial volume) is the interstitial volume per unit volume of a packed column ($\epsilon_i = V_0/V_L$).

The column should be designed so that the solutions passing through flow steadily through its whole cross-section. This is the reason why precision glass columns (especially of small diameter) are often drilled into glass rods or are made of precision capillary tubes.

The ion exchange resin layer, i.e., the resin bed, sits in the column on a flat support plate. It is recommended to use a fritted glass disk of appropriate porosity which is fused into the bottom part of the column. In cases when the fritted glass disk cannot be used, stainless steel or platinum net may be applied. In the case of small analytical columns, it is sufficient to seal their lower part with glass wool, porous Teflon®, polyurethane, or another suitable material.

The liquid flow is usually governed by a tap placed at the bottom part of the column. In the case of columns of small cross-section, the flow is controlled by the pressure necessary to overcome the hydrodynamic resistance of the ion exchange resin bed (backpressure of the resin bed). A tap is not necessary in columns of this type.

If the column is equipped with a tap, a minimum volume of the space between the support disk and the column mouth (outlet) should be secured.

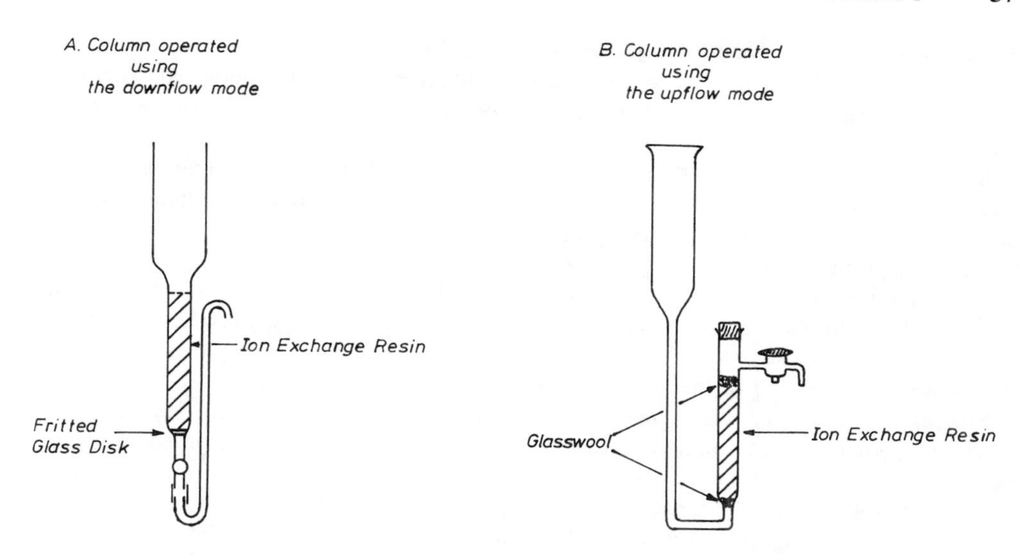

FIGURE 2. Ion exchange columns.

A certain space containing the liquid phase should be kept above the resin column. Its height is recommended to be 10 to 25% of the length (height) of the resin bed (column).

The ion exchange column is used most frequently in the form of a vertically fastened tube employing either the downflow or the upflow mode (i.e., down [descending] flow or reverse [ascending] flow technique). These two principles are illustrated by the column designs A and B shown in Figure 2. The funnel at the top of column A serves not only to pour in solutions, but also to receive the resin when it is "backwashed" before use. A stream of water is driven upward through the column, driving the resin into the upper part of the column. This releases air bubbles and classifies the resin beads so that they settle uniformly, the larger beads at the bottom and the smaller at the top, with spheres of roughly uniform size in any segment of the tube. This type of column can readily be equipped with a jacket for adjusting the temperature to higher or lower values as compared with the ambient temperature. Heating water for the jacket may be taken from a thermostatic bath.

Based on the two modes of flow a great variety of laboratory column types exists, and these have been illustrated in many publications dealing with the application of ion exchange resins to inorganic analytical chemistry. Therefore, it is not attempted here to give any recommendation as to which type of column is to be used for carrying out a certain analytical separation. This decision has to be made by the analyst or investigator. Because of the great popularity of ion exchange resins in analytical separation chemistry, it can be assumed that in all laboratories in which separations are carried out, ion exchange equipment of some sort will be available. If this should not be the case the reader is requested to consult the relevant, original references given in the tables of this handbook in which the separation procedures for the elements are presented. In these, the authors of the various methods usually give a description of the column type that has been used.

The same considerations are valid with respect to accessory equipment which is required in connection with certain chromatographic ion exchange separations. These accessories include the solution delivery system (pumps, etc.), fraction collectors, gradient elution devices, and detector systems allowing the continuous analysis of the effluent, e.g., by spectrophotometric, radiometric, or polarographic methods. The use of conductimetry has found extensive application in ion chromatography (see next chapter, Special Analytical Techniques Using Ion Exchange Resins).

Filling and Size of the Ion Exchange Column

The column is filled, as a rule, with a fully swollen resin. If the column contains a dry resin which swells in contact with water, the glass column may easily be ruptured due to the great volume change of the exchanger.

Before filling the column with a resin, the former is partially filled with deionized water which is free of dissolved gases. Furthermore, the porous support plate (or glass wool, etc.) carrying the resin bed should be completely freed from air by passing a small amount of water through the column.

If glass wool pads are used as supports for the resin bed the wool has to be thoroughly soaked in the solution from which the resin is filled into the column. Otherwise, air adhering to the glass wool will get into the resin bed. The time of soaking can be shortened considerably by occasional squeezing of the submerged glass wool, e.g., by means of a glass rod.

The resin suspension in water is poured along the column wall and the particles are allowed to settle in order to form a homogeneous layer (resin bed). The following procedure is also advantageous: a funnel with a broad stem is inserted into the top of the glass column which is completely filled with water. The funnel is partially filled with water. The resin, mixed in water, is then poured into the funnel and the tap in the bottom part of the column is opened. A homogeneous column of the exchanger (resin column; resin bed) can be rapidly obtained in this way.

This method of filling a column by gravitational settling has a disadvantage when the sizes of the resin particles vary over a wide range, because the larger particles will settle more rapidly and the resulting resin bed will have longitudinal zones of different particle sizes. If the column is not perfectly vertical during gravitational packing, there can also be radial variation in particle size. In ion exchange chromatography these variations affect the resolution of the column, and it will certainly make it difficult to pack several columns with reproducible properties. An additional problem exists when very small resin particles (less than 20 μm in diameter) are used in a chromatographic column. When the gravitational settling method is used for such particles, the settling velocity is very low and the time required to pack a column is prohibitive. To circumvent both of these problems a technique of dynamic packing can be used in which the ion exchanger particles are forced into the packed bed in a flowing fluid at a velocity much greater than their settling velocity. This method is suitable for packing high-pressure chromatographic columns.

For general analytical use, particles of 50-100 mesh are best. For difficult chromatographic separations a smaller size, say 200-400 mesh, should be chosen. The finer the particles, of course, the greater is the resistance to flow.

It is necessary to take care to exclude the entry of air in the resin column during filling as well as during further operations. The presence of air may lead to the formation of channels (''channeling'' causes irregular flow) in the bed or to its complete deterioration. When filling of the column is complete, water should always be present above the upper end of the resin bed. The column must never be allowed to drain; if it does, is must be backwashed to remove the air. A ''goose neck'' outlet tube is shown in Figure 2; this ensures that the solution level will not fall below the top of the bed. The ''goose neck'' can be omitted if surface tension will prevent the column from draining. If much air appears in the resin column, it is recommended to discharge the column and to fill it carefully again. If the column is to be filled with very small resin particles for high performance liquid chromatography (HPLC), special procedures are recommended.

For delicate chromatographic separations and when working at increased temperatures, it is necessary to deaerate the ion exchanger, i.e., to remove the adherent air before filling the column. The resin is mixed with water in a suitable beaker and placed in a dessicator. The dessicator is then evacuated. Another procedure is to place the resin mixed with water in a water bath, the temperature of which is about 15°C higher than the working temperature decided on. The flask is then evacuated. Deaeration is complete when bubbles released from the resin are no longer observed. If the column is to be used at increased temperature, it is necessary to carry out all operations including the preparation of solutions using boiled, deionized water.

The filled column is washed with water or with an appropriate solution. If further operations are carried out under pressure, it is necessary to wash the filled column under the same pressure.

In order to avoid formation of air bubbles in the resin column due to changes in the eluting solution composition (which may be a strong acid and/or contain a high concentration of an organic solvent) the filling of the ion exchange column and all subsequent rinsings and washings should be performed by use of a solution which has the same composition as the sorption solution. In other words, the resin is allowed to swell in such a solution and the entire procedure of filling the column and the subsequent washing process is carried out using this solution only. If this is not done a rupture of the resin bed due to formation of gas bubbles occurs, mainly caused by the heat evolved by mixing solutions of greatly varying composition. Thus, for example, when passing a sorption solution consisting of 90% organic solvent - 10% mineral acid through a resin bed that has been prepared by introducing the resin as aqueous suspension (e.g., in deionized water), excessive bubble formation will occur. This is not the case when the resin was filled into the column from the same organic solvent - acid mixture.

Since the solubility of air and other gases in liquids decreases with increasing temperature, formation of bubbles in a resin bed may also be caused by temperature changes in the environment. Thus, for example, when operating the column near a heater or in direct sunlight the compactness of the exchanger bed is often damaged. Similar negative effects are observed when passing solutions with temperatures differing considerably from that prevailing in the column, e.g., on percolating a solution that was kept in the refrigerator.

Gas bubbles may also be formed in the resin bed when passing samples of natural waters that have been acidified with an acid and from which the CO_2 formed has not been removed completely by boiling or prolonged standing.

After filling the glass column with an appropriate amount of resin a column of settled resin bed is obtained, which for simple nonchromatographic applications should have a diameter/height ratio in the range of 1:10 to 1:20. An empirical rule states that the diameter of this resin column should be more than 20 times greater than the diameter of the particles of the ion exchange resin forming this bed. Thus, when using a resin of 100-200 mesh (which corresponds to particle sizes of 0.08 to 0.16 mm) the bed diameter should be at least 1.6 to 3.2 mm. To obtain reasonable flow rates most analytical separations are performed in columns with diameters of 5 to 10 mm. Thin columns (<5 mm) are typical for work with carrier-free amounts of radioactive isotopes. A common mistake is to use a column that is too big. For most analytical work a resin bed 0.8 to 1.5 cm in diameter and 10 to 15 cm deep is suitable.

For difficult separations diameter/height ratios of 1:100 to 1:200 are chosen for the dimensions of the resin bed.

In chromatographic separations the height of the ion exchange column is determined by the separation factor of the ions under the given conditions. The approximate column height can be calculated using equilibrium data or results from a preliminary experiment.

In the tables of this *Handbook* describing separation procedures on ion exchange columns, the information given with respect to the size of the column is usually presented in the following abbreviated form: e.g., column — 10 × 0.6 cm; this means that the height (length) of the resin bed (resin column) is 10 cm and its diameter is 0.6 cm irrespective of the dimensions of the glass, quartz, or plastic column containing this resin bed.

For special separations coupled columns containing anion and cation exchange resins have been employed (for examples, see Tables 1, 5, and 10, 25, and 6 in the chapters on Titanium [Volume IV], Copper [Volume III], and Zirconium [Volume IV], respectively). Also, composite columns containing two different forms and/or types of ion exchange resins, separated by glass wool, have occasionally been utilized for analytical separations (see Tables 32, and 137 and 139 in the chapters on Copper [Volume III] and on Actinides [Volume II], respectively). The application of a rotating column has also been described (see Table 9 in chapter on Mercury, Volume IV).

The Flow Rate

The time required to pass a solution through an ion exchange resin bed depends on the cross-section area of the bed and the rate of flow per unit surface area. The latter is determined not only by the length of the resin bed and the height of the liquid column (with possible external pressure or vacuum), but also primarily by the particle size and form and also the viscosity of the solution. The particle size has the greatest effect; the form of particles is also significant, but to a lesser extent. Spherical grains produce greater resistance to flow than irregular grains. This is due to the fact that the void space is lower with spherical grains. A considerable decrease in the viscosity of aqueous solutions is obtained, as is known, with increased temperature. Working at elevated temperature may therefore be an advantage, especially when the resin particle size is low.

If a fast flow rate is desired (e.g., to shorten the separation time), coarse particles may be chosen, but is must then be remembered that the breakthrough capacity of the resin bed is considerably lowered and, therefore, a large excess of resin is required to obtain quantitative uptake in the sorption step. Higher flow rates can also be obtained by the application of pressure (e.g., with pumps) and by performing the separations at elevated temperatures (mainly used in ion exchange chromatography).

The rate of flow through an ion exchange column may be regulated, if desired, either by altering the variables just mentioned or by partly closing the stopcock in the outlet tube. Since a decrease in the particle size is advantageous, it is better to use smaller resin particles than to decrease the rate of flow by throttling the outlet or a similar regulatory device positioned between the reservoir and the column.

Usually flow rates in the range of 0.2 to 2 mℓ/min are employed for analytical nonchromatographic separations. Sometimes the flow rate is also given as mℓ/cm^2 min, where cm^2 refers to a column cross-section of 1 cm^2. Occasionally, flow rates are expressed in mℓ/hr and drops per minute or second.

In the procedures presented in the tables of this *Handbook* not only the overall flow rates through the columns are listed, but also those rates which are necessary for the elution of elements requiring different elution conditions.

Reduced flow rates through ion exchange columns during separations are obtained if Si has not been removed completely during the preliminary dissolution step of geological, environmental, and industrial materials containing this element. Silica precipitates on the resin and thus the rate is reduced. Similar reductions of flow rate can also be caused by precipitation of other insoluble compounds in the resin bed, as, for example, by hydroxides of Ti, Zr, Fe, Al, etc., sparingly soluble sulfates, phosphates, and also by insoluble organic material such as residual C particles after incomplete ashing of, e.g., coal or other C-rich materials. Also, insoluble inorganic and organic material contained in natural water, biological liquids, etc. which are directly passed through ion exchange resin beds may considerably reduce the flow rates through the columns. The formation of gas bubbles during the separation and elution steps may also entail a reduction of flow rate and eventually cause complete stoppage of the ion exchange operation. Consequently, great care has to be taken when preparing the sample solution for the ion exchange separation step.

Breakthrough Capacity of the Ion Exchange Column

This capacity (also called the effective capacity Q_B) gives the practical capacity of a known amount of the ion exchange resin in the column. A solution of the solute investigated is passed through the column (under chosen conditions) until the first traces of the solute appear in the effluent or until the concentration of the solute in the effluent reaches a chosen value. The breakthrough capacity is expressed in millimoles, milligrams, micrograms, or other appropriate units related to 1 g of dry resin or 1 mℓ of swollen resin. Thus, the effective capacity is determined by means of breakthrough curves. For this purpose the solution containing the ion in question at a concentration of c_0 is passed through the ion exchange column and the concentration c of this ion is determined in the individual fractions of the effluent. The results obtained are expressed in the form of the c/c_0 vs. effluent volume (V) plot. The curves obtained are called breakthrough curves or adsorption (exchange) isoplanes.

The breakthrough capacity depends on the particular conditions of the adsorption (grain size of resin, flow rate, solution composition, temperature, etc.). Thus, the breakthrough capacity of the column is greatly increased as the size of the particles is diminished. Where a quantitative uptake is desired, it is obviously of great importance to use relatively fine particles of the ion exchange resin in the column. The breakthrough capacity also becomes greater for increasing length of the column. With long and narrow columns a somewhat smaller amount of resin may be used than with short layers of large diameter. The breakthrough curves are sharper and the breakthrough capacity is higher at a lower flow rate than at a high rate. An increase in temperature causes a sharpening of the breakthrough curves and an increased breakthrough capacity.

For comparing effective capacities obtained under various conditions, it is necessary to make such a comparison at one and always the same c/c_0 value. The effective capacity is then defined as the amount of a substance retained by the exchanger column after reaching the chosen, comparative c/c_0 value.

Finally, it should be pointed out that the breakthrough capacity is not an exactly defined value, but depends on the sensitivity of the method used to determine the point where leakage of the solute begins.

Weight and Volume Capacity of the Ion Exchange Column

The weight capacity, Q_m, depends on the weight of the dry ion exchange resin placed in the column. It is expressed as the total number of millimoles of exchangeable ions. In addition to the weight capacity, the term volume capacity Q_V is also used. The relation between these two quantities is given by the following expression:

$$Q_V = (1 - \epsilon_i)\,\rho\,Q_m\left(\frac{100 - S}{100}\right)$$

where Q_V = the volume capacity of the column, Q_m = the weight capacity, ϵ_i = the relative interstitial volume, ρ = the density of the swollen resin, and S = the resin water content (in weight %).

The Separation Process

The column method of ion exchange processes enables one to carry out quantitative exchange of ions from the solutions as well as to separate ionic mixtures with maximum efficiency.

The following general procedure is used in ion exchange column separation (column operation):

1. Conditioning
2. Adsorption
3. Washing with a proper solution or solvent
4. Elution of retained ions (regeneration)

Conditioning of the Resin

After the resin has been transferred to the ion exchange column the resin bed must be prepared for the sorption step. This preparation is called conditioning, preconditioning, or pretreatment (the term "equilibration" is also used sometimes) and involves washing of the ion exchange resin with a suitable volume of a solution which has essentially the same composition as the sample solution, except that it does not contain the elements of the sample matrix and the ions which are to be separated from one another. This solution is usually identical with the so-called wash solution (see later) which is applied for rinsing and washing of the column after the sorption solution has passed through the resin bed.

Since appropriate pretreatment of the resin column is of great importance for the successful performance of the subsequent adsorption and separation steps, the solutions used for conditioning are in most cases included in the tables of this *Handbook* describing the separation procedures.

Adsorption (Sorption)

The method of selective adsorption is based on a choice of suitable adsorption conditions for one element or for a small group of elements present in the sample solution. In cation exchange, the adsorption conditions are chosen so that only elements of interest are retained, i.e., the undesirable components are converted to nonadsorbable forms. The most commonly used way is to convert the latter elements into sufficiently stable, nonadsorbable complexes by the addition of appropriate complex-forming agents (a very large number of separations of this type are presented throughout this *Handbook* using inorganic and organic complexing agents which include mineral acids and chelating agents). The method of selective adsorption is also used very frequently in anion exchange separation as is evident from the very numerous procedures presented in the various tables of this *Handbook*. As a matter of fact, selective adsorption of elements on anion exchange resins is by far a much more popular technique than that employing cationic resins.

In order to quantitatively separate a mixture of two elements by the method of selective adsorption, the distribution coefficients (K_d) of the retained and nonretained ions should be of the following approximate values: K_d of retained ions $>$ 100 to 300; K_d of nonretained ions $<$ 3 to 10. In other words, the separation factor should be greater than $\sim$100.

If the ions which are to be retained by the conditioned resin column (see above) show only weak to slight adsorption on the resin, or if an elution chromatographic separation is to be performed, the sample solution (sorption solution) should have a volume which is lower, comparable to or only somewhat larger than the volume of the resin bed. Otherwise, the adsorption zones of the ions will not be sharp and subsequent rinsing may further broaden the bands and even cause elution of weakly retained ions.

If, on the other hand, the ions to be determined are very strongly retained by the resin, the volume of sorption (sample) solution is only of little or no influence on the adsorption so that it may be larger by a factor of $>10^2$ to 10^3 than that of the resin bed. Examples are the quantitative adsorption of ions from liter volumes of natural waters as described in numerous procedures presented throughout this *Handbook*. The solution that has passed (was percolated) through the resin bed during the sorption process is called the effluent (or percolate).

In case that a chromatographic separation of the adsorbed ionic species is to be carried out, the "slurry column technique" of sorption is utilized occasionally. This procedure involves adsorption of the ions on a small batch of the resin (e.g., 0.1 to 2.0 mℓ), which after equilibration for a few minutes is transferred with a few milliliters of water or eluent solution to the column containing the bulk of the same resin (for examples see Tables 7, 31, 34 to 36, 45, 72, and 90 in the chapter on Rare Earth Elements). Thus, a resin bed with an upper layer is obtained in which the adsorbed ions are distributed homogeneously, so that on subsequent chromatographic elution well-defined bands are formed resulting in less broadening and tailing of the elution curves. The same slurry column technique, but using much larger amounts of resin (usually 50% of the entire quantity) for the preliminary batch process, is also employed, sometimes in connection with nonchromatographic separations, e.g., for the sorption of radioactive elements such as Pu (see Tables 54, 59, 60, 64, and 86 in the chapter on Actinides, Volume II) or to sorb metal ions directly from carbonate or other alkaline solutions obtained by the alkaline fusion of samples. In the latter case the adsorption is usually performed on a strongly acidic cation exchanger in the H form so that removal of carbonate (evolution of CO_2) and neutralization are effected simultaneously. For this purpose, also, a column flotation technique can be used in which an aqueous slurry of the powdered fusion products (e.g., of a silicate with a 1:1 mixture of K_2CO_3 and H_3BO_3) and the resin (fivefold excess) are agitated mechanically, e.g., by passing air or by magnetic stirring so that the resin particles are in constant flotation.[58-61] After dissolution of the sample melt the resin, on which all cations are adsorbed, is separated from the solution (containing Si and B), washed with water, and the elements are eluted with 2 *M* HCl or determined directly on the resin using X-ray fluorescence spectrometry[59,61] or emission spectrography.[60]

In the case of nonchromatographic separations up to 50% of the capacity of the resin column can be utilized for the adsorption of the ions from the sample solution. However, ion exchange chromatography requires that the volume of the resin layer retaining the elements to be separated should be about 3 to 5% of the total volume of the resin bed. When especially difficult chromatographic separations are to be carried out, this volume should not exceed 1% of the resin column.

Washing and Elution of Accompanying Elements

After the sample (sorption) solution has completely passed through the resin bed, the column reservoir and walls, tubings, etc. which have been in contact with this solution are repeatedly rinsed with small portions of a wash solution which usually has the same composition as the sample solution, but, of course, does not contain the elements to be separated. In most cases this wash solution is the eluent for elements which are not or only negligibly retained by the resin. With this washing step the removal of remaining (residual) amounts of these elements (which did not completely pass into the effluent during the sorption process) is effected before elution of more strongly adsorbed ionic species is started by application of suitable eluents. The latter serves the purpose of eluting foreign ions which were coadsorbed during sorption with the elements to be determined. If not removed these ions may accompany the elements to be assayed into the final eluate and possibly interfere with their determination. The flow rate during washing may conveniently be the same as in the preceding sorption step. Little time is saved by working at much higher speed, since the time required for complete displacement of ions is largely dependent upon their diffusion rates in the resin phase.

On the successive application of sorption, washing, and elution steps, cross-contamination of the various eluates may occur due to an upward or downward movement (diffusion) of the liquid phase remaining in the resin bed after each step. Thus, for example, a residual portion of the sorption solution (fraction of void volume next to the surface of the resin bed) may move (diffuse) into the wash solution; this effect is frequently observed when the flow rate is extremely low or zero, as it is the case when, e.g., the washing or elution process was stopped for a prolonged period of time (e.g., overnight) with wash or eluent solution remaining in the column reservoir. Mixing of various eluents also occurs whenever their compositions vary considerably, e.g., the sorption solution being a concentrated solution of salt or acid and the wash only very dilute acid.

All these adverse effects can be avoided if the resin bed is first treated portionwise with very small volumes of the respective mobile phase at a relatively high flow rate. Subsequently, the main portion of the liquid is transferred to the reservoir and passed at the required lower rate.

Elution and Quantitation

The desorption (elution) of the adsorbed ions is usually effected by the passage through the resin bed of suitable eluents in which the distribution coefficients of the elements to be eluted are very low so that only small volumes of the eluent are required for complete elution. The elution process may be started by passing the eluent through the column at a fairly high flow rate in order to replace the liquid in the void space so that the eluent will penetrate the entire resin column. Then the flow rate is preferably diminished to low values (e.g., 1 to 2 mℓ/cm² min). To achieve selective separation from interfering, coadsorbed elements this elution has to be performed by the use of special eluents, e.g., containing complexing ligands and reducing or oxidizing agents. Numerous separation methods based on this principle of selective elution are presented throughout this *Handbook*. In many cases this technique is employed following a procedure based on selective adsorption, so that separations of high selectivity can be achieved by this combination of methods, Occasionally it is advantageous to carry out the elution using the upflow mode (reverse flow elution) after the sorption and all other operations were preformed employing the downflow mode of column operation. This allows the rapid elution of very strongly adsorbed species with a minimum of eluent volume.

In some cases the elements adsorbed in the resin bed are desorbed by means of a batch elution process. For this purpose the resin is taken from the column and equilibrated with a suitable eluent, usually at elevated temperature to effect rapid batch elution (for examples see chapter on Platinum Metals, Volume III).

The volume of eluent necessary for an elution increases greatly when the particle size of the ion exchange resin is increased. Using a smaller particle size means a saving of eluent as well as of time. Increased concentration of the eluting agent, e.g., mineral acid, increases the elution rate because of a displacement of the ion exchange equilibrium (e.g., on a cation exchange resin). On the other hand, the resin shrinks at higher acid concentration, causing a diminished diffusion velocity of the ions in the matrix of the resin and, consequently, a lowered rate of elution. The increasing viscosity of the more concentrated acid has an influence in the same direction. Generally, it can be said that it is the "contact time" of the elution process that determines the efficiency, not the flow rate or the volume of acid used. It is useless to try to hasten the elution by increasing the flow rate over a certain value. For example, it was shown that if the flow rate of the eluent is raised from 2 to 10 mℓ/cm² min the only effect will be that the volume of eluent required for complete elution will increase fivefold. The efficiency of the elution will not be increased nor will the result be achieved in a shorter time.

After elution from the resin, the eluted elements, if possible, are determined directly, either in an aliquot portion or in the entire eluate using a suitable analytical method. If only trace quantities of the elements are present, the eluate is usually evaporated to dryness on a water bath, and organic material (from the resin and/or organic ligands used for the elution) is destroyed by wet or dry ashing. Subsequently, the elements are determined quantitatively using suitable analytical procedures.

Elution of the adsorbed ionic species is not required if their determination can be carried out directly on the resin, which in most cases is effected by the use of radiometric methods following neutron irradiation of the original sample or by postseparation irradiation of the resin containing the adsorbed elements. Other methods which allow the elements to be assayed directly on the resin utilize X-ray fluorescence spectrometry and the method of ion exchanger colorimetry (see next chapter, Special Analytical Techniques Using Ion Exchange Resins). Spectrographic methods are sometimes used following wet and/or dry ashing of the resins containing the adsorbed elements.

If X-ray fluorescence spectrometry is used for the determination of the adsorbed elements, the oven-dried resin beads are coated uniformly as a thin film (single layer of ~100 μm) on a strip of adhesive paper and then analyzed.[59,61,62] Occasionally, also the pellet technique

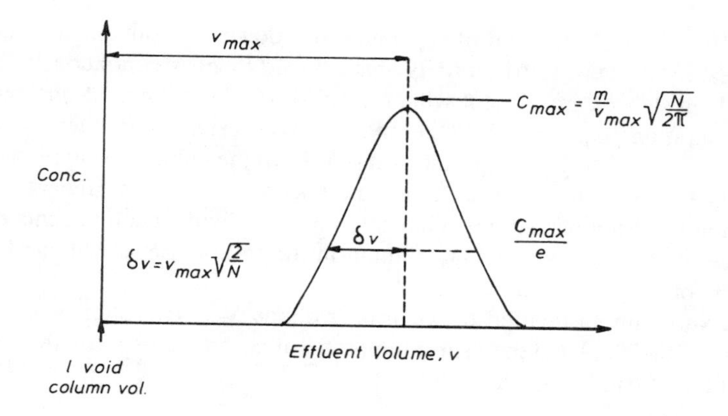

FIGURE 3. Shape of elution band.

has been used in connection with X-ray spectroscopy following column preconcentration.[63-66] In most cases, however, methods based on X-ray fluorescence are not used after column separation, but to measure metal ions that have been adsorbed on papers impregnated with ion exchange resins (e.g., Whatman® SA-2 and SB-2 resin-loaded filter papers).[67-75]

Regeneration of Resin

Following elution of the elements to be determined, ionic species still retained by the resin are eluted with suitable eluents. Then the resin is converted to the form in which it was originally present before passage of the sample (sorption) solution. However, this process of regeneration is in many cases very tedious and may require the application of substantial volumes of eluent solutions. Therefore, the use of a column with fresh resin is advocated in all those cases where only a few grams of the exchanger are involved, so that regeneration would be more expensive than its replacement.

Ion Exchange Chromatography

Ion exchange chromatography can be compared with conventional partition chromatography, but, in principle, the distribution of the species between the fixed and mobile phases is not controlled by solubilities, but rather by electrolytic equilibria, since the ions to be separated are bound electrostatically.

Chromatographic elution is used mainly for the mutual separation of ions showing very similar adsorption behavior on ion exchange resins, so that procedures based on selective adsorption and/or elution (see preceding sections) do not allow their fractionation into individual components. Examples are the chromatographic separations of the rare earth elements, trans-Pu elements, alkali metals, and alkaline earths (see in chapters treating these elements).

In elution chromatography a small amount of substance (e.g., Na ions) is adsorbed at the top of a column of ion exchange resin (e.g., strongly acidic cation exchange resin) and is gradually displaced down the column by passing an appropriate eluent or displacing solution (e.g., dilute HCl). If the concentration of adsorbed substance in the effluent, the solution emerging from the column, is plotted against the volume passed, a bell-shaped curve like Figure 3 is obtained. If the quantity of adsorbate is very small compared to the total capacity of the column, the curve approaches the shape of the standard curve of error (Gaussian curve). This is due to the nonideality of the system. Thus, a certain broadening of the individual bands (zones) occurs as a consequence of the difference of the concentration gradient within each solute band. The concentration of the solute on one side of the zone increases up to a maximum; on the other side of the exchanging zone it decreases to zero.

During the movement of the band of separated ions down the column, the maximum concentration steadily decreases. An initially sharp band (zone) is continuously broadened. If two adsorbed substances are present which travel down the column at different rates, two distinct concentration peaks are obtained, and the two substances are thereby separated; or, in other words, ions of lower affinity move faster down the column, while ions of relatively higher affinity to the exchanger used move more slowly. After a sufficient volume of the eluting solution has been passed down the column the individual components of the sample are distributed along the ion exchange column in the form of separated bands. In an ideal case they flow out separately from the column.

There is a very simple relation between the volume V_{max} (retention volume) needed to elute the peak of the band and the distribution ratio of adsorbate between the exchange resin and the solution. It can be stated as

$$\frac{V_{max}}{V_0} = D_g + 1$$

where V_0 = the void volume (dead, free, or interstitial volume) of the column (i.e., the volume of free solution in-between the resin beads) and D_g = a distribution coefficient = ratio of (adsorbate in the resin) to (adsorbate in the free solution) in any given resin segment. The derivation assumes that D_g is independent of the proportion of adsorbate on the resin and this will be strictly true only if the proportion is infinitesimally small. However, the relation is accurate enough for practical use. The peak volume (retention volume) V_{max} depends only on the distribution coefficient and not on the flow rate.

The width of the band is seen from Figure 3. The ratio ($\delta v/V_{max}$) depends inversely on the square root of the number of theoretical plates, N. Thus, from the equations shown in Figure 3 it is seen that an increase of the number of N results in a decrease of δv and an increase of c_{max}, i.e., a narrower (sharper) elution band is obtained.

The theoretical plate is a concept borrowed from fractional distillation. Its thickness is an experimental parameter, a number which expresses the closeness of approach to equilibrium as the solution flows past the resin. True equilibrium is never reached; changes in resin composition always lag behind changes in solution composition. To handle this delay mathematically we divide the resin column into imaginary segments called "theoretical plates" and suppose that the solution flowing into any given plate mixes completely with the resin in this plate, and comes to equilibrium with it, before discharging into the next plate, where it comes to equilibrium all over again. The narrower the "theoretical plates" the closer the approach to equilibrium.

The thickness of the theoretical plates has been calculated from diffusion coefficients, bead size, distribution coefficient, and flow rate. Under the conditions most often used in conventional ion exchange chromatography where diffusion within the beads is the rate-controlling step, the plate thickness is proportional inversely to the flow rate and directly to the square of the particle radius. It cannot, however, be smaller than the particle diameter and under typical operating conditions it is about 100 times the particle diameter.

The number of theoretical plates, N, in Figure 3 can be increased by decreasing the flow rate and the particle size and by increasing the length of the column. The resolution between bands, however, increases as the square root of N. To improve separations by a factor of two we must make the column four times as long, with other conditions, including flow rate, remaining the same. The flow rates normally employed in conventional ion exchange chromatography range from 0.2 to 2 mℓ/cm^2 of column area per minute.

One disadvantage of elution chromatography is that only a small amount of solute can be analyzed. It has been reported by many authors that the part of the column exchange capacity occupied by the separated ions should be less than 5% of the total exchange capacity of the column. If larger amounts of the substances are separated, complications may arise during the elution due to cross-contamination between the individual separated components.

For analytical separations usually elution chromatography is employed, although in a few instances also separations based on displacement chromatography have been used (see Table 53 in the chapter on Rare Earth Elements).

Gradient Elution Chromatography

The term gradient elution is used for procedures where the composition of the eluent is changed continuously during the elution. A certain concentration gradient is thus formed along the resin column. The following effects occur when the concentration of the eluting agent is increased: tailing is prevented because the elongated ends (tails) of the elution curves are shifted to the main zone as a consequence of the increasing concentration of the eluting agent; the elution curve becomes more symmetrical; elution of successive components becomes more rapid. The components are retained by the resin column for a shorter time and thus are shifted nearer to each other. A considerable shortening of the separation time can be obtained using gradient elution, especially in the case of an ion of low concentration.

However, the use of a gradient is unfavorable in some cases. Due to the increase of the eluting agent concentration the tightly following zones of the separated ion can be shifted such that they may overlap. Ions which were readily separated by normal elution cannot be separated using a gradient, especially a steep one.

Gradient elution in ion exchange chromatography is carried out either using a concentration gradient (concentration of the eluting agent is continuously changed) or by using a pH gradient (pH of the elution solution is continuously changed). However, procedures changing both the pH and the eluting agent concentration are also used. Practical examples of these various types of gradient elutions and their merits are discussed in the chapter on Rare Earth Elements and exemplified in Tables 7, 10, 17, 31 to 34, 36 to 45, 47, 64, 85, 90, 91, 93, 95, and 99 of the same chapter (see also Table 31 in chapter on Copper [Volume III] and Table 10 in chapter on Cadmium [Volume IV]).

Different forms of gradient can be applied. If the change of the eluting solution composition is constant in each time interval the gradient is linear. A convex form of the gradient occurs when the eluting solution composition is changed rapidly at the beginning of the elution and slowly at the end of it. A concentration change in the reverse order leads to the formation of a concave gradient.

The concentration gradient is prepared in the following way: the eluting solution (of concentration c_1) is transferred from a reservoir to a mixer containing the eluting solution at a lower concentration $c_2 (c_1 > c_2)$. The solution is led from the mixer to the column (see Figure 4). Its composition is continuously changed from concentration c_2 to concentration c_1. The formation of the gradient is affected not only by the composition of both solutions, but also by the volume ratio of the solutions in the reservoir and the mixer, as well as by flow rates of solutions flowing from the reservoir to the mixer and from the mixer to the column.

Continuous change in the elution solution composition can be obtained in the following three ways:

1. The volume of the mixer is constant, but the volume of the reservoir is changed.
2. The mixer and the reservoir form communicating vessels. The liquid level in both vessels decreases simultaneously.
3. The gradient is formed by a different rate of pumping of two eluting solutions of different composition into the mixer.

Effect of Temperature[76-80]

Raising the temperature in order to improve the quality of separation has often been

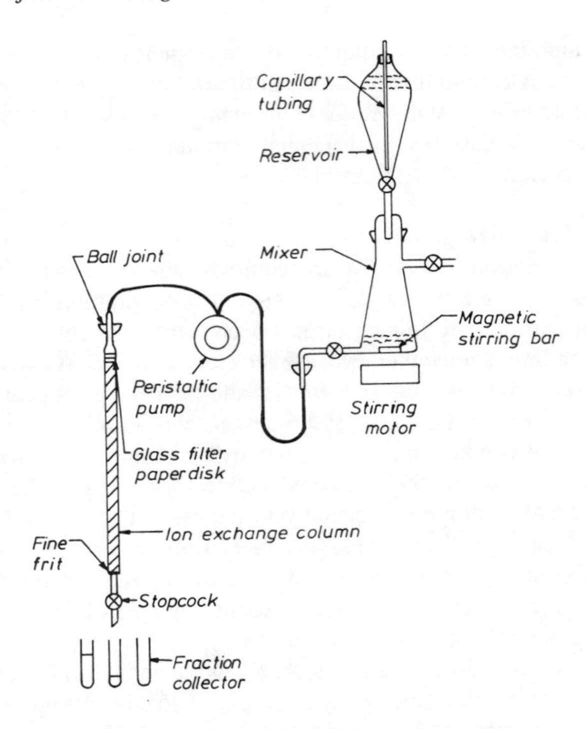

FIGURE 4. Gradient elution apparatus.

applied in ion exchange chromatography. The rise in temperature causes the diffusion coefficients to increase in the solution as well as in the resin phase and accelerates the stabilization of local equilibrium, and, hence, the height equivalent of the theoretical plate (HETP; H)(HETP[H] = L/N where L is the length of the column and N is the total number of plates in the column) diminishes. The smaller H is, the narrower are the elution curves of the separate components of the mixture, and the better the separation.

Although elevated temperatures are occasionally employed in separations based on selective adsorption on ion exchange resins to accelerate the ion exchange process, most work in which increased temperatures are applied is in the chromatography of rare earths and trivalent actinides on cation exchange resins with the use of organic acids as eluents (see in chapters on these elements).

In such systems the change of temperature only affects the absolute values of the distribution coefficients, while the ratio of the distribution coefficients, the separation factor (α), remains practically unchanged. In this case, the positive influence of the rise of temperature is exclusively concerned with decreasing H.

In systems in which the different rates of migration of the components of a mixture down the resin column are due to the differences in the individual affinity of the ions for the ion exchange resin, the change of temperature has a double effect:

1. H diminishes with rise of temperature as usual.
2. The values of the distribution coefficients of the individual ions undergo changes in accordance with the values and signs of the enthalpies of the ion exchange reaction, and, hence, the separation factor (α) changes as well.

Although the values of the enthalpies of ion exchange reactions are very small, i.e., of the order of several kilocalories (in the universally accessible range of temperatures of 0 to 100°C), the value of the distribution coefficient of a given ion can increase (or decrease) several times. At the same time the separation factor of two ions can increase several times, especially when the enthalpy sign of the two ion exchange reactions are opposite.

For example, on Bio-Rad® AG50W-X8 the separation factor for the Th-La pair increases from 4.12 at 20°C to 6.17 at 60°C in 4 *M* HCl. The peak for La is sharp and shows little tailing at 50°C, while at 20°C using 3 *M* HCl, which also provides a satisfactory separation, much more tailing is encountered and larger elution volumes are required for quantitative recoveries.[79]

It has also been shown that the separation factors for the Gd-Eu and Eu-Sm pair in cation exchange using EDTA solutions could be increased from 1.1 to 1.4 and from 1.4 to 1.8, respectively, by increasing the temperature from 25 to 92°C.

Rise of temperature still has another favorable effect, namely, it decreases the viscosity of solutions and thus diminishes the hydraulic resistance and permits the use of higher flow rates.

The Elution Process

A simple quantitative description of the elution process is possible via the elution constant (E) which is given by the relation:

$$E = \frac{\ell \cdot A}{V}$$

where A = the column cross-section (cm^2) and ℓ = the migration distance of a band (in centimeters) after the passage of volume V (milliliters) of eluting solution.

The relationship between the distribution coefficient $D_v(K_d)$ and the elution constant E is expressed by the equation:

$$E = \frac{\ell}{i + D_v}$$

where i = the fraction of the solvent in the resin bed.

The relationship between the elution constants of the first E_f and last, E_ℓ, traces of separated ion in the effluent is given by the equation:

$$E_f = \frac{\ell \cdot A}{V_f} \quad \text{and} \quad E_\ell = \frac{\ell \cdot A}{V_\ell}$$

where ℓ = column length (centimeters); V_f, V_ℓ = volumes of eluent (in milliliters) which are used to elute the first and last traces, respectively, of an ion from the column.

The separation of two ions M_1 and M_2 is quantitative if:

$$E_{\ell M_1} - E_{fM_2} > 0$$

E_{fM_2} = the elution constant of the first traces of element M_2, and $E_{\ell M_1}$ = the elution constant of the last traces of element M_1.

Theory of Ion Exchange

In the present *Handbook* the various theoretical aspects of ion exchange and ion exchange chromatography, such as the mathematical treatment of ion exchange equilibria, selectivity coefficients, and plate theory, are not discussed in detail. Relevant information is readily available from numerous monographs that have been written on these topics, so that the interested reader is requested to consult the literature treating this subject.[81-95]

REFERENCES

1. **Hajós, P. and Inczédy, J.,** Simple method for the determination of low cation-exchange capacity and the titration of cation exchangers by constant current coulometry, *J. Chromatogr.*, 201, 193, 1980.
2. **Sansoni, B.,** Coulometric micro and ultramicro determination of hydronium capacities of ion exchangers, *Angew. Chem. Int. Ed. Engl.*, 2(4), 218, 1963.
3. **Bunzl, K. and Sansoni, B.,** Determination of ion exchange capacity of ion-exchangers by difference weighings, *Anal. Chem.*, 48, 2279, 1976.
4. **Ficken, G. E. and Lane, E. S.,** Titration of basic groups in ion exchange resins with perchloric acid in glacial acetic acid, *Anal. Chim. Acta,* 16, 207, 1957.
5. **Fisher, S. and Kunin, R.,** Routine exchange-capacity determination of ion exchange resins, *Anal. Chem.*, 27, 1191, 1955.
6. **Ungar, J.,** Determination of ultimate capacity of weakly acidic and mixed-acid cation exchange resins, *Anal. Chem.*, 34, 413, 1962.
7. **Davankov, A. B. and Santo, I.,** Determination of the ion-exchange capacity of ionites with a high-frequency titrimeter, *Zavod. Lab.*, 29(11), 1304, 1963.
8. **Rachinskii, V. V. and Kolosov, I. B.,** Determination of the adsorption capacity of ion exchange resins by a gravimetric method, *Zavod. Lab.*, 29(8), 926, 1963.
9. **Polyanskii, N. G. and Shaburov, M. A.,** Rapid methods for the titrimetric determination of the capacity of anionites, *Zh. Anal. Khim.*, 18(3), 304, 1963.
10. **Vasil'ev, A. A. and Tsygankova, T. S.,** Determination of functional groups in sulfonated phenol-formaldehyde ion exchangers, *Zh. Prikl. Khim. (Leningrad),* 41(10), 2261, 1968.
11. **Pollio, F. X.,** Determination of moisture in ion exchange resins by Karl Fischer reagent, *Anal. Chem.*, 35, 2164, 1963.
12. **Heumann, W. R. and Rochon, F. D.,** Determination of water in cation-exchange resins by Karl Fischer reagent, *Anal. Chem.*, 38, 638, 1966.
13. **Sharma, H. D. and Subramanian, N.,** Determination of water in ion exchange resins by the Karl Fischer and drying methods, *Anal. Chem.*, 41, 2063, 1969.
14. **Sharma, H. D. and Subramanian, N.,** Determination of water in ion exchange resins: anion exchange resins, *Anal. Chem.*, 42, 1287, 1970.
15. **Van Acker, P., De Corte, F., and Hoste, J.,** Determination of water in a strong-base anion exchange resin by Karl Fischer titration, *Anal. Chim. Acta.*, 73, 198, 1974.
16. **Kristóf, J. and Inczédy, J.,** Application of continuous and selective water detector for quantitative measurements. Determination of water content of anion-exchange resins, *J. Therm. Anal.*, 19(1), 51, 1980.
17. **Blasius, E. and Schmitt, R.,** Determination of residual water in ion exchangers by means of tritium-labelled water and comparison of this method with the Karl Fischer titration, *Fresenius' Z. Anal. Chem.*, 241, 4, 1968.
18. **Heumann, W. R. and Rochon, F. D.,** Dehydration of cation exchange resins, *Can. J. Chem.*, 43, 3483, 1965.
19. **Marton, A., Kocsis, E., and Inczédy, J.,** Equilibrium and colorimetric study of the hydration of anion-exchange resins, *Talanta,* 30, 709, 1983.
20. **Grieser, M. D., Wilks, A. D., and Pietrzyk, D. J.,** Establishing water contents of hydrogen form cation resins by heats of immersion, *Anal. Chem.*, 44, 671, 1972.
21. **Webber, C. E., Garnett, E. S., and Godden, J. D.,** Exchangeable water in resins, *Clin. Chim. Acta,* 14(2), 282, 1966.
22. **Fricke, G. H., Rosenthal, D., and Welford, G. A.,** Wet weights of ion exchange resin beads by centrifugation, *Anal. Chem.*, 43, 648, 1971.
23. **Frankel, L. S.,** Hydrated porosity of macroreticular cation-exchange resins via nuclear magnetic resonance, *Anal. Chem.*, 43, 1506, 1971.
24. **Blasius, E. and Lander, H.,** Pyrolysis-gas chromatographic and pyrolysis-mass spectrometric determination of structural characteristics of exchangers with cyclic polyethers as anchor groups, *Fresenius' Z. Anal. Chem.*, 303, 272, 1980.
25. **Blasius, E. and Janzen, K. P.,** Analytical applications of crown compounds and cryptands, *Top. Curr. Chem.*, 98, 163, 1981.
26. **Blasius, E., Janzen, K. P., Adrian, W., Klautke, G., Lorscheider, R., Maurer, P. G., Nguyen, V. B., Nguyen, T. T., Scholten, G., and Stockemer, J.,** Preparation, characterisation and application of complex-forming exchangers with crown compounds or cryptands as anchor groups, *Fresenius' Z. Anal. Chem.*, 284, 337, 1977.
27. **Blasius, E., Adrian, W., Janzen, K. P., and Klautke, G.,** Preparation and properties of ion exchangers based on macrocyclic polyethers, *J. Chromatogr.*, 96(1), 89, 1974.
28. **Blasius, E., Janzen, K. P., Keller, M., Lander, H., Nguyen-Tien, T., and Scholten, G.,** Exchange materials having cyclic polyethers as anchoring groups. I. Preparation and characteristics, *Talanta,* 27, 107, 1980.

29. **Blasius, E., Janzen, K. P., Adrian, W., Klein, W., Klotz, H., Luxenburger, H., Mernke, E., Nguyen, V. B., Nguyen-Tien, T., Rausch, R., Stockemer, J., and Toussaint, A.,** Ion exchangers with cyclic polyethers as anchor groups. II, *Talanta,* 27, 127, 1980.
30. **Blasius, E. and Janzen, K. P.,** Preparation and application of polymers with cyclic polyether anchor groups, *Pure Appl. Chem.,* 54(11), 2115, 1982.
31. **Blasius, E., Janzen, K. P., and Neumann, W.,** Application of exchangers with cyclic polyethers as anchor groups in micro- and trace analyses, *Mikrochim. Acta,* 2, 279, 1977.
32. **Blasius, E., Janzen, K. P., Luxenburger, H., Nguyen, V. B., Klotz, H., and Stockemer, J.,** Application of exchangers with crown ethers as anchor groups in analytical and preparative chemistry, *J. Chromatogr.,* 167, 307, 1978.
33. **Blasius, E., Janzen, K. P., Klein, W., Klotz, H., Nguyen, V. B., Nguyen-Tien, T., Pfeiffer, R., Scholten, G., Simon, H., Stockemer, H., and Toussaint, A.,** Preparation, characterisation and application of ion exchangers with cyclic polyether anchor groups, *J. Chromatogr.,* 201, 147, 1980.
34. **Blasius, E. and Janzen, K. P.,** Analytical applications of crown compounds and cryptands, in *Topics in Current Chemistry, Springer-Verlag,* Berlin, 1981, 163.
35. **Sansoni, B. and Wiegand, W.,** Redox exchangers and their applications. XV. Reduction of iron (III) on redoxites and its subsequent oxidimetric determination, *Talanta,* 17, 973, 1970.
36. **Sansoni, B.,** Redox exchangers (redoxites and redox-ion exchangers), *Chem. Tech.,* 10(10), 580, 1958.
37. **Sansoni, B.,** Redoxites based on leucomethylene blue/methylene blue, *Naturwissenschaften,* 41(9), 212, 1954.
38. **Sansoni, B. and Sigmund, O.,** Ferrocenpolystyrene redoxite, *Angew. Chem.,* 73(9), 299, 1961.
39. **Ergozhin, E. E., Agibaeva, A. R., and Dzhandosova, K. D.,** Redox ionites based on diphenyl oxide, *Izv. Akad. Nauk Kaz. SSR Ser. Khim.,* (2), 65, 1977.
40. **Baistrocchi, R. and Grossi, F.,** Studies on electron exchange resins, *Ann. Chim. (Rome),* 54, 1218, 1964.
41. **Wanner, D. E. and Conrad, F. J.,** Purification of analytical grade cation exchange resin for spectrochemical applications, *Appl. Spectrosc.,* 21(3), 10, 1967.
42. **Edge, R. A.,** Removal of heavy metals from ion-exchange resins used for trace metal analysis, *Chemist Analyst,* 47(3), 72, 1958.
43. **Moody, G. J. and Thomas, J. D. R.,** Stability of ion-exchange resins. I. Mechanical and chemical stability, *Lab. Pract.,* 21(9), 632, 1972.
44. **Moody, G. J. and Thomas, J. D. R.,** Separations with ion exchange resins in organic and aqueous organic media. I. Stability and swelling of resins, *Lab. Pract.,* 19(3), 284, 1970.
45. **Moody, G. J. and Thomas, J. D. R.,** Stability of ion-exchange resins. II. Radiation stability, *Lab. Pract.,* 21(10), 717, 1972.
46. **Minto, M. A., Moody, G. J., and Thomas, J. D. R.,** Stability of ion-exchange resins. III. Solvent and thermal stability, *Lab. Pract.,* 21(11), 797, 1972.
47. **Kiseleva, E. D., Chmutov, K. V., and Krupnova, V. N.,** Influence of the exchanging ion and the degree of cross-linkage of DVB on the radiation stability of ion exchange resins, *Zh. Fiz. Khim.,* 36, 2707, 1962.
48. **Karyakin, A. V., Anikina, L. I., and Chirkova, T. S.,** Organic material liberated from ion exchange resins in contact with water, *Zh. Anal. Khim.,* 26(4), 816, 1971.
49. **Lenskaya, V. N.,** Interaction of certain cationites with solutions of oxidants and its analytical applications, *Tr. Kom. Anal. Khim. Akad. Nauk SSSR,* 6, 333, 1955.
50. **Armitage, G. M. and Lyle, S. J.,** Mass-spectrometric study of deterioration of polystyrene-based ion exchangers, *Talanta,* 20, 315, 1973.
51. **Specht, S., Dornov, V., Weinlaender, W., and Born, H. J.,** Variation in chromatographic system efficiency with transition to high activities. II. Comparision of the effects of alpha- and gamma-radiolysis of a cation exchanger on distribution coefficients, separation factors and plate heights, *J. Radioanal. Chem.,* 26(1), 17, 1975.
52. **Kubota, M., Konami, Y., Nakamura, H., and Amano, H.,** Radiation stability of macroporous and gel-type cation exchangers, *J. Radioanal. Chem.,* 45(1), 73, 1978.
53. **Kolditz, L. and Wendt, J.,** The radiation stability of ion exchange resins, *Z. Chem. Lpz.,* 6(10), 385, 1966.
54. **Kalkwarf, D. R.,** Safety evaluation of cation-exchange resins, Report BNWL-2391, U.S. Energy Research and Development Administration, 1977.
55. **Hoffmann, H. and Kauczor, H. W.,** Wet-ashing of ion exchange resins containing adsorbed metals, *Vom Wasser,* 51, 169, 1978.
56. **Bibler, N. E. and Orebaugh, E. G.,** Iron-catalyzed dissolution of polystyrenesulfonate cation-exchange resin in hydrogen peroxide, *Ind. Eng. Chem. Prod. Res. Dev.,* 15(2), 136, 1976.
57. **Vitkova, J., Jambor, J., and Vrchlabsky, M.,** Combustion of ion exchangers in high-frequency plasma. Spectroscopic determination of cadmium and lead adsorbed by cation exchangers, *Chem. Listy,* 72(4), 417, 1978.
58. **Govindaraju, K.,** Ion exchange dissolution method for silicate analysis, *Anal. Chem.,* 40, 24, 1968.

59. **Govindaraju, K. and Montanari, R.,** Routine performance of a matrix-correction free X-ray fluorescence spectrometric method for rock analysis, *X-Ray Spectrom.*, 7(3), 148, 1978.
60. **Govindaraju, K.,** New scheme of silicate analysis (16 major, minor and trace elements) based mainly on ion exchange dissolution and emission spectrometric methods, *Analusis*, 2(5), 367, 1973.
61. **Govindaraju, K.,** X-ray spectrometric determination of major elements in silicate rock samples, using a thin film technique based on ion exchange dissolution, *X-Ray Spectrom.*, 2, 57, 1973.
62. **Cesareo, R., Sciuti, S., and Gigante, G. E.,** XRF analysis improvement by a simple and fast pre-enrichment method, *Int. J. Appl. Radiat. Isot.*, 27, 58, 1976.
63. **Blount, C. W., Morgan, W. R., and Leyden, D. E.,** Preparation of pellets from ion-exchange resins for direct analysis for metal ions by X-ray spectroscopy, *Anal. Chim. Acta*, 53, 463, 1971.
64. **Blount, C. W., Channell, R. E., and Leyden, D. E.,** An improved technique for the preparation of pellets for X-ray spectrographic analysis on ion-exchange resins, *Anal. Chim. Acta*, 56, 456, 1971.
65. **Clark, P. J., Neal, G. F., and Allen, R. O.,** Quantitative multielement analysis using high energy particle bombardment, *Anal. Chem.*, 47, 650, 1975.
66. **Domel, G.,** The analysis of ion-exchange resins by X-ray fluorescence spectrometry, Report NIM-1607, National Institute for Metallurgy, Johannesburg, South Africa, 1974.
67. **Blasius, M. B., Kerkhoff, S. J., Wright, R. S., and Cothern, C. R.,** Use of X-ray fluorescence to determine trace metals in water resources, *Water Resour. Bull.*, 8(4), 704, 1972.
68. **Holynska, B., Leszko, M., and Nahlik, E.,** Applications of ion-exchanger foils in the determination of trace amounts of some metals in water by means of non-dispersive X-ray-fluoresence method, *J. Radioanal. Chem.*, 13(2), 401, 1973.
69. **Cesareo, R. and Gigante, G. E.,** Multi-element X-ray fluorescence analysis of natural waters by using a pre-concentration technique with ion exchange resins, *Water Air Soil Pollut.*, 9(1), 99, 1978.
70. **Van Espen, P., Nullens, H., and Adams, F. C.,** Automated energy-dispersive X-ray fluorescence analysis of environmental samples, *Fresenius' Z. Anal. Chem.*, 285, 215, 1977.
71. **Hooten, K. A. H. and Parsons, M. L.,** Practical limits of detection with ion exchange resin-loaded papers, *Appl. Spectrosc.*, 27(6), 480, 1973.
72. **Campbell, W. J., Spano, E. F., and Green, T. E.,** Micro and trace analysis by a combination of ion-exchange resin-loaded papers and X-ray spectrography, *Anal. Chem.*, 38, 987, 1966.
73. **Luke, C. L.,** Ultra-trace analysis of metals with a curved-crystal X-ray milliprobe, *Anal. Chem.*, 36, 318, 1964.
74. **Carlton, D. T. and Russ, J. C.,** Trace-level water analysis by energy-dispersive X-ray-fluorescence, *X-Ray Spectrom.*, 5(3), 172, 1976.
75. **Smith, F. C., Mathiesen, J. M., Orlando, M., and Spittler, T. M.,** Simultaneous analysis of dissolved metal pollutants in water, *Am. Lab.*, 7(12), 41, 1975.
76. **Dybczyński, R.,** Separation of rare earths on anion exchange resins. IV. Influence of temperature on anion behavior of the rare earth ethylenediamine tetraacetates, *J. Chromatogr.*, 14(1), 79, 1964.
77. **Dybczyński, R.,** Influence of temperature on tracer-level separations by ion exchange chromatography, *J. Chromatogr.*, 31(1), 155, 1967.
78. **Polkowska-Motrenko, H. and Dybczyński, R.,** Anion exchange separation of rare earth elements in orthophosphoric acid medium, *J. Radioanal. Chem.*, 59(1), 31, 1980.
79. **Strelow, F. W. E. and Gricius, A. J.,** Separation of thorium from lanthanum and other elements by cation exchange chromatography at elevated temperatures, *Anal. Chem.*, 44, 1898, 1972.
80. **Spiridon, S., Sabau, C., and Korkisch, J.,** Effect of temperature on anion exchange. Variation of the partition coefficients of zinc in organic solvent-hydrochloric acid mixtures, *J. Radioanal. Chem.*, 3(5-6), 377, 1969.
81. **Dorfner, K.,** *Ion Exchange Chromatography*, Akademie Verlag, Berlin, 1963.
82. **Fritz, J. S., Gjerde, D. T., and Pohlandt, C.,** *Ion Chromatography in Chromatographic Methods*, Bertsch, W., Jennings, W. G., and Kiser, R. E., Eds., Dr. Alfred Hüthig Verlag, Heidelberg, 1982.
83. **Griessbach, R.,** *Ion Exchange Adsorption in Theory and Practice*, Akademie Verlag, Berlin, 1957.
84. **Helfferich, F.,** *Ion Exchangers*, Vol. 1, Verlag Chemie, Weinheim, 1959.
85. **Helfferich, F. and Klein, G.,** *Multicomponent Chromatography*, Marcel Dekker, New York, 1970.
86. **Inczédy, J.,** *Analytical Application of Ion Exchangers*, Pergamon Press, London, 1966.
87. **Ionescu, T. D.,** *Ion Exchangers*, Technica, Bucharest, 1961.
88. **Kunin, R.,** *Elements of Ion Exchange, Reinhold Publishing*, New York, 1960.
89. **Marhol, M.,** Ion exchangers in analytical chemistry, their properties and use in inorganic chemistry, in *Comprehensive Analytical Chemistry*, Vol. 14, Svehla, G., Ed., Elsevier, Amsterdam, 1982.
90. **Marinsky, J. A., Ed.,** *Ion Exchange*, Vol. 1 and 2, Marcel Dekker, New York, 1966 and 1969.
91. **Nachod, F. C., Ed.,** *Ion Exchange. Theory and Applications*, Academic Press, New York, 1949.
92. **Rachinskii, V. V., Ed.,** *Theory of Ion Exchange and Chromatography*, Nauka, Moscow, 1968.
93. **Rieman, W. and Walton, W.,** *Ion Exchange in Analytical Chemistry*, Pergamon Press, London, 1970.
94. **Samuelson, O.,** *Ion Exchange Separations in Analytical Chemistry*, John Wiley & Sons, New York, 1963.
95. **Trémillon, B.,** *Separations by Ion Exchange Resins*, Gauthier-Villars, Paris, 1965.

SPECIAL ANALYTICAL TECHNIQUES USING ION EXCHANGE RESINS

METHODS BASED ON ION EXCHANGE CHROMATOGRAPHY WITH CONDUCTOMETRIC DETECTION (ION CHROMATOGRAPHY)

Ion chromatography uses the principles of high performance liquid chromatography (HPLC) and has been described in many reviews and monographs.[1-44c] The most comprehensive and complete treatment of the subject is presented by Fritz and collaborators.[44]

In the present chapter this topic is treated in two sections in which the conductometric analysis of ions by suppressed and unsuppressed ion chromatography is described. With the latter technique the separated ions are determined conductometrically in the presence of the eluent, while in the procedures based on suppressed ion chromotography the quantitative determination of the separated ions is performed after removal of the eluent.

In both cases the chromatographic separations are performed on ion exchange resins of low capacity, because ordinary ion exchange resins as, for example, Dowex® 1 or 50 have a high concentration of functional groups, i.e., high capacity, and therefore require eluents of high salt concentrations for chromatographic separation of sample ions. In this case conductometric measurements of the separated ions are not readily achieved due to the high background conductance.

It has been reported[45] that the background conductivity of the eluent, which varies with temperature, is cancelled effectively by use of a dual column and dual cell configuration. Also, with use of a highly sensitive conductivity meter, <1 nmol of common ions can be determined without use of a suppressor column.[46] For the same purpose the application of a bipolar-pulse conductivity detector has been suggested.[47]

Another drawback of conductivity measurements is that this type of detection requires accurate control of temperature (temperature coefficient of conductivity is $\simeq 2\%/°C$), which is especially important in ion chromatographic systems that use relatively high conductivity and nonsuppressed eluents. Therefore, the effect of fluctuating laboratory temperature necessitates thermal isolation of the active components of a chromatographic system from minor and rapid temperature gradients, so that maximum sensitivity and precision can be achieved.[48]

It has been shown that on cation exchange resins, the distribution ratio of metal cations decreases substantially when decreasing resin exchange capacity. Thus, from resins of low capacity, elution is possible with more dilute eluents. With respect to anion exchange resins of varying, low capacity, it was found that the selectivity coefficients for anions remained constant as the resin capacity was varied. This agrees with expected exchange behavior and has important implications for ion exchange chromatography.

Consider the exchange equilibrium for ions A^- and B^- on an anion exchange resin, R_s.

$$R_s - A^- + B^- \rightleftarrows R_s - B^- + A^- \tag{1}$$

The selectivity coefficient, K_A^B, is defined by the expression

$$K_A^B = \frac{(R_s - B^-)(A^-)}{(B^-)(R_s - A^-)} \tag{2}$$

where parentheses denote the concentration of an ion in either the solution or resin phase. In ion chromatography the amount of B^- is small compared to A^-. Therefore, the equation can be rewritten as follows:

$$K_A^B = \frac{D_B (A^-)}{Cap} \tag{3}$$

where D_B is the distribution coefficient for B^- (between the resin and solution) and Cap is the resin exchange capacity (equal to $[R_s - A^-]$). Simple chromatographic theory tells us that D_B is proportional to the adjusted retention time of sample anion B^- on an anion exchange column initially in the A^- form. The concentration of A^- in solution in column chromatography would be essentially that from the eluent (containing a salt such as $K^+ A^-$). Since K_A^B is constant with varying resin capacity and the adjusted retention time, t'_B,

$$\text{Constant} = \frac{t'_B (\text{Eluent})}{Cap} \tag{4}$$

$$t'_B = \frac{\text{Constant} \times Cap}{(\text{Eluent})} \tag{5}$$

Therefore, the final Equation 5 predicts that a significant reduction in resin capacity will permit a substantial reduction in eluent concentration while maintaining a desirable adjusted retention time.

Suppressed Ion Chromatography

This conventional type of ion chromatography was first introduced in 1975 by Small et al.[49] It is mainly used for the rapid determination of anions with $pK < 7$ and much less frequently also employed for the assay of some mono- and divalent cations. This dual-column chromatographic technique is a combination of ion exchange chromatography, with subsequent removal of the eluent by means of a suppressor and final determination of the separated ions by conductivity measurements. Peaks of eluted analyte ions are positive conductance signals relative to a very low conductivity eluent background. The technique is simple and has gained recognition as a useful analytical tool. Detection limits for common anions are in the range of 0.1 to 1 ppm. The determination of ultratrace (ppb) quantities of anions can be accomplished following their preconcentration on concentrator columns. Applications of the method range from the analysis of natural materials (geological samples including waters and air and biological materials) to the analysis of complicated industrial matrices.

Despite its many advantages, suppressed ion chromatography has some drawbacks:

1. The number of injections is restricted by the capacity of the suppressor.
2. The suppressor introduces extra band broadening which results in lower resolution.
3. Special equipment is needed for ion chromatography.
4. Only those eluents can be applied which, after passage through the suppressor, result in a low electrical background conductivity.

Anion Analysis

General

A flow diagram of an ion chromatograph based on the Dionex® system and which is suitable for anion analysis is presented in Figure 1. The sample solution (e.g., of pH 4 to 12), preferably neutral to weakly basic,[50] is injected into a flow of eluent via the sample loop and thus the sample is carried to the separator column (also called anion separator column, separation column, or analytical column) containing an anion exchanger in the

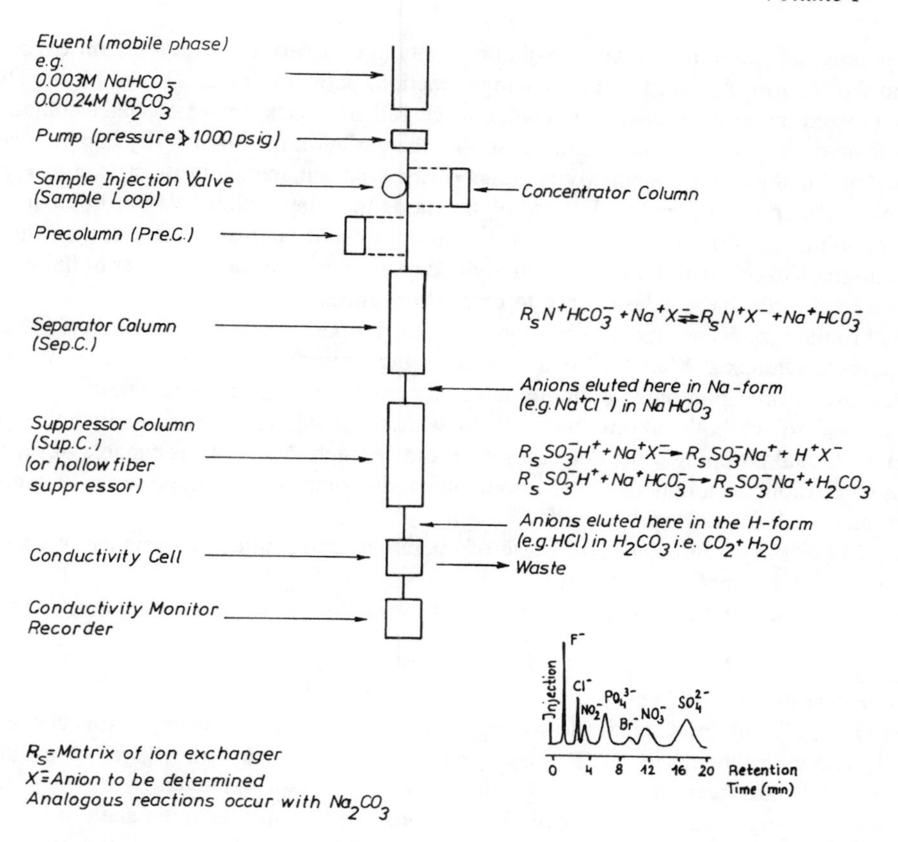

FIGURE 1. System for anion analysis by suppressed ion chromatography using conductometric detection (Dionex® system).

bicarbonate/carbonate form where selective adsorption and desorption of the anions take place. The adsorption and elution can be expressed by the reaction illustrated in Figure 1, and, as shown by the following selectivity scale, the greater the affinity of an anion for the anion exchanger the longer it will take to elute from the column.

Increasing affinity to anion exchange resin →

$F^- <$ formate $< OH^- < HCO_3^- < BrO_3^- < Cl^- < NO_2^- < PO_4^{3-} < Br^- <$

$< NO_3^- < SO_3^{2-} < SO_4^{2-} < S_2O_3^{2-} < I^-$

← decreasing retention time (elution time)

The elution position of orthophosphate relative to sulfate changes with pH. At higher pH levels it elutes after sulfate, while at lower levels it elutes before. These observations are compatible with the shift of orthophosphate from PO_4^{3-} to HPO_4^{2-} as the pH is reduced and the greater ease of elution of the divalent HPO_4^{2-} species compared to the trivalent PO_4^{3-}. This control over the elution position of orthophosphate ions is useful in the analysis of samples which contain phosphate, in addition to other ions of analytical interest.

The eluate from the separator column then passes to the suppressor column (also called anion suppressor column or stripper column) containing a strongly acidic cation exchange resin in the H form. In this, the ion exchange reactions shown in Figure 1 take place. Thus, highly ionized anion acids reach the conductance cell in a background of water containing carbonic acid (only very weakly ionized) so that the variation in conductivity can be readily detected and determined. Conductivity peaks of the type schematically illustrated in Figure 1 are obtained and compared with those of standards for quantitation.[51-53] For this purpose, standard solutions with known concentrations of the ions of interest are first run on the ion chromatograph to obtain a standard conductivity curve. Then, the conductivities of the sample are compared with the standard curve to obtain concentrations.

Ion chromatographs are manufactured by Dionex Corporation (Sunnyvale, Calif.). Thus, for example, Dionex® Model 10 is a complete single system to do either anion or cation separations. A dual pumping, a single detector system, the ion chromatograph Model 14 can be used to do both anions and cations without changing eluents, regenerants, and columns. A dual pumping, dual detector ion chromatograph, Model 16 is capable of running anions and cations simultaneously. If a computerized system is used, large series of samples can be analyzed unattended.[54]

The Dionex® system offers great ease of operation, and routine analysis can be carried out even by an inexperienced analyst.

Continuous concentration gradient elution was also employed in ion chromatography.[55,56]

Columns

Separator Column

The Dionex® columns contain low-capacity, pellicular anion exchange resins consisting of a styrene-divinylbenzene (DVB) copolymeric core on which a thin layer of an anion exchange latex has been deposited (agglomerated).[49,57] This is achieved by the superficial sulfonation of small beads of the copolymer (diameter: $\simeq$50 μm); the outer shell of the resin beads consists of a thin layer of sulfonic acid groups ($-SO_3^-\ H^+$). The resin is then converted into an anion exchange resin by treatment with a quaternary anion exchange material, e.g., containing methylammonium groups[58] which consist of spherical particles of uniform size in the range of 0.1 to 0.5 μm. (Simply stated, the procedure involves mixing macroparticles of surface sulfonated/DVB cationic resin with microparticles of strong base-type anion exchange resin.) These anion exchange particles can be synthesized by first polymerizing 5% DVB with 95% vinyl-benzyl chloride using emulsion polymerization techniques and then quaternized with dimethylethanolamine; as such the latex has the same anion functionality as the well-known Dowex® 2 anion exchange resin.[57] A latex of similar properties can be prepared from polystyrene-DVB by chloromethylation with chloromethoxymethane and subsequent amination with trimethylamine.[59] Due to electrostatic forces, the fine particles of the latex attach themselves to the resin surface to form a second thin shell surrounding the core of the beads. Consequently, the final resin consists of three distinct layers comprising the inner core, the sulfonated layer, and the layer of anion exchange particles.[58,60] A systematic illustration of such a surface agglomerated anion exchanger is presented in Figure 2. The advantage of this type of pellicular resin over conventional strongly basic resins of high capacity, e.g., Dowex® 2, lies in its much reduced diffusion path. This results in faster exchange, or mass transfer, of the eluent and sample ions and, consequently, in much improved column efficiency. In contrast to other pellicular resins with a silica[61-63] or glass core, the resin used in the Dionex® system is extremely stable even when strongly alkaline eluents are used.

Although the anion exchange material on the resin is of high capacity, the overall ion exchange capacity of the resin is low and typically in the range of 0.02 to 0.05 meq/g. This low capacity and the pellicular structure of the resin allow the separation of a variety of anionic species at relatively low eluent concentrations. Also, because of the short diffusion

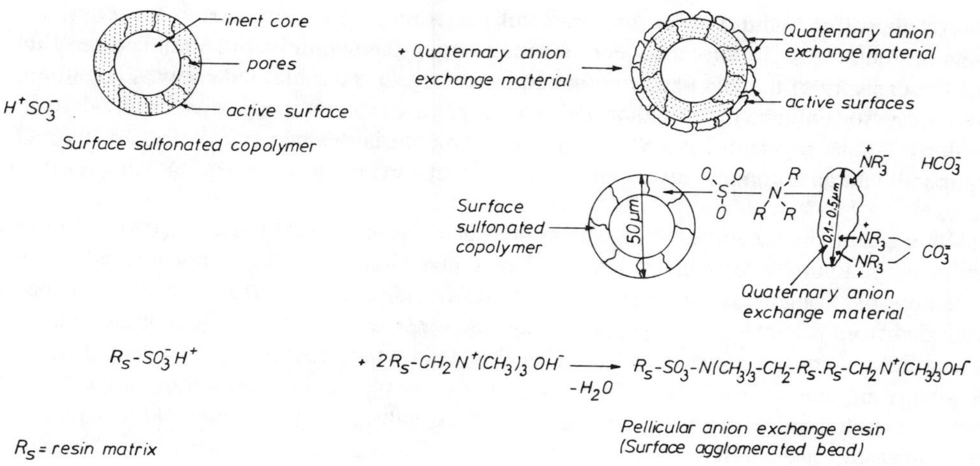

FIGURE 2. Surface agglomerated anion exchange resin bead.

path, the efficiency of the separator column is less affected by an increase of the flow rate. Therefore, relatively faster flow rates and, hence, shorter analysis time can be realized.

From the selectivity scale for anions presented in a preceding paragraph, it is seen that the selectivity of the Dionex® 2 separation column is comparable to that of conventional, high-capacity resins, e.g., Dowex® (see above). The order of affinity of the various anions is dependent not only on the nature of the active groups of the resin, but also on various factors such as their hydrated ionic radius, basicity, charge, and structure of the anion itself.

Interference occurring in the separator column may be due to the following reasons:

1. Excess of mineral acid or anionic metal species such as molybdate or tungstate in the injected sample solution may cause overloading of the separator.
2. Metals are likely to precipitate as hydroxides in an alkaline eluent environment. This results in pressure build-up and deteriorating performance of both the separator and suppressor columns.
3. Large amounts of hydroxide result in a substantial negative peak which may interfere with the determination of fluoride and chloride.

It has been shown[64] that successful separations of anions can also be achieved in separator columns (20 × 0.1 cm) containing 1:2 mixtures of Aminex® (13.5 μm; acetate form) and Dowex® 50W-X16 (200-400 mesh; H^+ form) using 0.001 *M* $NaHCO_3$-0.002 *M* Na_2CO_3 as eluent (flow rate: 1 mℓ/min).

Suppressor Column

From the reactions shown in Figure 1 it is seen that a suppressor column is used to exchange the cations of eluent and the sample for H ions. This removes the background conductance due to the eluent and results in the conversion of the sample anions to their highly conductive acids.

Any strongly acid H^+-form cation exchange resin of high capacity can be used in the suppressor column as long as it effectively accumulates the eluent cations and does not affect separations already achieved in the separator column.[60] The accumulation of eluent and sample cations continues until all active sites have been converted and the resin is almost completely in the Na form. Regeneration is accomplished by pumping dilute sulfuric acid solutions, e.g., 0.25, 1, or 2 *N* through the column. Due to a mass action effect of the H ions of the sulfuric acid, the Na ions are displaced from the resin and this is converted back to its original form. To avoid frequent regeneration of the suppressor column it is desirable that the capacity of the suppressor resin bed be as large as possible. Under most conditions, the suppressor column is not exhausted after 8 hr of continuous operation. In modern ion chromatographs replacement of the suppressor column is not necessary, because they are equipped with an automatic regeneration system that can be put into operation after working hours.

Other phenomena can occur in the suppressor column besides the conversion of eluent and sample anions into their acid form. This conversion results in highly ionized species for anions of strong acids. Because of the Donnan membrane effect, these ions are essentially excluded from the interior of the resin beads and emerge at the exit of the column with the interstitial volume. Weakly ionized species obtained by the conversion of weak acid anions, however, are able to enter the cation exchanger resin phase. This results in an increase in their retention times which is dependent on their pK_a values and the volume of the suppressor bed. Inorganic species likely to partition into the resin phase include nitrite, hypophosphite, phosphite, phosphate, cyanide, carbonate, borate, and fluoride. The retention of weak acid anions is complicated further by the place at which the anions are converted into their acid form, which in turn is dependent on the degree of exhaustion of the suppressor column.

The Dionex® suppressor column has been designed to reduce as far as possible effects that result in a loss of efficiency. Microporous gel-type cation exchangers with a DVB content of 8% and a particle size of 20 to 40 μm are typically employed as the suppressor material. The small size of the resin particles reduces the interstitial volume of the column bed and dispersion is kept to a minimum.

As a result of eluent dilution and the ion exclusion effect mentioned above, negative "dips" or deviations from the baseline conductance can interfere with fluoride and chloride analysis. A very small dip just before the fluoride peak is called the "water dip". A second dip, coinciding with the elution of chloride ion, is usually referred to as a "carbonate dip". Interference due to the negative deviations is negligible in the determination of large amounts of fluoride and chloride. However, when trace quantities (ppb) are analyzed serious interference can be expected.

Severe interferences in the determination of sulfite and nitrite from oxidation or interaction with the suppressor column resin were observed. Especially, the quantitation of sulfite by ion chromatography must be approached with caution because of the instability of sulfite toward oxidation, particularly when metal ions are present. It should be noted that complete removal of O from the system probably will not eliminate the catalytic oxidation of sulfite by metal ions. Therefore, the addition of formaldehyde was suggested to inhibit oxidation of sulfite.[65] Peak splitting was observed in the case of nitrite,[66] and is probably due to two factors. First, the nitrous acid enters the resin phase because of incomplete Donnan exclusion. This results in a later elution of the species and a consequent shift in retention time as a function of suppressor column exhaustion. Second, nitrite may oxidize partially to nitrate in the acidic environment of the suppressor column. The extent of this reaction also depends on suppressor column depletion.

Elution times and detector response for some ions vary depending on the percent exhaustion of the suppressor column, and peak band spreading and loss of resolution occur in the resin column (due to its void volume) which, as mentioned above, needs also to be regenerated periodically.

Many of these problems are eliminated through the utilization of a hollow fiber suppressor (also called cation exchange membrane suppressor)[55,67-73] which is used in place of the suppressor column. The effluent from the separator column is passed through the cores of a bundle of eight low density, sulfonated, polyethylene hollow fibers.[67] A solution of dilute (0.02 *N* or 0.025 *M*) sulfuric acid "regenerant" flows continuously through a tube that

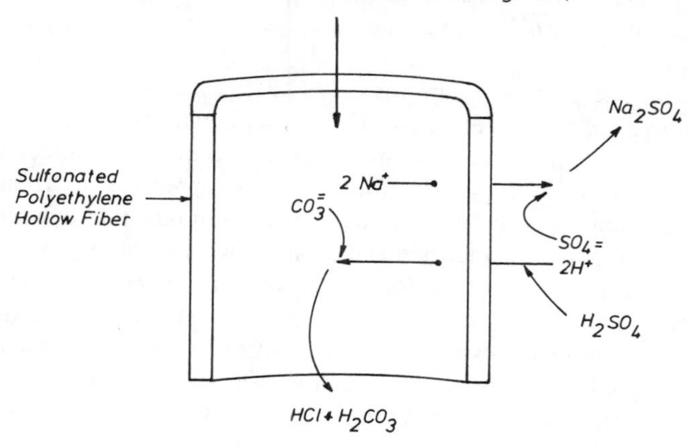

FIGURE 3. Operation of a hollow fiber.

contains the hollow fibers. As indicated in Figure 3, bicarbonate and/or carbonate eluent anions, and chloride or other sample anions do not permeate the fiber (membrane) wall because of Donnan exclusion forces, but cations do. Na bicarbonate and Na carbonate are converted to weakly conducting carbonic acid and Na chloride to highly conducting hydrochloric acid (HCl). The sulfuric acid regenerant is converted to Na sulfate and then replaced by fresh sulfuric acid. Thus, the hollow fiber suppressor is never exhausted; it can be operated for indefinite periods of time with little change in behavior. One limitation of the hollow fiber system is that the device can only suppress a maximum eluent flow of 80 mℓ/hr, because higher flows do not allow the diffusion of Na out of the eluent stream. Another disadvantage in the use of hollow fiber suppressors was found to be greater band spreading (e.g., of fluoride and chloride) than in a suppressor column packed with ion exchange resin. Reduced band spreading and better resolution were obtained by packing the tube of a hollow fiber suppressor with styrene-DVB copolymer beads.[72] A single tube (5 ft × 0.8 mm) of, for example, Nafion® 811-X packed with 500-μm-diameter beads was wound into a spiral and mounted in a cylindrical tube. A filament-filled helix (helically coiled Nafion® perfluorosulfonate cation exchange membrane tubing filled with monofilament nylon) was found to be more efficient than a packed-bead device of the same length and is superior with respect to dispersion and dead volume.[74,75] A low disperion suppressor has also been described[76] which contains one filament-filled membrane tube inserted inside another closely fitting membrane tube of the same type. The dual-membrane assembly is coiled as a small-diameter helix, the shape being retained by the filament. Column eluent flows in the annular space between the two membranes and regenerant flows through two separate channels, inside the inner membrane, and through a jacket which surrounds the whole device. Compared to simple filament-filled membrane helices, the device has a substantially larger available membrane surface area per unit dispersion of an injected band. One disadvantage is the need for a regenerant pump.

The dissolved gases, e.g., CO_2, contained in the effluent of a suppressor column (see Figure 1) or a hollow fiber system (see Figure 3) can be removed by use of gas-permeable tubings,[77,78] e.g., by a silicone rubber tubing (e.g., 3.6 m) enclosed in an evacuated jacket as a permeable membrane.[77] In other words, the tubing is utilized as a postsuppressor for removal of the background signal due to carbonic acid, which is decomposed to CO_2 and water; efficiency is increased by warming, e.g., to 79°C. Up to 90% suppression compared to normal suppressed ion chromatography can be achieved with such a postsuppressor; other advantages include enhanced sensitivity (≃40% for sulfate) and the ability to use gradient elutions with great differences in eluent strength without causing much shift in the baseline. The reduced background also improves the quantitative evaluation of ions coeluting with the carbonate dip.

If a column with a cation exchange resin in Hg(II) form is used as a postsuppressor, chloride ion is converted into unionized $HgCl_2$ so that conductance due to chloride can be eliminated.[79] Complete removal of chloride ion is also achieved by precipitation as AgCl using a cation exchange resin in the Ag^+ form as a postsuppressor. In Table 8 this column is called exclusion suppressor column. After passage through this column of a selected portion of the eluate (containing chloride and bromate which elute together).[80] the effluent is percolated through a concentrator column (see following paragraph). Subsequently, the latter is put in place of the sample loop of the ion chromatograph and eluent B is passed through the analytical and suppressor columns, and finally bromate is determined conductometrically (see Table 8). This principle of multidimensional ion chromatography has also been used for the determination of trace anions in natural waters.[81] Selenate, selenite, and arsenate were separated from the major anions chloride, nitrate, and sulfate by collecting the fraction of the ion chromatogram containing the former ions, after suppression, on a concentrator column and reinjecting at the original chromatographic conditions.

Concentrator Column

As indicated in Figure 1 sample loading can be performed with the concentrator column replacing the injection loop; the sample is then loaded by syringe or by pump.[17,18,58,82] The concentrator column may also be loaded off-line, disconnected from the ion chromatograph (e.g., for field use), and stored before analysis (storage for 1 week does not affect the results). In the former case, activation of the injection valve automatically places the concentrator column in-line with the separator column. In the latter case, the concentrator column must be manually connected to the flow path before the separator column. Since the concentrator column contains the same anion exchange resin as the separator column, the use of an additional precolumn (see below) becomes unnecessary. After injection into the concentrator column sample ions are separated and detected in the usual manner. Normal multispecies analysis often requires less than 5 min per ion; trace analysis requires the same time as routine analysis once the concentrator is loaded. Concentrator loading requires 5 to 15 min per sample of which usually 10 mℓ are passed, so that in most cases a 100-fold preconcentration of anions can be achieved (compared to the sample size of 100 $\mu\ell$ usually injected into the ion chromatograph) (see Tables 1 to 11). An example of use of a concentrator column is shown in Table 11.[58]

Preconcentration of anions, e.g., nitrate, can also be effected by Donnan dialysis using a Nafion® 811 cation exchange tubular membrane immersed in a slurry of Dowex® 50W-X4.[83]

Precolumn

This column (see Figure 1) contains the same resin as the more expensive separator column and its purpose is to protect the latter from inorganic and organic trace impurities present in the sample, which may gradually deteriorate the column performance.[50,71,84,85]

In the case that the precolumn contains a material which selectively removes an impurity in the sample solution, the column may be called a guard column. Examples of this type of column are the Ag^+ form of Dowex® 50-X8 (removes chloride from saline solutions, e.g., seawater[82]), a column packed with Nucleosil® $5NH_2$ (removes interfering humic substances contained in water samples[86]), and a MPIC-AG4 guard column.[87] The Ag chloride which precipitates in the guard column is finely dispersed so that the rate of flow through the system is only slightly reduced.

Table 1
DETERMINATION OF ANIONS IN SOLID GEOLOGICAL SAMPLES BY SUPPRESSED ION CHROMATOGRAPHY

Applications	Experimental conditions and remarks	Ref.
Determination of F, Cl, and S in geological materials (rocks, sediments, shales, etc.)	Dionex® Model 10 ion chromatograph Sample size (loop): 0.1 mℓ Pre.C: 15 × 0.3 cm Sep.C: 50 × 0.3 or 25 × 0.3 cm Sup.C: 25 × 0.8 cm Eluent: 0.0024 *M* Na_2CO_3 - 0.003 *M* $NaHCO_3$ Flow rate: 138 mℓ/hr Recoveries of F, Cl, and S were 78 to 88%, 100 ± 5%, and 85 ± 6%, respectively; detection limits averaged 1, 0.9, and 5 μg, respectively; before IC, the sample is ashed with V_2O_5, low-S Fe powder, Sn accelerator, and a porous silica wafer for 20 min in a stream of O saturated with H_2O (1.4 ℓ/min); the combustion products are absorbed in a solution of the same composition as that of the eluent; the resulting solution is subjected to IC	88
Determination of F in geological samples (sand, clay, pyrite, coal, and standard rocks)	Dionex® Model 12S ion chromatograph Sample size (loop): 0.1 mℓ Pre.C: 5 × 0.3 cm Sep.C: 25 × 0.3 cm Sup.C: 25 × 0.6 cm Eluent: 0.0024 *M* Na_2CO_3 - 0.003 *M* $NaHCO_3$ Flow rate: 138 mℓ/hr Before the separation, the sample (100 mg, containing 0.01 to 6% F) is decomposed by fusion with Na carbonate (200 mg); the detection limit is ≈8 ppm of F^- (90% confidence level)	89
Determination of fluoride and chloride in soils	Dionex® Model 10 ion chromatograph Sample size (loop): 0.1 mℓ Pre.C: 5 × 0.46 cm (fast run column) Sep.C: 2.5 × 0.4 cm (HPIC - AS3) Sup. C: hollow fiber suppressor Eluent: 0.0024 *M* Na_2CO_3 - 0.003 *M* $NaHCO_3$ Flow rate: 2.6 mℓ/min Before the separation, the sample (5 g; grain size: <250 μm) is stirred magnetically for 1 hr with 0.01 *M* NaOH and the suspension is removed by filtration Filtrate = sample solution	90
Determination of chloride in silicate rocks	Dionex® Model 10 ion chromatograph Sample size: 0.5 mℓ Pre.C: 15 × 0.3 cm Sep.C: 50 × 0.3 cm Sup.C: 25 × 0.9 cm Eluent: 0.0024 *M* Na_2CO_3 - 0.003 *M* $NaHCO_3$ Flow rate: 138 mℓ/hr Before the separation, the sample (≈0.2 g) is ashed with Fe chip and Cu accelerators for 5 min in a stream of O and the gaseous products are trapped in a solution (50 mℓ) of a composition corresponding to that of the eluent; the detection limit is 0.8 μg of Cl^- and the coefficient of variation is ≈5%	91
Determination of chloride in silicate rocks	Dionex® Model 12S ion chromatograph Sep.C: 25 × 0.3 cm Sup.C: 25 × 0.6 cm Eluent: 0.003 *M* Na_2CO_3 - 0.0015 *M* NaOH The coefficient of variation and detection limits were 8% and 7 ppm, respectively	159

Table 1 (continued)
DETERMINATION OF ANIONS IN SOLID GEOLOGICAL SAMPLES BY SUPPRESSED ION CHROMATOGRAPHY

Applications	Experimental conditions and remarks	Ref.
	The instrument is controlled by a microprocessor with a 3-min loading step followed by a 25-min recording interval; before the separation the sample is fused with Na_2CO_3	
Determination of F, Cl, N, and S in coal and oil shale[a]	Dionex® Model 14 ion chromatograph Sample size (loop): 0.1 mℓ Pre.C: 5 × 0.4 cm Sep.C: 25 × 0.4 cm Sup.C: 10 × 0.9 cm Eluent: 0.0024 *M* Na_2CO_3 - 0.003 *M* $NaHCO_3$ Flow rate: 2.3 mℓ/min This IC determination of F^-, Cl^-, NO_3^-, and SO_4^{2-} is performed after combustion of the sample in a Parr oxygen bomb	92
Determination of As in lake sediments	Dionex® Model 10 ion chromatograph Sample size: 10 mℓ Pre.C: 15 × 0.3 cm Sep.C: 50 × 0.3 cm Sup.C: 25 × 0.6 cm Eluent: 0.0014 *M* Na_2CO_3 - 0.003 *M* $NaHCO_3$ Flow rate: 137 mℓ/hr Cl^- and PO_4^{3-} are eluted before AsO_4^{3-}, while SO_4^{2-} is eluted afterwards Before the IC separation, the dried sample (500 mg) is extracted with 1 *M* HCl (5 mℓ) by shaking for 30 min at 30°C; an aliquot (1 mℓ) of the extract is diluted with 5% triethanolamine solution (5 mℓ), adjusted to pH 11.5 with NaOH, and all As is oxidized to As(V) by the addition of one to two drops of H_2O_2; finally, the solution is diluted with water to 100 mℓ to prepare the sample solution for IC; in case that only As(V) is to be determined, another 1-mℓ aliquot of the HCl extract is adjusted to pH 11.5 and diluted to 100 mℓ after 20 mℓ of 10% crotonaldehyde - 50% ethanol (prevents oxidation of As[III] to As[V]) and 5 mℓ of triethanol were added	128

Note: Pre.C = precolumn, Sep.C = separation column, Sup.C = suppressor column, and IC = ion chromatography.

[a] See also Reference 380.

Eluents

Solutions that can be used as eluent streams into which the sample is introduced must displace anionic species from the separator column and they must undergo an exchange reaction in the suppressor column to form water and another weakly ionized species that has very low conductance. In the Dionex® system the most widely use eluent is 0.0024 *M* Na carbonate-0.003 *M* Na bicarbonate (see Figure 1 and Tables 1, 2, 4, and 6 to 11).[30,58,80-120] This eluent is readily prepared by dissolving 0.5 g each of the carbonate and bicarbonate in 2 ℓ of water.

Mixtures containing similar concentrations of these two compounds have also frequently been employed as eluents (see Tables 1 to 5, 7, and 10).[50,60,68-70,82,87,121-145] All these buffered solutions which are prepared by mixing carbonate and bicarbonate have very good eluting characteristics because of the −2 valency state of carbonate and therefore can be used at relatively low concentrations. Changes in selectivity can be made simply by varying the

Table 2
DETERMINATION OF COMMON ANIONS IN NATURAL WATERS BY SUPPRESSED ION CHROMATOGRAPHY[a]

Applications	Experimental conditions and remarks	Ref.
Determination of chloride, nitrate, and sulfate in atmospheric precipitation	Dionex® Model 10 ion chromatograph Sample size: 0.3 mℓ Pre.C: 12.5 × 0.3 cm Sep.C: 25 × 0.3 cm (or 50 × 0.3 cm) Sup.C: 25 × 0.9 cm Eluent: 0.0024 M Na_2CO_3 - 0.003 M $NaHCO_3$ Flow rate: 165 and 180 mℓ/hr for 50- and 25-cm separation columns, respectively The method allows the analysis of 60 samples per day containing (per liter) 0.06—10, 0.022—2, and 0.028—1.5 mg of sulfate, nitrate, and chloride, respectively	85
Determination of anions in rain water	Dionex® Model 14 ion chromatograph Sample size (loop): 0.1 mℓ Sep.C: 50 × 0.3 cm Sup.C: 25 × 0.6 cm Eluent: 0.002 M Na_2CO_3 - 0.005 M NaOH Flow rate: 156 mℓ/hr F^-, Cl^-, NO_2^-, NO_3^-, SO_3^{2-}, and SO_4^{2-} are eluted in this order; limits of detection (100-μℓ sample) range from 0.1 ppm (for F^-) to 5 ppm (for SO_3^{2-})	160
Determination of fluoride, chloride, nitrate, phosphate, and sulfate in drinking, surface, and rain waters	Dionex® Model 12 ion chromatograph Sample size: 0.1 mℓ Sep.C: standard fast run anion separator Sup.C: standard anion suppressor Eluent: 0.0023 M Na_2CO_3 - 0.0048 M $NaHCO_3$ Flow rate: 115 mℓ/hr The time for analysis of one sample was ≃10 min	129
Determination of total N and P in river water and industrial waste water	Dionex® Model 10 ion chromatograph Sample size (loop): 0.1 mℓ Pre.C: 5 × 0.4 cm Sep.C: 25 × 0.4 cm Sup.C: 25 × 0.6 cm Eluent: 0.0015 M Na_2CO_3 - 0.005 M $NaHCO_3$ Flow rate: 2.5 mℓ/min The detection limits were 7 and 17 μg/ℓ for N and P, respectively Before IC, the sample (20 mℓ, containing <0.1 mg of N) is heated at 120°C for 30 min with 3.2% $K_2S_2O_8$ solution (3 mℓ) and 1 M NaOH (1.5 mℓ) to oxidize all N and P to nitrate and orthophosphate, respectively; the digest is treated with Bio-Rad® AG50W-X12 resin (100-200 mesh; H^+ form) (1.5 mℓ) and subsequently Ca and Mg are removed on a column (30 × 1 cm) of Chelex® 100 (100-200 mesh) using 0.05 NaOH as eluent; effluent = sample solution	130
Determination of anions in geothermal well water	Dionex® Model 10 ion chromatograph Sample size (loop): 0.1 mℓ Sep.C: 50 × 0.3 cm Sup.C: 25 × 0.6 cm Eluent; 0.0016 M Na_2CO_3 - 0.002 M $NaHCO_3$ Flow rate: 115 mℓ/hr The order of elution of the anions is F^-, Cl^-, Br^-, and SO_4^{2-}	131
Determination of fluoride, chloride, and bromide in mineral spring water	Dionex® Model 14 ion chromatograph Sample size (loop): 0.1 mℓ Pre.C: 15 × 0.3 cm	93

Table 2 (continued)
DETERMINATION OF COMMON ANIONS IN NATURAL WATERS BY SUPPRESSED ION CHROMATOGRAPHY[a]

Applications	Experimental conditions and remarks	Ref.
	Sep.C: 50 × 0.3 cm Sup.C: 25 × 0.6 cm Eluent: 0.0024 M Na_2CO_3 - 0.003 M $NaHCO_3$ Flow rate: 115 mℓ/hr The order of elution is F^-, Cl^-, and Br^-	
Determination of nitrate in drinking water	Dionex® Model 12 ion chromatograph Sample size (loop): 0.1 mℓ Sep.C: 25 × 0.6 cm Sup.C: 9 × 1 cm Eluent: 0.0024 M Na_2CO_3 - 0.0045 M $NaHCO_3$ Flow rate: ≃115 mℓ/hr	132
Determination of bromide in sea water	Dionex® ion chromatograph Sep.C: 21.5 × 0.3 cm Sup.C: 5.8 × 0.3 cm Eluent: 0.0024 M Na_2CO_3 - 0.003 M $NaHCO_3$ Flow rate: 2 mℓ/min Cl^- is eluted before Br^-; before IC the sample (10 mℓ) is mixed with CCl_4 (5 mℓ), KNO_3-HNO_3-buffer solution (pH 1.0) (2 mℓ) and 0.15% NaOCl solution (0.6 mℓ); elemental Br in the organic layer is reduced with H_2O_2 and the resulting mixture is acidified with acetic acid to prepare the sample solution	168

Note: Pre.C = precolumn, Sep.C = separation column, Sup.C = suppressor column, and IC = ion chromatography

[a] See also References 381 to 386 and 405.

ratio of bicarbonate to carbonate, i.e., by varying the pH value of the eluent. Also, the eluent concentration can be varied so that faster or slower elution is obtained without affecting the elution order of analyte ions. Other eluents consist of dilute solutions of Na carbonate (0.003 to 0.01 M; see Tables 3, 4, 9, and 11),[59,146-156] Na bicarbonate (see Table 4),[59,157,158] Na carbonate-Na (or K) hydroxide (see Tables 1 to 3, 5, and 10),[71,112,137,143,146-148,159,160] and Na tetraborate (0.0035 to 0.015 M $Na_2B_4O_7$) (see Tables 6 and 8),[80,161-164] in the absence or presence of Na bicarbonate[161] or Na carbonate.[164] Eluents containing Na tetraborate have a very low displacement potential and are useful if early eluting species such as fluoride, formate (see Table 6), and bromate (see Table 8) are to be separated. Na borate is converted into weakly ionized boric acid (H_3BO_3) in the suppressor column.

Carbonate solutions containing Na or K hydroxide are not only suitable eluents for common anions (see Tables 1, 2, 5, and 10),[137,143,159,160] but also for oxyanions of As, Cr, Mo, W, and Se (see Table 3).[146-148,165] The anionic metal species are also readily eluted with Na carbonate solutions (see Table 3).[146-148,152,154] Na hydroxide eluent would be ideal with respect to suppression, because it is converted into water. A disadvantage is its low affinity for the separator resin, which results in the displacement of slightly retained anions only. Increasing the concentration of the Na hydroxide solution increases its displacement potential, but this results in rapid exhaustion of the suppressor column.

Other eluents that may be employed in suppressed ion chromatography include Na phenoxide,[49] Na or K benzoate,[166,167] K hydrogen phthalate,[167] tripotassium citrate,[167] disodium glutamate,[21] and 2-morpholinoethane-sulfonic acid (zwitter ionic eluent).[62] All these compounds are converted into weakly ionized organic acids in the suppressor column.

Table 3
DETERMINATION OF B, AS,[a] CR, MO, W, AND SE IN WATERS BY SUPPRESSED ION CHROMATOGRAPHY

Applications	Experimental conditions and remarks	Ref.
Determination of B in water	Dionex® Model 14 ion chromatograph Sep.C: 50 × 0.3 cm brine anion separator Sup.C: 25 × 0.6 cm Eluent: 0.004 *M* Na_2CO_3 - 0.003 *M* $NaHCO_3$ Flow rate: 138 mℓ / hr With a standard separation column (15 × 0.3 cm) the eluent is 0.0024 *M* Na_2CO_3 - 0.003 *M* $NaHCO_3$. Before the IC determination the neutral or alkaline solution of the sample (containing 5—100 μg of B) is passed through a column (4 × 0.75 cm) containing ≃1 mℓ of Amberlite® XE-243 resin (40-80 mesh; free-base form) using a flow rate of 1 mℓ/min; then, the resin on which B is adsorbed as borate is washed in this order with water (2 mℓ), 3 *M* ammonia solution (3 mℓ), water (2 mℓ), and finally with 10% HF solution (2 mℓ) to convert borate to tetrafluoroborate; after 10—15 min the column is rinsed with water (2 mℓ) and 0.3 *M* ammonia solution and again with water (2 mℓ); finally, BF_4^- is eluted with 1 *M* NaOH (5 mℓ) followed by water (5 mℓ) and the eluate is diluted with water to 50 mℓ to prepare the sample solution which is then injected into the IC The same technique can be used to determine B in nuclear fuel dissolvent solutions A modified version of this IC procedure as well as a method based on the use of an IC exclusion separator column has been employed in connection with the determination of B in borax, howlite, kernite, silt, urine, and blood;[376] the ion exclusion process is superior when a Dionex® IC exclusion separator column is used, with 0.1 *M* mannitol - 0.001 *M* HCl as eluent	133, 134
Determination of As, Mo, W, Se, and Cr in river or main water and sewage	Dionex® Model 10 or 14 ion chromatograph Flow rate: 230 mℓ/hr Sep.C: 50 × 0.3 cm (Dionex® low capacity or brine column) Sup.C: 25 × 0.6 or 10 × 0.9 cm Eluent A: 0.003 *M* Na_2CO_3 (separates SeO_4^{2-}, AsO_4^{3-}, WO_4^{2-}, MoO_4^{2-}, and CrO_4^{2-}) Eluent B: 0.005 *M* Na_2CO_3 (is used in the absence of SeO_4^{2-}; analysis time is reduced from 45 to 25 min) Eluent C: 0.01 *M* Na_2CO_3 (if the sample contains 50 μg/mℓ of SO_4^{2-} and NO_3^-; further reduction of analysis time to 15 min) Eluent D: 0.003 *M* Na_2CO_3 - 0.005 *M* KOH (for determining As, Mo, W, and Cr in the presence of large amounts of SO_4^{2-} and NO_3^-) Eluent E: 0.0025 *M* Na_2CO_3 - 0.002 *M* KOH (this eluent is used for the determination of Se in waters) With increase in pH of the eluent, the retention times for SeO_4^{2-}, MoO_4^{2-}, WO_4^{2-}, and CrO_4^{2-} decrease, while that for AsO_4^{3-} increases; the detection limits of W, Mo, As, Se, and Cr are 0.35, 0.25, 0.10, 0.01, and 0.05 μg/ℓ, respectively; there is no interference from 5000-fold or greater amounts of SO_4^{2-} or other inorganic ions	146—148
Determination of Mo and W in natural waters	Dionex® Model 14 ion chromatograph Sample size: 1 mℓ	152

Table 3 (continued)
DETERMINATION OF B, AS,[a] CR, MO, W, AND SE IN WATERS BY SUPPRESSED ION CHROMATOGRAPHY

Applications	Experimental conditions and remarks	Ref.
	Sep.C: 25 × 0.3 cm Sup.C: 25 × 0.6 cm Eluent: 0.006 *M* Na_2CO_3 Flow rate: 90 mℓ/hr WO_4^{2-} is eluted before MoO_4^{2-}; coadsorbed vanadate is eluted ahead of these two anions; the limit of determination is ≃1 μg/ℓ; before IC, the sample (1 ℓ) is adjusted to pH 1—2 and Mo plus W are adsorbed on a 1-cm column containing 2 g of Chelex® 100 (H^+ form) using a flow rate of ≯ 5 mℓ/min; subsequently, Mo and W are eluted with conc ammonia solution (7 mℓ) followed by water (10 mℓ); this eluate is evaporated until no ammonia fumes are detected and then diluted to 10 mℓ with water to prepare the sample solution About 90% of the W and Mo is recovered	
Determination of Se in waters	Dionex® Model 16 ion chromatograph with conductivity detector and connected to a graphite furnace atomic absorption spectrometer (AAS) Sample size: 1 mℓ Pre.C: 5 × 0.3 cm Sep.C: 15 × 0.3 cm Sup.C: 25 × 0.3 cm Eluent: 0.008 *M* Na_2CO_3 (elution of selenite, selenate, Cl^-, NO_2^-, NO_3^-, Br^-, SO_4^{2-}, PO_4^{3-}, and other anions) Flow rate: 0.46 mℓ/min The eluate is passed through the conductivity detector and is then routed to the AAS which is used as a Se-specific detector (the detection limit is 20 ng for Se for each of the species); in the presence of greater amounts of Cl^- and/or SO_4^{2-}, Se cannot be determined conductometrically	154

Note: Pre.C = precolumn, Sep.C = separation column, Sup.C = suppressor column, and IC = ion chromatography.

[a] See also Reference 387.

Applications

Geological Materials

Suppressed ion chromatography has very frequently been employed in connection with the determination of anions in rocks (see Table 1),[88,89,91,159] sediments (see Table 1),[88,128] and other solid geological materials which include soils (see Table 1),[90,112,126] shales and coal (see Table 1),[50,88,89,92,117] and mineral such as pyrite (see Table 1) and scheelite.[167] Even more often, this technique has been utilized for the assay of anions in waters (see Tables 2 and 3),[11,12,30,59,62,69,79,82,85,86,93,119,121,122,129-134,141,146-148,152,154,160,167-174] and air and other gases (see Tables 4 to 6).[1,49,95,96,99,114,118,125,135-138,141,145,149,157,158,161,175-184]

Biological-Organic Materials

In addition to the methods presented in Tables 7 to 9 suppressed ion chromatography has also been utilized for the assay of anions in plasma,[185] human saliva and sweat,[186] beer,[87] fruit juice,[69] foods[164] including meat and meat products,[71] oyster,[14] plants,[187] fuel oil,[116,188] waste oil,[14] and organic compounds[70,111,189] other than those listed in Table 9.

Table 4
DETERMINATION OF HALOGENS AND OTHER ELEMENTS IN AIR AND AIR-PARTICULATE MATTER BY SUPPRESSED ION CHROMATOGRAPHY[a]

Applications	Experimental conditions and remarks	Ref.
Fluoride, chloride, carbonate, nitrate, and sulfate in coal fly ash	Dionex® Model 10 ion chromatograph Sample size (loop): 0.1 mℓ Sep.C: 25 × 0.3 cm (Chromex® resin) Sup.C: 50 × 0.6 cm (Chromex® resin) Eluent: 0.0024 *M* Na_2CO_3 - 0.003 *M* $NaHCO_3$ Flow rate: 92 mℓ/hr The order of elution is F^-, Cl^-, CO_3^{2-}, NO_3^-, and SO_4^{2-} Before IC the sample (22.5 g) is leached for 3 hr with water (90 mℓ) using sonic agitation The precision of the method is ±3% and the accuracy ±5%; detection limits are 0.05, 0.07, 0.1, 0.1, and 150 μg/mℓ for F^-, Cl^-, SO_4^{2-}, NO_3^-, and CO_3^{2-}, respectively	94
Determination of fluoride, chloride, nitrite, nitrate, and sulfate in atmospheric aerosol	Dionex® Model 10 ion chromatograph Sample size (loop): 0.1 mℓ Pre.C: 1.5 × 0.3 cm Sep.C: 50 × 0.3 cm Sup.C: 25 × 0.6 cm Eluent: 0.0024 *M* Na_2CO_3 - 0.003 *M* $NaHCO_3$ Flow rate: ~3 mℓ/min Order of elution: Cl^-, NO_3^-, SO_4^- Before IC the atmospheric aerosol components are extracted from the collection filter by the same solution which is used as eluent Extract = sample solution	95
Determination of As and S in flue dust	Dionex® Model 10 ion chromatograph Sample size (loop): 0.1 mℓ Pre.C: 15 × 0.3 cm Sep.C: 25 × 0.3 cm Sup.C: 25 × 0.6 cm Eluent: 0.0035 *M* Na_2CO_3 - 0.0026 *M* NaOH Flow rate: 1.15 mℓ/min The anions are eluted in the order F^-, Cl^-, NO_2^-, SO_3^{2-}, SO_4^{2-},PO_4^{3-}, and AsO_4^{3-} Before IC, a 5-mℓ extract of the sample is mixed with 30% H_2O_2 (5 mℓ), 50% NaOH solution (0.1 mℓ), and MnO_2 (5 mg) to oxidize all As to AsO_4^{3-}	135
Determination of gaseous and particulate fluoride in air	Dionex® Model 10 ion chromatograph Sample size (loop): 0.1 mℓ Pre.C: 25 × 0.3 cm Sep.C: 50 × 0.3 cm Sup.C: 25 × 0.6 cm Eluent: 0.003 *M* $NaHCO_3$ This method makes it possible to detect down to 2 ppb of F^-; analysis time; ≃15 min	157
Determination of I_2 in industrial atmospheres	Dionex® Model 14 ion chromatograph Sample size (loop): 0.1 mℓ Pre.C: 12.5 × 0.3 cm Sep.C: 25 × 0.3 cm Sup.C: 25 × 0.6 cm Eluent: 0.01 *M* Na_2CO_3 Flow rate: 184 mℓ/hr For a 15-ℓ sample, the detection limit for I_2 was 0.17 mg/m³; before IC the air is drawn through charcoal impregnated with alkali metal hydroxide; the charcoal is then extracted with	149

Table 4 (continued)
DETERMINATION OF HALOGENS AND OTHER ELEMENTS IN AIR AND AIR-PARTICULATE MATTER BY SUPPRESSED ION CHROMATOGRAPHY[a]

Applications	Experimental conditions and remarks	Ref.
	0.01 *M* Na_2CO_3 in an ultrasonic bath, and the extract is filtered and analyzed for I_2 by IC	

Note: Pre.C = precolumn, Sep.C = separation column, Sup.C = suppressor column, and IC = ion chromatography.

[a] See also References 388 to 392.

Industrial Products

From the procedures outlined in Table 10 it is seen that suppressed ion chromatography has been used for the determination of anions in a variety of industrial products. Additional applications include analyses of photographic materials (silver halides),[123,190] glasses,[124] cosmetic products,[156] color additives,[120] smelter flue dust,[165] concentrated electrolytes,[191] electrolyte from the halogen tin-plating process,[127] utility waste liquors and leachates,[10] U leach liquors,[59] and pulping and bleaching liquors.[4,192]

Synthetic Mixtures

To determine anions in synthetic mixtures the procedures outlined in Table 11 and other methods can be used.[49,58,110,153,193]

Cation Analysis

General

A flow diagram of an ion chromatograph based on the Dionex® system, and which is suitable for cation analysis, is presented in Figure 4. The sample solution in very dilute nitric acid or HCl (e.g., 0.0025 to 0.01 *M*) is injected via the sample loop into a flow of eluent and thus carried to the separator column containing a cation exchange resin in the H form where selective sorption and elution of the cations occurs. This adsorption and desorption can be illustrated by the reaction shown in Figure 4 and, as is evident from the following selectivity scale, the greater the affinity of a cation for the cation exchanger the longer it will require to elute it from the separator column.

Increased affinity to cation exchange resin →

$$Li^+ < Na^+ < NH_4^+ < K^+ < Cs^+ < Mg^{2+} < Ca^{2+} < Sr^{2+} < Ba^{2+}$$

← decreasing retention time (elution time)

The effluent from the separator column then passes to the suppressor column containing a strongly basic resin in the hydroxide form. In this, the ion exchange reactions shown in Figure 4 take place. Thus, highly ionized metal hydroxides reach the conductance cell in a background of water so that the variation in conductivity can be readily detected and determined by the use of standards (see preceding section).

Columns

Separator Column

This column contains a low-capacity cation exchange resin which is obtained by superficial sulfonation of styrene-DVB copolymer beads. The resin beads are treated with concentrated sulfuric acid and a thin layer of sulfonic acid groups ($-SO_3H$) is formed on the surface. The

Table 5
DETERMINATION OF S COMPOUNDS IN AIR BY SUPPRESSED ION CHROMATOGRAPHY[a]

Applications	Experimental conditions and remarks	Ref.
Determination of S dioxide in air	Dionex® Model 14 ion chromatograph Sample size (loop): 0.1 mℓ Pre.C: 12.5 × 0.3 cm Sep.C: 50 × 0.3 cm Sup.C: 25 × 0.6 cm Eluent: 0.003 M Na_2CO_3 - 0.003 M $NaHCO_3$ Flow rate: 138 mℓ/hr Before IC, air containing SO_2 is passed for 4 hr at a rate of up to 200 mℓ/min through a glass tube (70 × 0.4 cm) containing 100- and 50-mg portions of impregnated charcoal (No. 580-19; 20 × 40 mesh; Barneby-Cheney Co., Columbus, Ohio) in adsorption and backup sections, respectively; on the charcoal SO_2 is oxidizing to SO_4^{2-} which is extracted, with a solution of the same composition as the eluent, using ultrasonic agitation; subsequently, the coal is filtered off to prepare the sample solution	136
Determination of sulfuryl fluoride in air	Dionex® Model 10 ion chromatograph Sample size (loop): 0.1 mℓ Sep.C: 50 × 0.3 cm Sup.C: 25 × 0.6 cm Eluent: 0.0035 M Na_2CO_3 - 0.004 M NaOH Flow rate: 1.8 mℓ/min The range of determination is 1—10 ppm The average recovery is 95.8%, with a coefficient of variation of 11%; the sampling period is 4 hr Before IC the air is pumped through a tube containing activated charcoal to adsorb SO_2F_2; subsequently, this compound is extracted into NaOH solution in which it is hydrolyzed according to the reaction: $SO_2F_2 + 2NaOH \rightarrow NaF + NaSO_3F + H_2O$; the fluorsulfonate is then further hydrolyzed by evaporation of a portion of the hydrolysis mixture, and NaF and Na_2SO_4 (formed by the reaction: $NaSO_3F + 2NaOH \rightarrow Na_2SO_4 + NaF + H_2O$) are determined by IC	137
Determination of sulfuric acid and ammonium sulfates in the atmosphere	Ion chromatograph Sample size (loop): 0.1 mℓ Pre.C: 5 × 0.46 cm Sep.C: 25 × 0.46 cm Sup.C: hollow fiber suppressor Eluent: 0.004 M Na_2CO_3 - 0.004 M $NaHCO_3$ Flow rate: 2.5 mℓ/min The detection limit of sulfate is 0.005 μg/m³ at 40 m³ air sample volume Before IC, the sulfate aerosol is collected on a PTFE (Teflon®) filter and extracted with benzaldehyde and 2-propanol to separate sulfuric acid from the ammonium sulfates and NH_4HSO_4 from $(NH_4)_2SO_4$, respectively	138

Note: Pre.C = precolumn, Sep.C = separation column, Sup.C = suppressor column, and IC = ion chromatography.

[a] See also References 392 and 413.

final capacity of the resin is related to the thickness of the layer and is dependent on the bead diameter, the type of resin, and the temperature and time of contact with the acid. Typical capacities range from 0.005 to 0.1 meq/g as compared to 5 meq/g for conventional,

Table 6
DETERMINATION OF N COMPOUNDS IN AIR BY SUPPRESSED ION CHROMATOGRAPHY[a]

Applications	Experimental conditions and remarks	Ref.
Determination of N oxides in exhaust gases	Dionex® Model 16 ion chromatograph Sample size (loop): 0.1 mℓ Sep.C: 65 × 0.3 cm Sup.C: 25 × 0.6 cm Eluent: 0.0024 *M* Na_2CO_3 - 0.003 *M* $NaHCO_3$ Flow rate: 2 mℓ/min Anions are eluted in the order F^-, Cl^-, NO_2^-, NO_3^-, and SO_4^{2-} One determination takes ≃15 min; recovery of >95% of NO is achieved without interference Before IC, the sample (150 mℓ, e.g., of boiler or waste incinerator exhaust gas) is drawn into a 200-mℓ syringe, and 30 mℓ of O_2 is added; the gas mixture is shaken for 5 min with 0.5 *M* NaOH (5 mℓ) and 0.3% H_2O_2 solution (20 mℓ) to oxidize N oxides to nitrate A similar method can be used for the determination not only of N oxides, but also of S oxides and HCl in exhaust gases;[378] in this case the elution of Cl^-, NO_3^-, and SO_4^{2-} in this order is effected on a smaller separation column, i.e., 25 × 0.46 cm using 0.004 *M* Na_2CO_3 - 0.004 *M* $NaHCO_3$ as eluent (injection volume: 10 μℓ)	96, 377
Determination of HCN in air	Dionex® Model 16 ion chromatograph Sample size (loop): 0.1 mℓ Pre.C: 15 × 0.3 cm Sep.C: 50 × 0.3 cm Sup.C: 25 × 0.6 cm Eluent: 0.005 *M* $Na_2B_4O_7$ (elutes formic acid) Flow rate: 138 mℓ/hr Before IC, air (40 ℓ; filtered if necessary) is drawn at 1.5 ℓ/min through 0.2 *M* NaOH (15 mℓ) in a small impinger; this solution and rinse (2 mℓ of 0.2 *M* NaOH) is heated at 110°C for 24 hr to convert NaCN into Na formate (according to the reaction $NaCN + 2H_2O \rightarrow NH_3 + HCOONa$) and then diluted with water to 50 mℓ to prepare the sample solution	161

Note: Pre.C = precolumn, Sep.C = separation column, Sup.C = suppressor column, and IC = ion chromatography.

[a] See also Reference 413.

high-capacity resins of the sulfonic acid type. The superficially sulfonated resins show ion exchange properties similar to those of the pellicular resins (see preceding section) with respect to diffusion path and speed of mass transfer. Also, because of the rigidity of the resin core, there is less tendency for the bead to compress. This means that higher flow rates (at relatively low back pressures) can be used than would be possible with conventional resins. Superficially sulfonated resins are stable over the pH range of 1 to 14 and swelling is minimal. The selectivity of the resin is similar to that observed for strongly acidic resins of the sulfonic acid type. In place of a low-capacity resin, the separator column (13 × 0.1 cm) can be packed with a 1:2 mixture of Aminex® A-5 (13 μm; Na^+ form) and Dowex® 1-X8 (200-400 mesh; Cl^- form) using 0.005 *M* HCl and 0.0012 *M* *m*-phenylenediamine dihydrochloride-0.00125 *M* HCl as eluents for monovalent cations (alkalies) and divalent cations (earth alkali metals), respectively.[64]

Table 7
DETERMINATION OF COMMON ANIONS IN BIOLOGICAL MATRICES BY SUPPRESSED ION CHROMATOGRAPHY[a]

Applications	Experimental conditions and remarks	Ref.
Determination of sulfate in plasma, cerebrospinal fluid, and hepatic tissue	Dionex® Model 10 ion chromatograph Sample size (loop): 1—1.5 mℓ Pre.C: 5 × 0.3 cm Sep.C: 50 × 0.3 cm Sup.C: 25 × 0.6 cm Eluent: 0.0024 *M* Na_2CO_3 - 0.003 *M* $NaHCO_3$ Flow rate: 140 mℓ/hr The anions are eluted in the order Cl^-, PO_4^{3-}, and SO_4^{2-}; the average recovery of SO_4^{2-} is ±101 ± 3% Before IC, the sample of blood, serum, or plasma is diluted to 1—1.5 mℓ with 0.001 *M* NaOH and to a sulfate concentration of 5—25 μ*M;* liver tissue is homogenized in 0.001 *M* NaOH, the mixture is centrifuged, and the supernate is diluted tenfold with 0.001 *M* NaOH	97
Determination of chloride, phosphate, bromide, nitrate, and sulfate in human serum and urine	Dionex® Model 10 ion chromatograph Sample size (loop): 0.1 mℓ Pre.C: 5 × 0.4 cm (fast run) Sep.C: 25 × 0.4 cm (fast run) Sup.C: 6 × 0.6 cm Eluent: 0.002 *M* Na_2CO_3 - 0.003 *M* $NaHCO_3$ Flow rate: 2.6 mℓ/min For each anion the coefficient of variation is ≤1.3%	139
Determination of phosphate, bromide, nitrate, and sulfate in human serum	Dionex® Model 14 ion chromatograph Sample size (loop): 0.1 mℓ Pre.C: 5 × 0.4 cm Sep.C: 25 × 0.4 cm Sup.C: 25 × 0.6 cm Eluent: 0.0024 *M* Na_2CO_3 - 0.003 *M* $NaHCO_3$ Flow rate: 180 mℓ/hr	98
Determination of phosphate in urine	Dionex® Model 14 ion chromatograph Sample size (loop): 0.1 mℓ Pre.C: 15 × 0.3 cm Sep.C: 50 × 0.3 cm Sup.C: 25 × 0.3 cm Eluent: 0.0024 *M* Na_2CO_3 - 0.003 *M* $NaHCO_3$ Flow rate: 115 mℓ/hr Before IC, the urine sample is diluted 300-fold and then filtered	100

Note: Pre.C = precolumn, Sep.C = separation column, Sup.C = suppressor column, and IC = ion chromatography.

[a] See also References 393, 394, and 406.

Suppressor Column

This column contains a strongly basic anion exchange resin which is in the hydroxide form, provided that the eluent is a dilute acid, the dihydrochloride of an organic diamine, or a dilute solution of zinc nitrate (see Tables 12 and 13). Eluents consisting of dilute solutions of Ba or Pb nitrate are suppressed on anion exchange resins in the sulfate and iodate forms, respectively (see Table 13). From Figure 4 it is seen that on a resin in the hydroxide form highly conducting metal hydroxides are formed in a background consisting essentially of water. When using as eluents the solutions of dihydrochlorides of phenylenediamine or other diamines, the effluent from the suppressor column also contains the free amines (chloride is retained) which, however, do not interfere with conductometry.

Table 8
DETERMINATION OF BROMATE IN BREAD[80,162,163] AND FISHPASTE[80,162] BY SUPPRESSED ION CHROMATOGRAPHY

Experimental Conditions and Remarks

Dionex® Model 10 ion chromatograph
Sample size (loop): 0.1 mℓ; flow rate: 115 mℓ/hr
Pre.C: 15 × 0.3 cm; Sep.C: 50 × 0.3 cm
Sup.C: 25 × 0.3 cm; exclusion Sup.C: 14 × 0.4 cm (Ag^+ form)
Concentrator column: 5 × 0.3 cm
Eluent A: 0.0024 *M* Na_2CO_3 - 0.003 *M* $NaHCO_3$
Eluent B: 0.005 *M* $Na_2B_4O_7$ - 0.003 *M* $NaHCO_3$
Cl^-, PO_4^{3-}, and SO_4^{2-}, which interfere with the determination of BrO_3^-, are removed by the initial four columns (Cl^- by precipitation on the Ag^+-form resin) using eluent A; bromate is determined after elution from the concentrator column with eluent B
The detection limit is 1—2 ppm of BrO_3^-

Dionex® Model 14 ion chromatograph
Sample size: 0.3 mℓ; flow rate: 117 mℓ/hr
Sep.C: 100 × 0.3 cm; Sup.C: 25 × 0.6 cm
Eluent: 0.0035 *M* $Na_2B_4O_7$ (bromate is eluted before chloride)
Before IC of bromate, the freeze-dried, powdered sample (2 g) is extracted ultrasonically at 20°C for 10 min with water (20 mℓ); the aqueous extract is diluted to 50 or 100 mℓ with water and an aliquot is ultrasonically mixed with Dowex® 50W-X8-10 (Ag^+ form) to remove chloride
The detection limit is 0.5 μg BrO_3^- per milliliter

Note: Pre.C = precolumn, Sep.C = separation column, Sup.C = suppressor column, and IC = ion chromatography.

Table 9
DETERMINATION OF COMMON ANIONS AND CHROMATE IN ORGANIC MATERIALS BY SUPPRESSED ION CHROMATOGRAPHY

Applications	Experimental conditions and remarks	Ref.
Determination of halogens and S in organic compounds	Ion chromatograph Sep.C: 5 × 0.3 cm (MCI Gel CAO4S/Hitachi® Gel 3011-0 [4:9]) Sup.C: 15 × 0.8 cm (Hitachi® Custom 2612) Eluent: 0.01 *M* Na_2CO_3 Flow rate: 2 mℓ/min The order of elution is F^-, Cl^-, Br^-, NO_3^-, PO_4^{3-}, and SO_4^{2-} and these anions can be determined within 15 min with detection limits of 0.01—0.8 ppm	150
Determination of sulfate in antibiotics	Dionex® Model 14 ion chromatograph Sample size (loop): 0.1 mℓ Sep.C: 15 × 0.3 cm Sup.C: 25 × 0.6 cm Eluent:0.0024 *M* Na_2CO_3 - 0.003 *M* $NaHCO_3$ Flow rate: 2.25 mℓ/min	101

Note: Sep.C = separation column and Sup.C = suppressor column.

Because the function of the suppressor column in cation analysis is similar to that described for anions (see preceding section) the same considerations apply with respect to the choice of resin. Any type of anion exchange resin of high capacity can be used as long as it does not affect separations that have been achieved in the separator column. To inhibit most of the side effects on the analyte cations during their passage through the suppresssor column,

Table 10
DETERMINATION OF ANIONS IN INDUSTRIAL PRODUCTS BY SUPPRESSED ION CHROMATOGRAPHY

Applications	Experimental conditions and remarks	Ref.
Determination of fluoride and chloride in Ta	Dionex® Model 14, ion chromatograph Sample size (loop): 0.1 mℓ Sep.C: 50 × 0.3 cm Sup.C: Dionex® No. 0.30015 Eluent: 0.0024 *M* Na_2CO_3 - 0.003 *M* $NaHCO_3$ Flow rate: 150 mℓ/hr Fluoride is eluted before chloride With 1-g samples the detection limit is 1 ppm for each element and the coefficient of variation is 6% for chloride and 2% for fluoride Before IC, F^- and Cl^- are isolated from the sample (0.5 g) by means of a pyrolytic technique	102, 103
Determination of nitrite in metal cutting fluids	Dionex® Model 16 ion chromatograph Sample size (loop): 500 μℓ Sep.C: 25 × 0.4 cm Sup.C: 7 × 0.6 cm Eluent: 0.0024 *M* Na_2CO_3 - 0.003 *M* $NaHCO_3$ Flow rate: 3 mℓ/min The sample is diluted 1000-fold with water for analysis and bromide (as KBr) is added as internal standard Down to 0.04 ppm of nitrite can be determined with a recovery of >97%	104
Determination of hydrofluoric, hydrochloric, nitric, and sulfuric acids in alloy pickling baths	Dionex® Model 10 ion chromatograph Sample size: 25.6 or 100 μℓ Pre.C: 15 × 0.3 cm Sep.C: 25 × 0.3 cm or 50 × 0.3 cm Sup.C: 25 × 0.6 cm Eluent: 0.0024 *M* Na_2CO_3 - 0.003 *M* $NaHCO_3$ Flow rate: 180 mℓ/hr Fluoride is eluted before chloride and nitrate	105
Determination of chloride and sulfate in chloridedoped Cd sulfide	Dionex® Model 10 ion chromatograph Sample size: 0.1 mℓ Pre.C: 15 × 0.3 cm Sep.C: 50 × 0.3 cm Sup.C: 25 × 0.6 cm Eluent: 0.0018 *M* Na_2CO_3 - 0.003 *M* $NaHCO_3$ (pH 9.75) Flow rate: 2 mℓ/min Chloride is eluted before sulfate Before the separation, the sample (100 mg) is dissolved in a 1:1 mixture (20 mℓ) of the eluent and 30% H_2O_2 and then diluted to 1 ℓ to obtain the sample solution; for the determination of soluble chloride and sulfate the sample (100 mg) is leached ultrasonically for 20 min with the eluent (100 mℓ)	140
Determination of chloride in liquid bromine	Dionex® Model 10 ion chromatograph Sample size (loop): 0.1 mℓ Sep.C: 45 × 0.3 cm Sup.C: 50 × 0.3 cm Eluent: 0.002 *M* Na_2CO_3 - 0.002 *M* $NaHCO_3$ Flow rate: 1.5 mℓ/min Chloride is eluted before bromide Before the separation, the sample (0.2 mℓ) is shaken for 1.5 min with 0.01% KBr solution (20 mℓ) and free Br is removed by extraction with CCl_4 (4 × 20 mℓ)	60

Table 10 (continued)
DETERMINATION OF ANIONS IN INDUSTRIAL PRODUCTS BY SUPPRESSED ION CHROMATOGRAPHY

Applications	Experimental conditions and remarks	Ref.
Determination of chloride in Na hydroxide and sulfuric acid	Dionex® Model 16 ion chromatograph Sample size (loop): 0.1 mℓ Sep.C: 50 × 0.3 cm Sup.C: 25 × 0.6 cm Eluents: 0.0024 M Na_2CO_3 - 0.003 M $NaHCO_3$ (for H_2SO_4 samples) and 0.01 M $Na_2B_4O_7$ (for NaOH samples) Flow rate: 180 mℓ/hr Before IC, sulfuric acid is diluted to ≤600 ppm, while NaOH is diluted to ≃0.8 M	106
Determination of chloride and sulfate in chromic acid	Dionex® ion chromatograph Sep.C: 5 × 0.4 cm Sup.C: 25 × 0.6 cm Eluent: 0.0024 M Na_2CO_3 - 0.003 M $NaHCO_3$ Detection limits are 0.1 ppm for Cl^-, F^-, and PO_4^{3-} and 1 ppm for SO_4^{2-} in a 0.1% solution of H_2CrO_4; adsorbed CrO_4^{2-} is eluted with 0.0054 M Na_2CO_3	107
Determination of As in material from a Cu smelter	Ion chromatograph Sample size: 0.1 mℓ Sep.C: 50 × 0.3 cm Sup.C: 25 × 0.6 cm Eluent: 0.0024 M Na_2CO_3 - 0.003 M $NaHCO_3$ Flow rate: 1.5 mℓ/min The anions are eluted in the order F^-, Cl^-, NO_3^-, AsO_2^-, and SO_4^{2-} The method was applied to aqueous solutions containing 2.5—12.8 μg/mℓ of As(V)	115
Determination of fluoride, chloride, sulfate, and phosphate in waste water precipitate	Ion chromatograph Sample size: 60 μℓ Sep.C: 50 × 0.3 cm Sup.C: 25 × 0.6 cm Eluent: 0.0017 M Na_2CO_3 - 0.0021 M $NaHCO_3$ Flow rate: 2 mℓ/min Recoveries of the anions are 91 to 96% and detection limits are 8, 4, 5, and 10 μg for F^-, Cl^-, PO_4^{3-}, and SO_4^{2-}, respectively Before IC, the sample (~40 mg of precipitate plus membrane on which it was collected) is fused with Na_2CO_3 (1 g) and the aqueous solution of the melt is passed through a column (100 mℓ Nalgene buret) containing Bio-Rad® AG50W-X12 (H^+ form) to convert Na_2CO_3 into CO_2; the effluent and washings (≃150 mℓ water) are evaporated to ≃5 mℓ and this concentrated solution plus 0.21 M $NaHCO_3$ - 0.17 M Na_2CO_3 (0.1 mℓ) is diluted with water to 100 mℓ to prepare the sample solution	142
Determination of chloride and bromide in process streams of nuclear fuel reprocessing plants	Dionex® Model 10 ion chromatograph Sample size (loop): 0.1 mℓ Pre.C: 15 × 0.3 cm Sep.C: 50 × 0.3 cm Sup.C: 25 × 0.6 cm Eluent: 0.0024 M Na_2CO_3 - 0.003 M $NaHCO_3$ Flow rate: 138 mℓ/hr Chloride is eluted before the bromide Before IC, the aqueous sample (1.7 M in H^+ and containing, e.g., Zr, Al, Sn, and anionic constituents) is passed through a column (3 × 0.4 cm) of Bio-Rad® AG50W-X8 (Ag^+	108

Table 10 (continued)
DETERMINATION OF ANIONS IN INDUSTRIAL PRODUCTS BY SUPPRESSED ION CHROMATOGRAPHY

Applications	Experimental conditions and remarks	Ref.
	form); the precipitated AgCl and AgBr are dissolved in ammonia solution and Ag is removed by passage of the solution through a Jones reductor; the ammoniacal effluent is the sample solution in which Cl and Br are determined by IC	
Determination of anions in boiler blowdown water	Ion chromatograph Sample size: 0.5 mℓ Sep.C: 100 × 0.28 cm Sup.C: 30 × 0.28 cm Eluent: 0.005 *M* Na_2CO_3 - 0.004 *M* NaOH Flow rate: 105 mℓ/hr Determined were Cl^-, PO_4^{3-}, SO_3^{2-}, and SO_4^{2-}	143

Note: Pre.C = precolumn, Sep.C = separation column, Sup.C = suppressor column, and IC = ion chromatography.

Table 11
DETERMINATION OF ANIONS IN SYNTHETIC MIXTURES BY SUPPRESSED ION CHROMATOGRAPHY

Applications	Experimental conditions and remarks	Ref.
Analysis of synthetic mixtures of chloride, phosphate, nitrate, and sulfate	Dionex® Model 14 ion chromatograph Sample size: 10 mℓ Concentrator column: 5 × 0.3 cm Sep.C: 50 × 0.3 cm Sup.C: 25 × 0.3 cm Eluent: 0.0024 *M* Na_2CO_3 - 0.003 *M* $NaHCO_3$ Flow rate: 138 mℓ/min Cl^-, PO_4^{3-}, NO_3^-, and SO_4^{2-} are eluted in this order	58
Separation of hypophosphite, phosphite, and phosphate	Dionex® Model 10 ion chromatograph Sample size (loop): 0.1 mℓ Pre.C: 5 × 0.3 cm Sep.C: 15 × 0.3 cm Sup.C: 25 × 0.6 cm Eluent: 0.0024 *M* Na_2CO_3 - 0.003 *M* $NaHCO_3$ (pH 10.3) Flow rate: 112 mℓ/hr The order of elution is $H_2PO_2^-$, HPO_3^{2-}, and HPO_4^{2-} The method is applicable to 0.2—0.3 mg/ℓ of P and was applied to determine hypophosphite and phosphite in plating solutions	110
Separation of tungstate, molybdate, chromate, tellurate, and tellurite	Dionex® Model 14 ion chromatograph Sample size (loop): 0.1 mℓ Eluent: 0.006 *M* Na_2CO_3 Flow rate: 2.3 mℓ/min Tungstate is eluted first and then MoO_4^{2-} and CrO_4^{2-}; TeO_4^{2-} is eluted sooner than TeO_3^{2-} With 0.0048 *M* Na_2CO_3 - 0.006 *M* $NaHCO_3$ as eluent, EDTA and PO_4^{3-} could be separated similarly	153

Note: Pre.C = precolumn, Sep.C = separation column, and Sup.C = suppressor column.

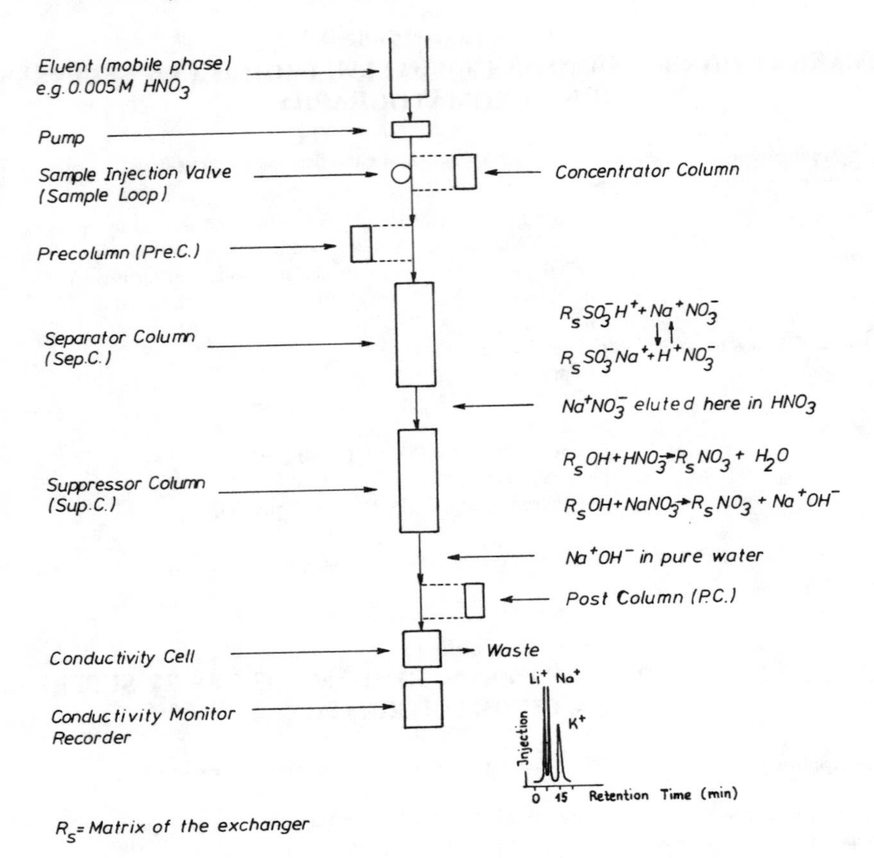

FIGURE 4. System for cation analysis by suppressed ion chromatography using conductometric detection (Dionex® system).

microporous resins of the type Dowex® 1-X10 are usually employed. These moderately cross-linked resins minimize incomplete Donnan exclusion of weakly basic solutes, such as the ammonium ion, which are converted into their free bases by the hydroxide-form suppressor column. The larger the pore volume of the resin (the lower the cross-linkage), the more pronounced is the tendency of such species to enter the resin phase. Retention times and consequently peak heights of weak bases can be further influenced by the degree of exhaustion of the suppressor column, i.e., the extent to which resin hydroxide has been neutralized.

When the column is exhausted (its active sites have been completely used up and eluent anions are not longer removed), it is regenerated by pumping 0.1 *M* Na hydroxide through the column.

Postcolumn

This column contains a strongly acidic cation exchange resin in the H form (e.g., Dowex® 50W-X12 or Bio-Rad® AG50W-X16) and is placed between the suppressor column and detector (see Table 13 and Figure 4).[194-196] It serves to minimize the effect of pH on the conductometric base line and amplifies divalent cation peak heights by a factor of approximately 5, thus greatly increasing the sensitivity of the technique. In other words, the cations, e.g., Mg and Ca, contained in the effluent from the suppressor column, when adsorbed on this postcolumn, liberate equivalent amounts of H ions which show much higher conductivity than the metal ions themselves.

Table 12
DETERMINATION OF AMMONIA AND ALKALI METALS BY SUPPRESSED ION CHROMATOGRAPHY[a]

Applications	Experimental conditions and remarks	Ref.
Determination of ammonia and lower amines in the atmosphere	Dionex® Model 14 ion chromatograph Sample size (loop): 0.1 mℓ Sep.C: 25 × 0.6 cm Sup.C: 25 × 0.9 cm (Dowex® 1-X8; OH^- form) Eluent: 0.0025 *M* HNO_3 Flow rate: 110 mℓ/hr Before IC, air is passed at 6 ℓ/min through a glass fiber filter (area 19.6 cm^2) impregnated with 8% oxalic acid - 10% glycerol solution; the NH_3 and lower amines (methyl-, dimethyl-, and trimethylamine) collected are extracted with water and the solution is subjected to IC Similar methods have been described for the determination of NH_3 and amines in aqueous solutions[379] and in air;[205] the latter was passed (at 1 ℓ/min for 10 min or 100 mℓ/min for ≤7.5 hr) through sampling tubes containing 780 mg of silica gel (some tubes were pretreated by injecting 200 μℓ of 25% H_2SO_4) Samples collected with the pretreated tubes could be stored at room temperature for ≤21 days and samples collected by untreated tubes could be kept in a refrigerator for up to 32 days; after extraction with 0.1 or 0.2 *N* H_2SO_4 in aqueous 90% methanol (for pretreated or untreated samples, respectively), the extract was analyzed by IC with use of a Sep.C (23 × 0.9 cm) packed with surface sulfonated divinylbenzene-styrene (230—325 mesh) and a suppressor column (18 × 0.9 cm) packed with Bio-Rad® AG1-X10 (200-400 mesh; OH^- form) with 0.015 *M* HCl as eluent using a flow rate of 3.5 mℓ/min; recoveries of NH_3, mono-, di-, and trimethyl-amine were 91, 93, 92, and 84%, respectively	197
Determination of alkalies in geothermal well water	Dionex® Model 10 ion chromatograph Sample size (loop): 0.1 mℓ Sep.C: 25 × 0.6 cm Sup.C: 25 × 0.9 cm Eluent: 0.003 *M* HNO_3 Flow rate: 184 mℓ/hr The order of elution is Li^+, Na^+, NH_4^+, K^+, and Rb^+	131
Determination of Na and K in antibiotics	Dionex® Model 14 ion chromatograph Sample size (loop): 0.1 mℓ Pre.C: 15 × 0.3 cm Sep.C: 25 × 0.6 cm Sup.C: 25 × 0.9 cm Eluent: 0.005 *M* HNO_3 Flow rate: 3.0 mℓ/min	101

Note: Pre.C = precolumn, Sep.C = separation column, Sup.C = suppressor column, and IC = ion chromatography.

[a] See also Reference 395.

Table 13
DETERMINATION OF ALKALINE EARTH ELEMENTS BY SUPPRESSED ION CHROMATOGRAPHY[a]

Applications	Experimental conditions and remarks	Ref.
Determination of Mg and Ca in waters, soils, and serum	Dionex® Model 10 ion chromatograph Sample size (loop): 0.1 mℓ Sep.C: 15 × 0.3 cm (Ba^{2+} form) (conditioned with 3 × 0.1 mℓ of 0.1 *M* $BaCl_2$ followed by water) Sup.C: 25 × 0.6 cm (Bio-Rad® AG1-X10; 200-400 mesh; SO_4^{2-} form) (conditioned with 1.2 *M* Na_2SO_4 followed by water) PC: 15 × 0.3 cm (Bio-Rad® AG 50W-X12; 100-200 mesh; H^+ form) Eluent: 0.001 $Ba(NO_3)_2$ at pH 4 Flow rate: 2.3 mℓ/min Mg, Ca, and Sr are eluted in this order and then adsorbed on the PC liberating equivalent amounts of H^+ ions which are determined conductometrically; on the Sup.C the eluent is removed by the reaction: $(R_s)_2SO_4 + Ba(NO_3)_2 \rightarrow 2R_sNO_3 + BaSO_4$ Transition metals such as Mn(II), Fe(II), Co, Ni, Zn, Cd, and Cu are eluted together with the alkaline earths Regeneration of the Sup.C is effected by treatment with 0.5 *M* EDTA (tetrasodium salt) followed by water As eluent, also, 0.001 *M* $Pb(NO_3)_2$ can be used; this is removed on the Sup.C by precipitation as $PbSO_4$; when using a Sup.C in which the resin is in the IO_3^- form the Pb is precipitated as insoluble iodate;[195] in this case the Pb eluent is 0.0001 *M* in HNO_3	194
Determination of alkaline earth metals in synthetic mixtures	Dionex® Model 16 ion chromatograph Sample size (loop): 0.1 mℓ Pre.C: 15 × 0.3 cm (anion exchange resin) Pre.C: 15 × 0.3 cm (cation exchange resin) Sep.C: 25 × 0.6 cm (cation exchange resin) Sup.C: 50 × 0.3 cm (OH^--form anion exchanger) PC: 25 × 0.9 cm (H^--form resin) Eluent: 0.0025—0.005 *M* $Zn(NO_3)_2 \cdot H_2O$ - 0.003—0.004 *M* HNO_3 Flow rate: 40% Mg, Ca, Sr, and Ba are eluted in this order Na^+, K^+, and NH_4^+ are eluted, as a group, before the Mg; in the Sup.C the following reactions occur: $2R_sOH + Zn(NO_3)_2 \rightarrow 2R_sNO_3 + Zn(OH)_2$ and $R_sOH + HNO_3 \rightarrow R_sNO_3 + H_2O$ The Sup.C is regenerated by treatment with 0.5 *M* NaOH	196

Note: Pre.C = precolumn, Sep.C = separation column, Sup.C = suppressor column, and PC = post column (cation suppressor column).

[a] See also References 395, 396, and 407.

Precolumn

This column, which usually contains the same resin as the separator column, serves the same purpose as the analogous columns used in anion analysis (see preceding section).

Eluents

The choice of eluent is dependent on the affinity of the cations to be analyzed for the resin in the separator column. Thus, dilute solutions of mineral acids such as 0.0025 to

0.005 *M* nitric acid (see Table 12)[101,131,197-200] or 0.0025 to 0.01 *M* HCl[49,201-203] as well as 0.01 *M* HCl - 40% methanol[204] are effective eluents for the separation of the alkali metals.

Divalent cations such as the alkaline earths and transition metals exhibit a much stronger retention on the cation exchange resin and they can no longer be eluted with the eluents used for monvalent sample ions. The eluent recommended by the Dionex Corporation for the elution of the alkaline earths is 0.001 *M* *m*-phenylenediamine dihydrochloride - 0.0025 *M* nitric acid. This solution should be prepared fresh every day, because it turns brown in a few days and it deposits a dark-colored material on the precolumn. For the same elution, also, 0.0025 *M* *m*-phenylenediamine dihydrochloride - 0.0025 *M* nitric acid or HCl,[198,200,201] or 0.0025 to 0.005 *M* ethylene-diamine dihydrochloride[199] can be used. Effective elutions and mutual separations of alkaline earth elements, e.g., Mg and Ca, can also be achieved by the use of solutions containing the nitrates of Ag,[49] Ba,[194] Pb,[194,195] and Zn.[196] Examples for separations of this type are illustrated in Table 13, from which it is evident that when solutions of these metal nitrates are used as eluents the suppressor column has to be in a form which takes up the cation and anion of the eluent. Thus, for example, on the sulfate form of Bio-Rad® AG1-X10, Ba is precipitated as the sulfate and simultaneously nitrate ion is retained by the resin sites originally occupied by the sulfate ion. In other words, the entire eluent $Ba(NO_3)_2$ is taken up by the sulfate resin (see Table 13) so that the effluent contains only the nitrates of the analyte cations, i.e., those of Mg and Ca. With Ag nitrate solution as eluent the suppressor column contains an anion exchange resin in the chloride form on which Ag is precipitated as Ag chloride.[49]

Applications

The technique of suppressed ion chromatography has been utilized in connection with the determination of alkali metals (mainly, Na and K) and ammonium in water (see Table 12),[131,198] air (see Table 12),[197,205] urine,[25,49,203] serums,[25,49] grape and orange juice,[49] plants,[201] and antibiotics.[101] Mg and Ca were determined in waters,[194,198] soils,[194] serum,[194] plants,[201] and synthetic mixtures (see Table 13).[194-196]

Unsuppressed Ion Chromatography

Ion chromatographic systems which are designed to work without a suppressor column were first developed by Fritz and co-workers in 1978. This technique, which is called unsuppressed ion chromatography, uses a specially designed conductivity detector that tolerates some conductivity of the eluent. The term "single column" ion chromatography is also used when it is necessary to distinguish it from the Dionex® system of ion chromatography, which requires two columns in series (see preceding section).

The apparatus used in unsuppressed ion chromatography is a conventional high performance liquid chromatograph and consists of the following components (arranged in essentially the same sequence as those shown in Figures 1 and 4; see section on suppressed ion chromatography):

1. A pump to force the eluent through the sample loop, column, and conductivity cell at a flow rate of 0.8 to 2.0 mℓ/min
2. A sample injection valve, which makes use of a loop and enables samples of 100 μℓ to be injected
3. A separator column, usually 2.0 to 3.0 mm ID and 500 mm in length, packed with a low capacity ion exchange resin of small uniform particle size
4. A fluid conductivity cell assembly, coupled to a conductivity detector to monitor continuously the conductance of the column effluent
5. A recorder to trace the output of the conductivity detector

In unsuppressed ion chromatography, the first peak to appear on the chromatogram is an "injection peak", which is analogous to the "void volume peak" encountered in other chromatographic methods. Each of the subsequent peaks represents one individual ion, the position of the peaks being determined by the retention times of the specific ions on the separator column. By use of the conductivity detector, the signals of the normal chromatographic peaks are a measure of the difference between the conductivities of the particular ions and that of the eluent (the "background conductivity"). The conductivity of the injection peak, on the other hand, is related to the total ionic content of the injected sample.

Anion Analysis

General

By using an anion exchange resin of low capacity and employing an efficient eluent of low conductance, anions can be separated chromatographically and detected by a conductivity detector. Under these conditions the equivalent conductance of chloride, nitrate, sulfate, and other common anions is appreciably higher than that of the eluent anions (e.g., benzoate or phthalate). Thus, when an adsorbed sample anion passes into solution (by exchange with an eluent anion) and ultimately is eluted from the separator column, there is an increase in conductance and a peak is recorded for the anion on the chromatogram (plotted as conductance vs. elution time). The sensitivity attainable with this technique varies with the anion separated, but generally is around 1 ppm (100 ng in a 100-$\mu\ell$ sample). Use of a short column to concentrate sample anions improves sensitivity to the point that 2 ppb of some anions can be determined.

Separator Column

This column is usually packed with an anion exchange resin of the XAD type[206-213] or with surface-coated silica.[37,214-229]

An anion exchanger based on Rohm and Haas® XAD-1, which is a highly cross-linked, macroreticular, styrene-DVB copolymer consisting of beads with a network of hard microspheres with pores and channels in-between (surface area of 100 m^2/g and average pore diameter of 205 Å), can be prepared by chloromethylation, followed by amination with a tertiary amine.

$$-C_6H_5 \xrightarrow[ZnCl_2]{ClCH_2OCH_3} -C_6H_4-CH_2Cl \xrightarrow{(CH_3)_3N} -C_6H_4-CH_2N(CH_3)_3^+Cl^-$$

XAD-1 Chloromethylation Amination Anion exchange resin

Anion exchange resins of variable, but low, exchange capacities are produced by using mild conditions and short reaction times in the chloromethylation reaction. Conditions for the amination are chosen to convert as much as possible of the chloromethyl group to the quaternary ammonium chloride. Highly suitable procedures for preparing anion exchange resins of this type have been described by Fritz and collaborators.[44] Depending on the conditions selected, resins based on XAD-1 with ion exchange capacities from 0.001 to 0.09 meq/g are obtained. This range includes all of the capacities that one would be likely to use in anion chromatography.

Silica beads (porous silica spheres) coated with a thin shell of an organic material containing quaternary ammonium functional groups are commercially available under the trade names Vydac® SC and Vydac® I.C.302 (produced by Wescan Instruments, Inc., Santa Clara, Calif.).[210-212,214,216,218,219,221-223,226,227,229-235] The former has a bead diameter of 30 to

44 μm and an exchange capacity of ≈0.1 meq/g. Vydac® I.C.302 is of high mechanical strength with a particle diameter of ≈15 μm. The particles have an area of 86 m^2/g and an average pore diameter of 330 Å. Coated silica beads, available from Toyo Soda (Japan) under the trade name TSK-gel®, have also been used for anion analysis.[37,215,217,220,224,225,228,236] Silica gel coated with a polyamide crown resin has also been prepared.[237,238] The order in which anions are eluted from these coated silica beads is somewhat different from that observed with all-organic anion exchangers such as on the XAD-based resins. For example, nitrate is eluted later than sulfate and bromide on Vydac® SC. No fluoride peak is obtained, probably because of reaction with the silica matrix.

In HPLC a column packing of small particle size and a narrow size distribution is necessary to achieve a small H value (height equivalent to a theoretical plate). An average particle size of ~5 μm diameter appears to be the lower practical limit. Small, uniform particles are also necessary for efficient separations. One of the best column packings is obtained with the XAD-1 resin which is available as 20-50 mesh beads which must be ground and sized to obtain material useful for chromatography.

Elution and Eluents

Before a series of separations, the eluent to be used is pumped through the anion exchange column (separator column) until an equilibrium is attained in which the resin exchange sites contain the eluent anion (preconditioning of the resin bed). When a sample containing salts of various anions is injected, the sample anions will be taken up by the resin and exchanged for an equivalent amount of the eluent anion. The sample volume is rather small (usually 100 μℓ). If there is no mixing, a zone corresponding to the sample volume travels down the separator column at a rate equal to the eluent flow rate. This zone contains the cations of the original sample and an eluent anion concentration equivalent to that of the sample anion. If the conductance of the cations and anions in this zone is greater than that of the eluent, a positive pseudo peak will be observed when this zone passes through the conductivity detector. However, if the conductance of ions in this zone is lower than that of the eluent, a negative pseudo peak will result. This pseudo peak causes no difficulty provided elution of the sample anions is delayed after the pseudo peak has passed. After the sample plug has passed, the baseline is quickly restored to that with the eluent alone. However, the solute anions gradually move down the column, pushed by the mass-action effect of the eluent anion on the ion exchange equilibrium. The total cation concentration in solution is fixed by the eluent anion concentration, because a solute anion can only enter the solution phase by uptake of an equivalent number of eluent anions. Therefore, the change in conductance when a sample solute band passes through the detector results from replacement of some of the eluent anions by solute anions, although the total ion concentration remains constant. This change is directly proportional to the sample concentration and the difference in equivalent conductance of the eluent anion and sample anion.

The most widely used eluents are solutions containing phthalate (Na^+, K^+, or NH_4^+ as cations),[206,207,210-214,216,219,221-223,225,227,229-232,239,240] benzoate,[210-213,225] and citrate[37,212,213,225] at concentrations ranging from approximately 5×10^{-5} to 4×10^{-3} *M*. These large organic anions are less mobile than most inorganic anions and therefore have lower equivalent conductances. For example, benzoate anion has an equivalent conductance of 32 mho cm^2/eq, while chloride, sulfate, and other typical sample anions have higher equivalent conductances (≈70 mho cm^2/eq). This difference in conductance is vital to the success of unsuppressed anion chromatography with a conductivity detector. When the equivalent conductance of a sample anion is significantly higher than that of the eluent anion, a sample peak will occur on a baseline of lower background conductance.

Usually the eluent pH is between 5.8 and 6.8, which is close to the natural pH of a benzoate or phthalate salt solution. If the eluent pH is raised substantially, the eluent now contains OH^- as a second major anion. Use of such an eluent is often complicated by extra positive or negative peaks before or after the sample anion peaks. If the eluent pH is made more acidic than the usual pH 5.8 to 6.8, the eluting power is diminished, because some of the eluent anion is converted either into a molecular acid (e.g., benzoate to benzoic acid) or into an anion of lower charge (e.g., phthalate^{2-} to H phthalate$^-$).

Other eluents that can effectively be used for anion analysis include solutions of tartrate,[37,225] tartaric acid,[220] orthosulfobenzoate,[211] methanesulfonate,[233] succinic acid,[206,209] malic acid,[220] nicotinic acid,[209] EDTA,[236] acetoacetic acid,[226] Na-ethylenediaminetetra-acetatocobaltate(III),[218] tris(ethylenediamine) Co(III)-iodide,[218] boric acid-tetraborate-gluconate,[218a] and K hydroxide solutions,[215,217,224,228] Acetoacetic acid can be thermally decarboxylated (to acetone and CO_2) so that considerable eluent conductivity suppression is obtained.[226]

Even though the pH of a sample may be different from that of the eluent used, the sample volume is quite small (usually 100 $\mu\ell$) and so the sample anions are quickly brought to the pH of the eluent. Some anions will exist in different protonated forms, depending on the eluent pH. Thus, phosphate can be present as $H_2PO_4^-$ or as HPO_4^{2-} according to the eluent pH. Carbonate can be converted to bicarbonate or to carbonic acid. Cyanide can only exist as the anion in a basic environment and therefore requires a basic eluent for separation and detection (at the usual eluent pH of 5.8 to 6.8 cyanide is present as HCN, which is too weakly ionized to be detected). Control of pH can be very useful in adjusting the retention time of certain anions relative to most other anions.

With all the eluents mentioned above the sequence of elution of common inorganic anions is similar to that observed in the case of suppressed ion chromatography (see preceding section), i.e., the retention times (elution times) for fluoride, chloride, nitrite, bromide, nitrate, and sulfate increase in this order.

With K or Na hydroxide solutions and other strongly basic eluents the peaks for the sample anions appear as negative peaks (decreased conductance), because the hydroxide ion is more mobile than other anions. However, the peak height (or area) is still a function of the amount of sample anion and the sensitivity is even better than with the more acidic eluents where positive peaks are obtained. The theory of negative peaks for sample anions is easy to understand. Suppose we use 0.001 *M* Na hydroxide as the eluent. The background conductance will be the sum of the Na and hydroxide conductances and will be relatively high. Injection of a sample will result in the uptake of the anions by the resin column, with an equivalent amount of resin hydroxide ion passing into solution. A "pseudo" peak will occur when the sample zone passes through the conductivity detector, because the conductance of this zone will be either higher or lower than the eluent background conductance. Once the pseudo peak is through, the anion concentration of the column effluent will be constant, as fixed by the 0.001 *M* eluent concentration. When a sample anion, A^-, is eluted from the column and passes through the detector, the eluent hydroxide ion will be decreased because a constant anion concentration must be maintained $[(A^- + (OH)^- = 0.001]$. The equivalent conductances of most anions range from about 30 to 80 mho cm^2/eq, while the hydroxide ion has an equivalent conductance of 199. Thus, the conductance will decrease when a sample anion is being eluted and the height (or area) of this negative peak will be proportional to the concentration of the anion.

K and Na hydroxide solutions are effective eluents for cyanide, acetate, arsenite, fluoride, chloride, nitrate, and other easily eluted anions, but multivalent anions such as carbonate and sulfate are eluted very slowly.

Applications

Conductometric quantitation of inorganic anions, after separation by unsuppressed ion chromatography, has in most cases been used in connection with their determination in natural waters[211,213,215,217,223,228,230-232,234,236,241] and other environmental materials (see Table

Table 14
CONDUCTOMETRIC DETERMINATION OF INORGANIC ANIONS AFTER UNSUPPRESSED ION CHROMATOGRAPHIC SEPARATION ON SURFACE-COATED SILICA OF THE VYDAC® TYPE[a]

Applications	Experimental conditions and remarks	Ref.
Determination of chloride, nitrate, sulfate, and total S in environmental samples (natural waters, soils, marine sediments, and plants)	Separator column: Vydac® 302 IC Sample size (loop): 250 μℓ Eluent: 0.004 *M* phthalic acid buffer adjusted to pH 5.0 with Na borate Flow rate: 2.0 mℓ/min Pretreatment with Dowex® 50W-X8 (200-400 mesh; Pb form) on which sulfate is precipitated as $PbSO_4$ permits the analysis of samples high in sulfate, as in the determination of available soil NO_3^- The determination of the anions can also be effected by use of an indirect UV technique	223
Determination of phosphate, chloride, nitrite, sulfate, thiosulfate, and bicarbonate in natural waters	Separator column: 25 × 0.46 cm, Vydac® 3021C Sample size (loop): 0.2 mℓ Eluent: 0.004 *M* NaH-phthalate of pH 4.5 Flow rate: 2 mℓ/min The limit of detection was 0.1—0.2 ppm (4—5 ppm for $H_2PO_4^-$ and HCO_3^-); with use of a preconcentration column and an injection volume of 2—10 mℓ the detection limit was improved by a factor of 10—50	234
Determination of nitrate, nitrite, phosphate, chloride, and sulfate in salad and vegetables	Separator column: 25 × 0.46 cm, Wescan® 269-001 Sample size (loop): 100—200 μℓ Eluent: 0.0044 *M* phthalate buffer of pH 3.9 Flow rate: 2 mℓ/min Run time was ≃30 min; detection limits were 0.1 ppm for NO_3^-, NO_2^-, and SO_4^{2-}, 0.04 ppm for Cl^-, and 0.4 ppm for PO_4^{3-} Before the separation, the sample (10 g) was homogenized rapidly with water (40 mℓ), the slurry was filtered, and the filtrate diluted to 1 ℓ; a 2-mℓ portion of this solution was mixed with 2 mℓ of 0.0088 *M* KH-phthalate and filtered and a 0.1—0.2-mℓ aliquot was passed through a precolumn containing the same ion exchange material	219
Determination of azide (N_3^-) in human serum	Separator column: 25 × 0.46 cm, Vydac® 302 Sample size (loop): 100 μℓ Eluent: 0.001 *M* phthalic acid adjusted to pH 3.5 with pyridine Flow rate: 2.5 mℓ/min The limit of detection is 0.2 μg of N_3^-	222
Determination of chloride, nitrate, and sulfate in rain water	Separator column: 25 × 0.46 cm, Wescan® 269-001 Sample size (loop): 2 mℓ Eluent: 0.002 *M* KH-phthalate (pH 4.5) Detection limits are <0.1 ppm for Cl^- and NO_3^- and 0.25 ppm for SO_4^{2-}; one analysis takes <25 min	231

[a] See also References 397 to 401, 408, and 409.

14),[223] which include atmospheric aerosols[227] and biological materials (see Table 14).[219,222,223] For these separations low capacity anion exchangers of the Vydac® (see Table 14), TSK-gel® (see Table 15), and XAD types (see Table 16) have been utilized.

Other applications using this technique include the determination of anions in toothpaste,[209] paint-bath solution,[209] curing salt,[211] and boric acid.[229]

Table 15
CONDUCTOMETRIC DETERMINATION OF INORGANIC ANIONS AFTER UNSUPPRESSED ION CHROMATOGRAPHIC SEPARATION ON SURFACE-COATED SILICA OF THE TSK-GEL® TYPE[a]

Applications	Experimental conditions and remarks	Ref.
Determination of anions in lake water	Separator column: 5 × 0.4 cm, TSK-gel® IC-anion-PW (grain size: 10 μm) Eluent: 0.002 *M* KOH Flow rate: 1 mℓ/min The coefficients of variation for 8 ppm of Cl^- and 670 ppb of nitrate-N were 0.17 and 3.51%, respectively The peak for SO_4^{2-} overlapped with that for CO_3^{2-}, and that for F^- with those for low molecular weight organic acids A similar procedure using 0.0005 KOH as the eluent (flow rate: 1.2 mℓ/min) has been used to determine silicic acid (soluble silicate) in natural waters;[217,228] interference from F^- was eliminated by the addition of H_3BO_3 and that from Mg and Ca was reduced by pretreatment with Dowex®50W-X2 (H^+ form); the detection limit was 22 ppb (as Si), and the coefficient of variation was 2.1% at the 1-ppm level of silicic acid	215, 217, 228
Determination of anions and cations in natural waters	Separator column: 5 × 0.46 cm, TSK-gel® IC-anion-SW (particle size: 5 ± 1 μm; capacity: 0.1 meq/g) Eluent: 0.001 *M* EDTA of pH 6.0 Flow rate: 1 mℓ/min Detection limits (ppm) were 0.04 for Cl^-, 0.05 for NO_2^-, NO_3^-, Mg, and Ca, and 0.1 for $H_2PO_4^-$ and SO_4^{2-}; the precedure was used to determine Cl^-, NO_3^-, SO_4^{2-}, Mg, and Ca in rain water and tapwaters at ppm levels and Mg, Ca, and sulfate in seawater diluted 100-fold	236
Determination of anions in synthetic mixtures	Separator column: 5 × 0.4 cm; TSK-gel® 6205A (grain size: 9 ± 1 μm; capacity 0.1—0.3 meq/g) Sample size (loop): 0.1 mℓ Eluent: 0.001—0.006 *M* KOH Flow rate: 1 mℓ/min Detection limits for F^-, Cl^-, Br^-, NO_2^-, and NO_3^- were 1.5, 2.5, 15, 15, and 15 ppb, respectively	224

[a] See also References 402, 403, and 410.

Cation Analysis

General

The separator column contains a cation exchange resin of low capacity and the eluent is a dilute solution of a mineral acid (e.g., nitric acid) or of an ethylenediammonium salt. The background conductance is sufficiently low so that a conductivity detector can be placed immediately after the separation column. When a sample peak is eluted, the conductance is decreased (due to the presence of low conducting metal cations in place of highly conducting eluent cations, e.g., H^+) and the peaks are in the negative direction. Consequently, the amount of sample ion is proportional to the height of the decreasing conductance peak. Less than 1 ppm of many cations can be determined.

Resins and Eluents

The separator column is packed with lightly (superficially) sulfonated beads of a polymer, as, for example, obtained by sulfonating polystyrene - 4% DVB (Benson Co., BN-X4 beads of 20 μm diameter) under mild conditions.[242,243]

Table 16
CONDUCTOMETRIC DETERMINATION OF INORGANIC ANIONS AFTER UNSUPPRESSED ION CHROMATOGRAPHIC SEPARATION ON ANION EXCHANGE RESINS OF THE XAD TYPE[a]

Applications	Experimental conditions and remarks	Ref.
Indirect determination of cyanide	Precolumn: 7.5 × 0.2 cm with unfunctionalized XAD-4 resin Separator column: 50 × 0.2 cm packed with functionalized XAD-1 resin (325-400 mesh; capacity: 0.025 meq/g) Sample size (loop): 0.1 mℓ Eluent: 0.0002 *M* Na (or K) phthalate Flow rate: 1.5 mℓ/min Before the separation, the solution is diluted, if necessary, so that the concentration of CN^- is in the range of 0.5—10 ppm; then the solution (40 mℓ) is mixed with 0.05 *M* acetate buffer (pH 4.75) (1 mℓ) and 0.1 *M* I_2 in 95% ethanol (0.3 mℓ) (excess of I_2) and the mixture is diluted to 50 mℓ; under these conditions the following reaction occurs: $I_2 + HCN \rightleftharpoons H^+ + I^- + ICN$ After 5 min, an aliquot of the solution (2—3 mℓ) is applied to the precolumn to remove excess I_2; then the iodide formed by the above reaction is separated on the separator column and determined conductometrically; after a series of injections the precolumn is flushed with 0.05 *M* HNO_3 in acetone to desorb I_2 Interference is caused by Hg(II), Co, $S_2O_3^{2-}$, and SCN^-	207
Determination of chloride and sulfate in steam condensate and boiler feed water	Concentrator column: 7 × 0.2 cm, functionalized XAD-1 resin (grain size: 45—57 μm; capacity: 0.0175 meq/g) Separator column: 5 × 0.3 cm filled with the same resin Eluents: 0.0002 *M* K-benzoate of pH 6.2 (for chloride determination); 0.0002 *M* K-citrate of pH 6 (for sulfate determination) To determine chloride and sulfate simultaneously, the technique was similar but the concentrator column (3 × 0.25 cm) contained functionalized XAD-4 resin (capacity 0.5 meq/g), and three coupled separator columns (each 5 × 0.2 cm), each containing XAD-1 resin (capacity: 13 μeq/g), were used; the eluent was 0.0002 *M* K-phthalate of pH 6.2; detection limits were 3—5 ppb; a precolumn of Dowex® 50-X8 or -X16 (H^+ form) in the sample loading line was required for the boiler feed samples To determine chloride or sulfate the sample was pumped directly from its container to the concentrator column	213

[a] See also Reference 404.

The eluent concentration needed for a given chromatographic separation will vary somewhat according to the exchange capacity of the resin. A resin of higher capacity will require an eluent of higher concentration. Resins with approximately 0.06 meq/g exchange capacity are probably near the upper practical limit for some types of separations because of the higher background conductance of the eluent. The effective capacity of a resin can be decreased by blending it with up to 1.5 times as much unfunctionalized resin. This results in a lowering of the eluent concentration needed and, thus, a lowering in the background conductance.

An eluent-containing H ion, such as a dilute solution of nitric,[242] HCl, or perchloric or picolinic acid,[242a] is effective in separating mixtures of alkali metal ions and the ammonium cation. The acid eluents should be pure and not contain divalent metal ions, such as the alkaline earths. Eluents containing *m*-phenylenediamine salt have been recommended for separation of the alkaline earths.[242a]

Eluents containing a salt of ethylenediamine give even better separations and do not discolor the column.[242] Even when diluted, eluents containing ethylenediammonium salts give poor separations of the alkali metal ions. Thus, ethylenediammonium nitrate solutions are satisfactory for separation of mixtures of divalent metal ions, and a dilute nitric acid solution works well for separation of monovalent metal ions. It is also possible to incorporate a complexing anion such as tartrate,[243,243a] citrate,[243a,243b] or hydroxyisobutyrate in an eluent containing the ethylene-diammonium cation.

Applications

The technique outlined above has been applied for the mutual separation of alkali metals and alkaline earth elements.[242,243] Separation of the alkali metals was also achieved on a crown ether-polyamide resin and on silica gel treated with $SOCl_2$ and substituted with 4-hydroxymethylbenzo-15-crown-5.[243c]*

CHROMATOGRAPHIC METHODS WITH NONCONDUCTOMETRIC DETECTION

Although conductometric detection is the most general technique for monitoring separation by ion exchange (see preceding section on ion chromatography), a wide variety of other detectors has also been used for the automatic detection of ions in ion exchange chromatography. These include UV-visible spectrophotometric, electrochemical, flame emission spectroscopic, and atomic absorption spectroscopic detectors which are selective techniques, because they respond only to certain ions that may be present in the detector cell. Spectrophotometric detectors can be made into almost universal or general-type detectors by the postcolumn reaction of sample ions with certain color-forming reagents.[244-247]

In detectors of this type the selectivity can be changed simply by choosing different wavelengths. Furthermore, the versatility of the detector can be increased by adding a color-forming reagent to the eluent or to the column effluent. Finally, high quality instrumentation already exists because of the use of the detector in the determination of organics by HPLC.

The postcolumn method of derivatization of eluate fractions has been well established from older ion exchange separations. An appropriate reaction is done on each fraction to determine the metal ion concentration in that fraction. The automatic addition of a color-forming reagent to an ion exchange column effluent and analysis by flow-through cell detection are more recent; however, many of the color-forming reagents and buffers are the same as those used in the classical fraction method determinations. The most important parameters in the design and operation of this type of detection system are

1. Fast color-forming reaction (a typical holdup time between the mixing chamber outlet and detector cell might be 1 to 100 msec)
2. Low (or zero) background absorption of the color-forming reagent
3. High absorption of the metal complex
4. Rapid mixing of the reagent and column effluent, yet the mixing chamber (postcolumn reactor) should be as small as possible to reduce peak broadening

Besides the above criteria, the ideal color-form reagent should be a general reagent that reacts with a large number of metals ions. Examples for general reagents are PAR [4-(2-pyridylazo)-resorcinol],[245,248-248b] arsenazo,[249-249b] and xylenol orange. Examples for specific

* For other applications see References 411 and 412.

reagents are Chrome Azurol S (for Al),[249c] 1,10-phenanthroline and derivatives (for Fe),[249d] Ba-chloroanilate (for sulfate),[249e] and iodide (for BrO_3^-).[249f]

Most anions and also some cations are best detected by means of photometric detectors, e.g., operating in the range of 190 to 300 nm.[250-271f]* Non-UV absorbing anions and also cations may be detected by monitoring the decrease in refractive index of an eluent containing an aromatic counter ion.[272-274a] Detection methods based on postcolumn fluorescence and phosphorescence measurements have also been described.[275-277]**

Electrochemical detection comprises techniques based on polarography, potentiometry, amperometry, and coulometry.[109,260,278-294c]*** These detectors are highly selective and sensitive. They are selective because they operate on the principles of oxidation or reduction of substances at an electrode. The ability of a substance to be oxidized or reduced is different for each substance and is measured by the potential required to induce the electrolysis. Detector selectivity is controlled simply by controlling the magnitude of the potential applied to the cell.

Secondary electrochemical detection (postcolumn derivatization) is useful when the sample species of interest is not electroactive. A reagent is mixed with the column effluent to convert the sample into an electroactive species.

Other detectors that have been used to monitor chromatographic separations automatically include detectors based on atomic absorption,[155,295,296]† flame photometry,[297] and radiometry.[298,298a]

Because all the detectors mentioned above are selective, not only very low but also relatively high eluent concentrations can be employed for many separations using the principles of conventional HPLC. High eluent concentrations are more compatible with high-capacity resins and separations can be designed by using numerous ion exchange distribution data already collected. Another advantage of selective detection is that the strength of an eluent can be programmed. This allows sample ions that differ greatly in selectivity behavior to be eluted in the same chromatogram.

Applications

Numerous procedures based on this type of chromatographic separation are presented throughout this *Handbook* (for examples, see in chapter on Rare Earth Elements (Tables 15, 32, 34, 36 to 42, 45, and 46), Actinides (Volume II) (Tables 37, 40, 145, 157, and 161), Copper (Volume III) (Tables 41, 59, 61, 63, 64, and 79), Platinum Metals (Volume III) (Tables 13 and 15), and Mercury (Volume IV) (Tables 6 and 14).

ION EXCLUSION CHROMATOGRAPHY (IEC)

This type of separation method (also called Donnan exclusion chromatography), which was first described by Wheaton and Bauman,[299] is based on the principle by which anions are separated using a cation exchanger or cations using an anion exchanger. In other words, coions of an ion exchanger (coions = mobile ionic species with a charge of the same sign as the fixed ions, e.g., with $R_sSO_3^-H^+$ coions, can be fluoride, phosphate, sulfate, etc.) are excluded from the ion exchanger (e.g., strongly acidic or basic resin) by electrostatic repulsion with their fixed charged groups. In both cases ions of higher charge are eluted earlier according to charge exclusion by the fixed ionic groups of the ion exchanger.

Separations by IEC of various acids and acid salts are based on the ability of nonionic

* See also References 414 and 415.

** See also Reference 416.

***See also Reference 417.

† See also References 418 to 420.

weak acids to enter the resin phase, whereas strong, highly ionized acids are excluded and pass through a cation exchange column unretained.

Among other applications this technique has been employed for the separation of phosphorus oxy anions,[300-309] nitrate, nitrite, and ammonium ions,[310-312] fluoride,[313,314] bicarbonate,[315] and sulfate.[316]

INORGANIC AFFINITY CHROMATOGRAPHY

An example for an anlytical separation based on affinity chromatography is shown in Table 6 in the chapter on Fluorine, Volume VI.

NONCHROMATOGRAPHIC METHODS

The Concept of Combined Ion Exchange - Solvent Extraction (CIESE)

Separations carried out in systems in which both ion exchange and liquid-liquid extraction are operative simultaneously have been investigated by Korkisch and collaborators,[317-320a] who have coined the term CIESE, which stands for combined ion exchange - solvent extraction (sometimes also called CIESE effect). In media of this type the organic solvent, as such or an extracting agent dissolved in it, is acting as an extractant for one element or a group of elements, while for other metal ions it causes an increase of selectivities toward the ion exchanger, so that selective separations can be achieved on ion exchange columns.

Compared to most other separation procedures on columns, CIESE can be regarded as being bifunctional, because with the other techniques (e.g., ion exchange in pure aqueous systems, adsorption chromatography, and partition chromatography) either ion exchange or adsorption or partition are the single separation parameters.

The CIESE systems which have been investigated by Korkisch and various other research groups[317-325] using both anion and cation exchange resins can be roughly divided into three groups.

Ethers or ketones in the presence of mineral acids — The CIESE-active components are ethers, ketones, and mineral acids.

Organic P compounds in the presence of mineral acids and organic solvents — The CIESE-active components are organic P compounds and mineral acids (sometimes the organic solvent is also CIESE active).

Chelating organic compounds in the presence of organic solvents — The CIESE-active components are the chelating organic compounds (e.g., dithizone and cupferron).

In these systems the following equilibria are assumed to exist in the liquid phase when taking acetone and tetrahydrofuran (THF) as examples:

$$(CH_3)_2C{=}\overline{\underline{O}} + \text{aq. HCl} \rightleftarrows \left[(CH_3)_2C{=}\overline{O} \rightarrow H\right]^+ Cl^-$$

Acting as a Lewis base — Oxonium salt

$$\left[(CH_3)_2C{=}\overline{O} \rightarrow H\right]^+ Cl^- + \text{e.g. } FeCl_4^- \rightleftarrows \left[(CH_3)_2C{=}\overline{O} \rightarrow H\right]^+ FeCl_4^- + Cl^-$$

anionic chloride complex — Ion association complex

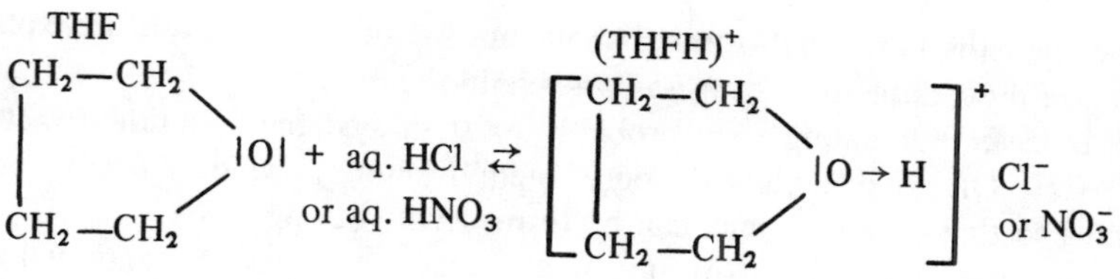

$$\text{THF: } \begin{matrix} CH_2-CH_2 \\ | \\ CH_2-CH_2 \end{matrix} \rangle \overline{O}| + \text{aq. HCl or aq. } HNO_3 \rightleftarrows \left[\begin{matrix} CH_2-CH_2 \\ | \\ CH_2-CH_2 \end{matrix} \rangle |O \rightarrow H \right]^+ Cl^- \text{ or } NO_3^-$$

Acting as a Lewis base — $(THFH)^+$ Oxonium salt

$$(THFH)^+Cl^- + \text{e.g. } FeCl_4^- \rightleftarrows (THFH)^+FeCl_4^- + Cl^-$$

anionic chloride complex

$$(THFH)^+NO_3^- + \text{e.g. } [UO_2(NO_3)_3]^- \rightleftarrows (THFH)^+UO_2(NO_3)_3^- + NO_3^-$$

anionic nitrate complex — Ion association complex

Equilibria of a similar nature very probably also exist in systems containing organic P compounds such as triocytlphosphine oxide (TOPO). The latter reacts according to the following equations:

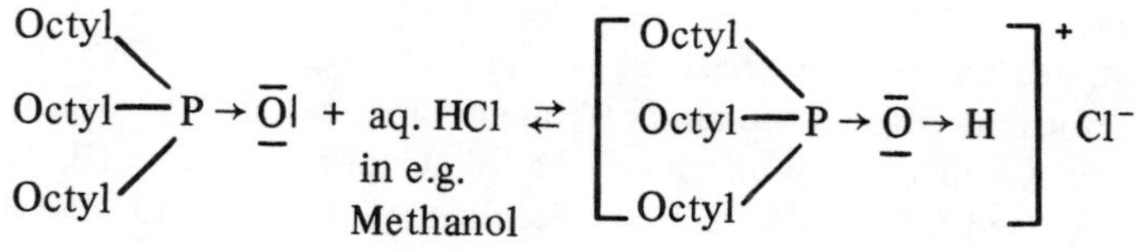

$$(Octyl)_3P \rightarrow \underline{O}| + \text{aq. HCl (in e.g. Methanol)} \rightleftarrows [(Octyl)_3P \rightarrow \underline{O} \rightarrow H]^+ Cl^-$$

Acting as a Lewis base — Oxonium salt (for short $TOPOH^+Cl^-$)

$$TOPOH^+Cl^- + ScCl_4^- \rightleftarrows (TOPOH)^+ScCl_4^- + Cl^-$$

Ion association complex

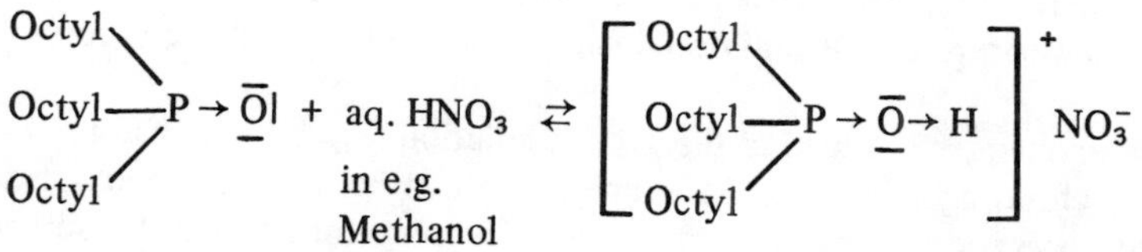

$$(Octyl)_3P \rightarrow \underline{O}| + \text{aq. } HNO_3 \text{ (in e.g. Methanol)} \rightleftarrows [(Octyl)_3P \rightarrow \underline{O} \rightarrow H]^+ NO_3^-$$

Acting as a Lewis base — Oxonium salt (for short $TOPOH^+NO_3^-$)

$$2\,TOPOH^+NO_3^- + Th(NO_3)_6^{2-} \quad (TOPOH^+)_2Th(NO_3)_6^{2-} + 2NO_3^-$$

Ion association complex

Similar mechanisms to those illustrated above are frequently employed to explain liquid-liquid extraction procedures involving ion association complexes.

Equilibria of a similar nature very probably exist in systems containing methyl glycol ($CH_3O–CH_2–CH_2–OH$) and hexone (isobutyl methyl ketone), which can also be regarded as Lewis bases so that oxonium salts can be formed with strong acids.

At high concentrations of THF (or any other of the Lewis bases mentioned above) the oxonium salt is formed to a higher percentage (according to the law of mass action) and, hence, its reaction with the anionic metal complex will be more complete. As a consequence, the ion association complex will be present at a much higher concentration than at low percentages of this organic solvent. Therefore, the metal ions which are readily extractable by ethers and/or ketones (such as trivalent Fe, Ga, trivalent Au, and Mo from HCl solutions and, e.g., the uranyl ion from nitric acid solutions) will, at high concentrations of these organic solvents, show a behavior toward cation and anion exchange resins which deviates considerably from that observed in pure aqueous systems containing HCl or nitric acids. This different behavior which manifests itself by highly decreased adsorption of these metal ions on ion exchangers is used frequently to isolate various elements from matrices which contain large quantities of the weakly adsorbed elements.

Similar equilibria as those shown above for TOPO very probably exist in systems containing tributyl phosphate (TBP) or bis-(2-ethyl)-hexyl phosphoric acid (HDEHP).

At high concentrations of TOPO (or TBP or HDEHP) the oxonium salt is formed to a higher percentage, and as a consequence metal ions which form very strong complexes of the oxonium salt type will not be retained on columns of cation or anion exchange resins, but will pass into the effluent as extracted species. On the other hand, metal ions, which under these conditions do not readily form ion association complexes, will be adsorbed on the ion exchanger.

Separations based on CIESE can be effected on strongly basic and acidic cation exchange resins as is illustrated by the flow sheets shown in Figures 5 and 6. This and similar types of CIESE separations have been used rather frequently for analytical purposes; for examples, see in chapters on Actinides (Volume II) (Table 3), Copper (Volume III) (Tables 14, 33, 40, 46, 50, 54, 56, and 62), and Rare Earth Elements (Tables 15, 18, and 29).

Ion Exchanger Phase Absorptiometry

This technique, which is also called ion exchanger colorimetry or ion exchanger phase spectrophotometry, is based on the direct measurement of the degree of light absorption by an ion exchanger resin phase which has sorbed a sample component.[326,327] Since light absorption measurements in this method are made directly on the ion exchanger phase of small volume to which light-absorbing components are concentrated from a sample solution of large volume, much higher sensitivity is attained than with an ordinary solution spectrophotometry. There are three ways of developing the color in the ion exchanger phase, depending on the nature of the individual sample component and the chromogenic agent:

1. The ion exchanger is added to the sample solution (e.g., 1 ℓ of natural water) together with the chromogenic agent. This procedure can be applied when the color reaction is highly specific for the analyte and the complex formed can be sorbed on the ion exchanger (for examples, see in chapters on Chromium [Volume IV], Iron [Volume V], Cobalt [Volume V], Copper [Volume III], Bismuth [Volume VI], Actinides [Volume II], Phosphorus [Volume VI], and Sulfur] [Volume VI]).
2. The chromogenic reagent, presorbed (loaded) on the ion exchanger, is added to the sample solution. This procedure is applied when the colored complex cannot be directly sorbed from the sample solution. The chromogenic agent should be sorbed irreversibly on the ion exchanger under conditions such that it is retained on the exchanger during

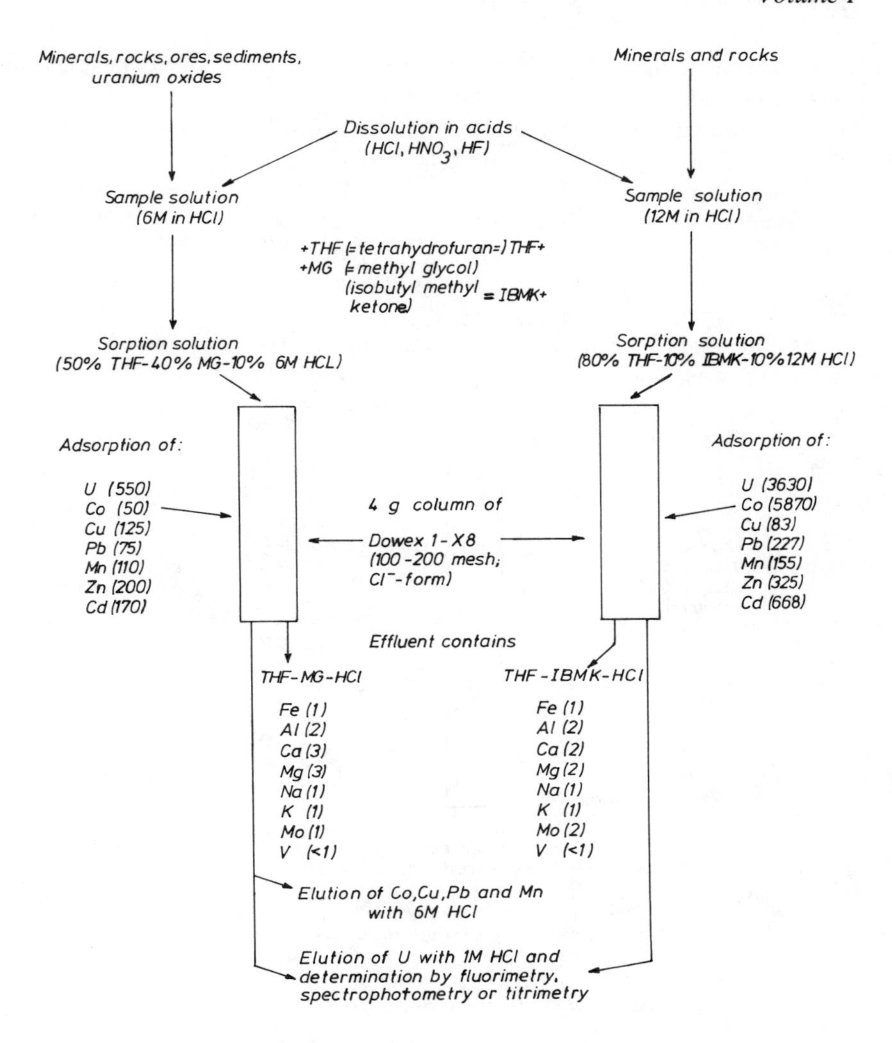

FIGURE 5. Flow sheet of CIESE separations on the strongly basic anion exchange resin Dowex® 1.

the equilibration with the sample solution. In this case the ion exchanger loaded (impregnated) with the reagent behaves like a chelating resin (for examples, see in chapters on Copper [Volume III], Nickel [Volume V], Silicon [Volume VI], and Phosphorus [Volume VI]).

3. The analyte is first sorbed on the ion exchanger from solution and then the chromogenic agent is added. This procedure is applied when the chromogenic agent has poor selectivity for the analyte (for examples, see in chapters on Zinc [Volume IV], Cadmium [Volume IV], and Iron [Volume V]).

Following equilibration with the sample solution some of the ion exchanger containing the colored sample species is packed as a slurry into a quartz cell (usually a 1-mm cell) (after most of the equilibrated solution has been decanted or filtered). The sedimentation and collection of the resin may be accelerated by adding finely powdered oppositely charged ion exchanger as coagulant (e.g., see in chapter on Copper [Volume III] Tables 8, 44, and

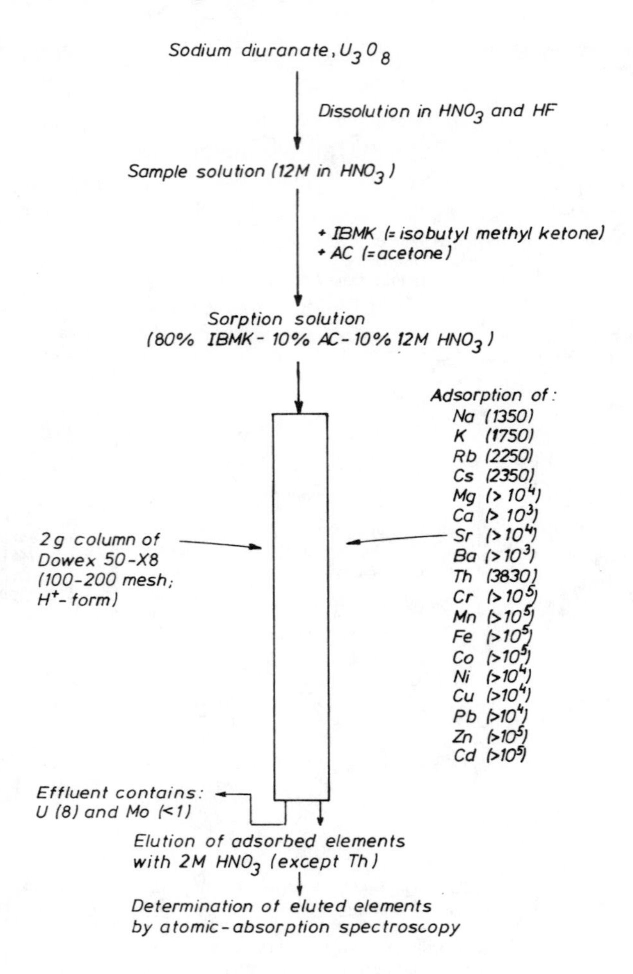

FIGURE 6. Flow sheet of a CIESE separation on the strongly acidic cation exchange resin Dowex® 50.

61).[328,329] In this case the coagulated particles are collected on a filter paper under suction, to make a uniform thin layer. The cell containing the ion exchanger is set in a spectrophotometer and the light transmitted through the cell is measured.

Reactive Ion Exchange (RIEX)

According to Janauer et al.,[330-335] RIEX encompasses all such ion exchange processes which are accompanied by reactions resulting in the production of new chemical species and/or the consumption or transformation of initially present species (particularly counter ions) in any number of participating phases. Consequently, RIEX includes *in situ* redox and precipitation reactions which usually are performed on columns containing common cation and anion exchange resins in ionic forms which can be oxidized or reduced[333,336-340] or contain counter ions which form insoluble precipitates with ions present in the sample (sorption) solution.[335,341-344] Numerous examples for reactions of this type are presented throughout this *Handbook*. The process involving separations based on the *in situ* precipitation of insoluble compounds is sometimes called precipitation chromatography.

To RIEX also belong complex and chelate formation and neutralization reactions on ion exchange resins, of which a very large variety and number are presented in the present *Handbook*.

Chemical Amplification Reactions with Ion Exchange Resins

According to Weisz and collaborators,[345,346] chemical amplification reactions can be carried out by passage of the solution through ion exchange columns in two alternate ionic forms. For example, if a solution containing H^+ was applied to serial columns of Na^+- and H^+-form resins (e.g., Dowex® 50-X8; 20-50 mesh), the H^+ were retained on the first column; the Na^+ displaced were subsequently retained on the second column, thereby displacing H^+. The original H^+ could be eluted with aqueous NaCl and added to the percolate from the second column, so giving twice the original concentration of H^+. By successive use of ten columns, a fivefold multiplication was achieved, and the method was applied to determination of micromole amounts of cations and anions. Ca, Mg, and Al were determined after passage through columns of the cation-exchange resin (H^+ and Na^+ forms), and Cl and F were determined with use of columns of Dowex® 1-X4 anion-exchange resin (Cl^- and OH^- forms). A variation of this multiplication procedure utilizes an inversion tube containing two different forms of the resin (e.g., in bags made from nylon net) which must be suitably separated so that they do not come in contact with each other (but the sample solution must have free access).[346] Thus, for example, a Na^+-form resin is placed in one half of the glass tube (18 cm × 18 mm with a short 9-mm-diameter central section) and the same resin in H^+ form is placed in the other half. To determine H^+, for example, a defined amount of acid is placed in the half of the tube containing the resin in Na^+ form, whereby Na^+ are liberated. The tube is then inverted and the Na^+-containing solution exchanges with the H^+-form resin to liberate more H^+; each two-step inversion cycle releases more H^+ and, after a chosen number of cycles, the H^+ accumulated on the Na^+-form resin are eluted with NaCl and titrated.

Application of the Ion Exchange Resins to the Determination of Free Acid and Total Salt Concentration

In solutions containing hydrolyzable cations the concentration of free acid is defined as the acid concentration that would be determined if the hydrolysis of the cations could be suppressed. During an acid-base titration, hydrolyzable cations form basic salts or hydroxides, and extra protons are liberated. The liberated protons are indistinguishable from protons of free acid, and estimates of the acid concentration in the original sample based on such titrations are anomalously high. This adverse effect of hydrolyzable cations can be eliminated by their removal with ion exchange resins. For example, if a solution of uranyl nitrate containing free acid is passed through a column containing a strongly acidic cation exchange resin, e.g., Amberlite® IR-120 in the H form (R_sH), and the column is washed with water, the effluent contains the free acid present in the solution, together with the acid liberated by exchange of the uranyl ions for the H ions of the resin as shown by the reaction:

$$2R_sH + UO_2(NO_3)_2 + \text{free } HNO_3 \rightleftarrows (R_s)_2UO_2 + \overbrace{2\ HNO_3 + \text{free } HNO_3}^{\text{total acid}}$$

The total acid can be found by titrating the effluent against standard alkali. The free acid can then be calculated by deducting the acid equivalent to the uranyl ions in the solution. Thus, the difference between the total acidity and that equivalent to the ion exchange of the U gives the free acidity. The uranyl ions are determined in a separate aliquot of the sample solution using a suitable analytical method.

This separation technique has been employed in connection with the determination of free acid in solutions containing U,[347-351] Th,[347,350] Pu,[349] Fe,[352,353] Al,[350] Ce,[354] and La.[350]

For the assay of free acid in uranyl nitrate solutions, a method based on the removal of U by anion exchange on Amberlite® IRA-400 (sulfate form) has also been described.[355]

The principle of total salt content determination (e.g., of NaCl in water) depends on the following reactions:

$$R_sSO_3H + NaCl \rightarrow R_sSO_3Na + HCl$$

$$R_s\overset{+}{N}(CH_3)_3OH + NaCl \rightarrow R_s\overset{+}{N}(CH_3)_3Cl + NaOH$$

The acid or base liberated can be easily determined by titration.

Ion Exchange Resins in Speciation Studies

Ion exchange is an attractive technique to study the trace metal speciation in environmental materials such as natural waters, because separations can be carried out with little manipulation and little opportunity for contamination, a most important consideration when working with very low metal concentrations. Conventional cation and anion exchange resins have been used for investigations involving metal speciation (see in chapters on Mercury [Volume IV], Arsenic [Volume VI], and Chromium [Volume IV]) but the iminodiacetate chelating resin Chelex® 100 has found much more application (see chapter on Copper, Volume III).*

Isotopic Ion Exchange

Isotopic ion exchange involves the simultaneous operation of ion exchange and isotopic exchange to effect a separation. Thus, for example, when the component M* (radionuclide to be determined) of a radioactive mixture has a high selectivity for a cation exchange resin, the concentration of this component is adjusted so that the distribution coefficients of the other trace radioactive components are greatly suppressed. Upon passing the mixture through a column previously saturated with the same cation in the stable form (R − M), the radioactive component M will be retained on the column by isotopic exchange while the other radioactive components are easily eluted.

Thus, for example, for the effective neutron activation analysis of Th interfering elements, e.g., U, Mn, Ba, Cs, Co, and lanthanoids, were eliminated from irradiated samples by passing them in a solution containing 0.5 mg/mℓ of Th and 0.5 *M* HCl through columns of Bio-Rad® AG50W resins in the Th form; such a column retained only the ^{233}Th, by a process of isotopic exchange.[355a]

Other analytical separation procedures based on isotopic ion exchange are outlined in the chapter on Copper (Tables 25 and 32), the chapter on Gold (Table 14), and the chapter on Rubidium-Cesium-Francium (Table 5).

Analytical Applications of Individual Beads of Ion Exchange Resins

Resin Spot Tests

Numerous methods for the microchemical detection of traces of elements by means of resin spot tests have been reported by Fujimoto and collaborators,[356-362] as well as by other investigators.[363-365]

The high sensitivity of resin spot tests is based mainly on the fact that the strongly colored reaction products (e.g., formed between a metal ion and a suitable reagent) are not uniformly distributed in the liquid phase (few drops), as is the case with the normal spot tests, but can be adsorbed on a few beads of a colorless or essentially transparent ion exchange resin (cation, anion, or chelating resin). This results in a considerable increase of the local concentration of the colored species (10- to 100-fold increases are usually observed).

* See also Reference 421.

Many of these techniques are described in the corresponding chapters and tables of this *Handbook* in which applications of ion exchange resins to the analysis for the various elements are treated. Therefore, no further details will be presented here.

In addition to resin spot tests the microchemical detection of elements can also be achieved on microcolumns of resins and on ion exchange resin paper.[365] Furthermore, columns of ion exchange resins have been used in a qualitative analysis scheme for cations based on ion exchange properties.[366]

Indicator Resins[367-369]

Acid-base indicator forms of ion exchange resins can act as indicators and may be used in conventional acid-base titrations. The indicator form of an anion exchange resin, e.g., Amberlite® IRA 400 (20-50 mesh), is prepared by shaking the resin with a 0.1% ethanol solution of the indicator (e.g., phenolphthalein, bromophenol blue, cresol red, or bromocresol purple). To be used as an indicator, a few beads (20 to 30) of this impregnated resin are added to the titration vessel. Titration is done in the normal manner and the end point is indicated by a color change in the resin beads, e.g., from colorless to a pink hue in the case of beads impregnated with phenolphthalein when titrating an acid with a standard base. The indicator resin beads, once used, may be filtered off, washed with water, and reused (at least a dozen acid-base cycles without loss in efficiency). As little as one resin bead may be used as indicator. Indicator beads may be dried and stored for future use. The prime advantage of this technique is that such indicators are in a phase separate from the solution whose pH they indicate. Cation exchange resins can be loaded with weak base indicators such as thymol blue.

Resin beads can also be used as indicators in precipitation and redox titrations.[369] Thus, for example, Zn, Cd, Pb, and Ag are titrated with $K_4Fe(CN)_6$ solution, in the presence of a few beads of Dowex® 50W-X8 resin in the Fe(III) form (at the end point the beads assume a blue color due to the formation of $Fe[III]K[Fe(CN)_6]$). Bivalent Fe and thiosulfate are titrated with K bichromate, ceric sulfate, or K permanganate solutions in the presence of the resin in the diphenylamine form. Chloride, bromide, and iodide can be titrated with Ag nitrate solution in the presence of resin beads containing *p*-dimethylamino-benzylidenerhodanine.

Resin Bead Technique

This procedure, which is also called the single-bead method, is described in Tables 65 and 71 in the chapter on Actinides (Volume II) and in Table 10 in the chapter on Zirconium and Hafnium (Volume IV). In all cases a few beads of an anion exchange resin are loaded with the element, i.e., U, Pu, Zr, etc., and a single bead is introduced into the filament of a mass spectrometer for quantitation.

Resin Bead Micro Standards

The preparation of standards of this type has been described by several investigators.[370-375] They have been used in connection with activation analysis[372-374] and with the electrothermal atomic absorption spectrometry of metals.[375] The standards are suitable for calibration purposes in the submicrogram range and for the determination of chemical detection limits.

REFERENCES

1. **Jackson, C. J. and Ferrett, J.,** The role of automation as applied to ion chromatography, *Anal. Proc. (London),* 21(11), 420, 1984.
2. **Lyle, S. J. and Pearson, C. H. G.,** Studies on surface anion exchangers for ion chromatographic determinations, *Anal. Proc. (London),* 22(1), 22, 1985.
3. **Lindgren, M.,** Ion-exchange chromatography in water analyses, *Vatten,* 36(3), 249, 1980.
4. **Franklin, G. O.,** Ion chromatography provides useful analysis of the chemistry of pulping and bleaching liquors, *Tappi,* 65(5), 107, 1982.
5. **Jansen, K. H.,** Chromatography even in inorganic analysis by means of ion chromatography, *GIT Fachz. Lab.,* 22(12), 1062, 1978.
6. **Shpigun, O. A.,** Ion chromatography in the analysis of natural waters, *TrAC Trends Anal. Chem. (Pers. Ed.),* 4(1), 29, 1985.
7. **Edwards, P.,** Ion chromatography. A valuable analytical tool for the food chemist, *Food Technol. (Chicago),* 37(6), 53, 1983.
8. **Rich, W. E.,** Ion chromatography: new technique for automated analysis of ions in solution, *Anal. Instrum.,* 15, 113, 1977.
9. **Masters, M. B.,** Use of ion chromatography in surfactant analysis, *Anal. Proc. (London),* 22(5), 146, 1985.
10. **Holcombe, L. J. and Meserole, F. B.,** Applications of ion chromatography for analysis of utility waste liquors and leachates, *Water Qual. Bull.,* 6(2), 37, 1981.
11. **Resch, G. and Gruenschlaeger, E.,** Ion chromatography — an analytical method for studying water and waste water, *VGB Kraftwerkstechnik,* 62(2), 127, 1982.
12. **Mosko, J. A.,** Automated determination of inorganic anions in water by ion chromatography, *Anal. Chem.,* 56, 629, 1984.
13. **Mulik, J. D. and Sawicki, E.,** Ion chromatography, *Environ. Sci. Technol.,* 13(7), 804, 1979.
14. **Koch, W. F.,** Sample preparation in ion-chromatography, *J. Res. Natl. Bur. Stand. (Washington, D.C.),* 84(3), 241, 1979.
15. **Small, H.,** Modern inorganic chromatography, *Anal. Chem.,* 55, 235A, 1983.
16. **Johnson, E. L.,** Modern ion chromatography, *Int. Lab.,* 14(3), 110, 1982.
17. **Pensenstadler, D. F. and Fulmer, M. A.,** Pure steam ahead, *Anal. Chem.,* 53, 859A, 1981.
18. **Jansen, K. H.,** Ion chromatography — a new technique in analytical chemistry, *Labor Praxis,* No. 3, 1, 1978.
19. **Kempf, T. and Sonneborn, M.,** Use of ion-exchangers in the determination of trace elements in waters, *Mikrochim. Acta,* 1(3-4), 207, 1981.
20. **MacDonald, J. C.,** Ion chromatography, *Am. Lab. (Fairfield, Conn.),* 11(1), 45, 1979.
21. **Jardy, A. and Rosset, R.,** Coupling ion exchange and conductometric detection: ion chromatography, *Analusis,* 7(6), 259, 1979.
22. **Lipski, A. J. and Vairo, C. J.,** Applications of ion chromatography in an analytical services laboratory, *Can. Res.,* 13(1), 45, 1980.
23. **Nickless, G.,** Trace metal determination by chromatography, *J. Chromatogr.,* 313, 129, 1984.
24. **Bag, S. P.,** Ion chromatography: a novel monitoring method for pollution, *Chem. Age India,* 36(1), 85, 1985.
25. **Anderson, C.,** Ion chromatography: new technique for clinical chemistry, *Clin. Chem. (Winston Salem, N.C.),* 22(9), 1424, 1976.
26. **Jupille, T., Burge, D., and Togami, D.,** Ion chromatography uses only one column to get all the ions, *Res. Dev.,* 26(3), 135, 1984.
27. **Westwell, A.,** Industrial applications of ion chromatography, *Anal. Proc. (London),* 21(9), 320, 1984.
28. **Mu, S. and Chen, L.,** Ion chromatography, *Fenxi Huaxue,* 11(3), 232, 1983.
29. **Masters, M. B.,** Latest developments in ion chromatography, *Anal. Proc. (London),* 21(9), 322, 1984.
30. **Coggan, C. E.,** Use of ion chromatography in water analysis, *Anal. Proc. (London),* 19(12), 567, 1982.
31. **Cooke, M.,** Ion chromatography, *Anal. Proc. (London),* 21(9), 321, 1984.
32. **Basili, N.,** Ion chromatography, *Com. Naz. Energ. Nucl. (Rapp. Tec.) CNEN-RT/CHI(Italy),CNEN-RT/CHI(82),* 3, 25, 1982.
33. **Mulik, J. D. and Sawicki, E.,** Ion chromatography, *Environ. Sci. Technol.,* 13(7), 804, 1979.
34. **Wetzel, R.,** Ion chromatographs and columns give better separations, *Ind. Res. Dev.,* 24(4), 92, 1982.
35. **Franklin, G.,** Ion chromatography permits rapid process analysis, trouble shooting, *Pulp Pap.,* 56(2), 91, 1982.
36. **Vialle, J.,** Modern ion-separation techniques in inorganic analysis: ion chromatography and isotachophoresis, *TrAC Trends Anal. Chem. (Pers. Ed.),* 3(2), 61, 1984.
37. **Matsushita, S., Tada, Y., Baba, N., and Hosako, K.,** High performance ion chromatography of anions, *J. Chromatogr.,* 259(3), 459, 1983.

38. **Girard, J. E. and Glatz, J. A.,** Ion chromatography with conventional h.p.l.c. instrumentation, *Int. Lab.*, 11(8), 62, 1981.
39. **Smith, R. E.,** Applications of ion chromatography, Report BDX-613-2865, U.S. Department of Energy, 1983.
40. **Johnson, E.,** Modern ion chromatography, *GIT Fachz. Lab.*, 26(3), 241, 1982.
41. **Haddad, P. R. and Cowie, C. E.,** Computer-assisted optimization of eluent concentration and pH in ion chromatography, *J. Chromatogr.*, 303(2), 321, 1984.
42. **Heisz, O.,** Detection methods in ion chromatography, *GIT Fachz. Lab.*, 29(2), 113, 1985.
43. **Hershcovitz, H., Yarnitzky, Ch., and Schmuckler, G.,** Quantitative interpretation of the injected peak in ion chromatography, *J. Chromatogr.*, 244, 217, 1982.
44. **Fritz, J. S., Gjerde, D. T., and Pohlandt, C.,** *Ion Chromatography*, Hüthig Verlag, Heidelberg, 1982.

44a. **Franklin, G. O.,** Development and applications of ion chromatography, *Int. Lab.*, 15(6), 56, 1985.

44b. **Colenutt, B. A. and Trenchard, P. J.,** Ion chromatography and its application to environmental analysis: a review, *Environ. Pollut. Ser. B*, 10(2), 77, 1985.

44c. **Cox, D., Harrison, G., Jandik, P., and Jones, W.,** Application of ion chromatography in the food and beverage industry, *Food Technol. (Chicago)*, 39(7), 41, 1985.

45. **Sato, H.,** Practical application of ion-exchange chromatography with a high-sensitivity differential conductivity meter, *Jpn. Analyst*, 31(3), T23, 1982.
46. **Sato, H.,** Ion-exchange chromatography-conductivity detection with a low capacity column and a low-concentration eluent, *Jpn. Analyst*, 31(2), 97, 1982.
47. **Keller, J. M.,** Bipolar-pulse conductivity detector for ion chromatography, *Anal. Chem.*, 53, 344, 1981.
48. **Jenke, D. R. and Pagenkopf, G. K.,** Temperature fluctuations in nonsuppressed ion chromatography, *Anal. Chem.*, 54, 2603, 1982.
49. **Small, H., Stevens, T. S., and Bauman, W. C.,** Novel ion exchange chromatographic method using conductimetric detection, *Anal. Chem.*, 47, 1801, 1975.
50. **Chakraborti, D., Hillmann, D. C. J., Zingaro, R. A., and Irgolic, K. J.,** Ion-chromatographic determination of fluorine in Texas lignite, *Fresenius' Z. Anal. Chem.*, 319, 556, 1984.
51. **Goebl, M.,** The rectilinearity problem of conductivity detection in ion chromatography. II. Behavior of the detector in measurement of the conductivity of ionic species in aqueous solution, *GIT Fachz. Lab.*, 27(5), 373, 1983.
52. **Jenke, D. R. and Pagenkopf, G. K.,** Modelling of analyte response to changing eluent composition in suppressed ion chromatography, *J. Chromatogr.*, 269, 202, 1983.
53. **Van Os, M. J., Slanina, J., De Ligny, C. L., and Agterdenbos, J.,** Linear calibration in ion chromatography by calculating total amounts of sample from measured conductivity data, *Anal. Chim. Acta*, 156, 169, 1984.
54. **Slanina, J., Bakker, F. P., Jongejan, P. A. C., Van Lamoen, L., and Möls, J. J.,** Fast determination of ions by computerized ion chromatography coupled with selective detectors, *Anal. Chim. Acta*, 130, 1, 1981.
55. **Tarter, J. G.,** Gradient elution ion-chromatographic determination of inorganic anions using a continuous gradient, *Anal. Chem.*, 56, 1264, 1984.
56. **Sundén, T., Lindgren, M., Cedergren, A., and Siemer, D. D.,** Separation of sulfite, sulfate and thiosulfate by ion chromatography with gradient elution, *Anal. Chem.*, 55, 2, 1983.
57. **Stevens, T. S. and Langhorst, M. A.,** Agglomerated pellicular anion exchange columns for ion chromatography, *Anal. Chem.*, 54, 950, 1982.
58. **Wetzel, R. A., Anderson, C. L., Schleicher, H., and Crook, G. D.,** Determination of trace level ions by ion chromatography with concentrator columns, *Anal. Chem.*, 51, 1532, 1979.
59. **Pohlandt, C.,** Assessment of pellicular anion exchange resins for the determination of anions by ion chromatography, Report NIM-2132, National Institute for Metallurgy, Randburg, South Africa, October 1981.
60. **Reigler, P. F., Smith, N. J., and Turkelson, V. T.,** Solvent extraction-ion chromatography determination of chloride in liquid bromine, *Anal. Chem.*, 54, 84, 1982.
61. **Burns, I. W.,** Separation and determination of anions using reversed phase h.p.l.c. columns, *Anal. Proc. (London)*, 21(6), 200, 1984.
62. **Ivey, J. P.,** Zwitter-ionic eluents for suppressed-ion chromatography, *J. Chromatogr.*, 287, 128, 1984.
63. **Van Os, M. J., Slanina, J., De Ligny, C. L., Hammers, W. E., and Agterdenbos, J.,** Determination of traces of inorganic anions by means of high-performance liquid chromatography on Zipa-SAX columns, *Anal. Chim. Acta*, 144, 73, 1982.
64. **Sanemasa, I., Mizoguchi, T., Ohtsuka, J., Deguchi, T., and Nagai, H.,** Simple preparation of separation columns for ion chromatography, *Jpn. Analyst*, 32(7), 420, 1983.
65. **Lindgren, M., Cedergren, A., and Lindberg, J.,** Conditions for sulfite stabilisation and determination by chromatography, *Anal. Chim. Acta*, 141, 279, 1982.

66. **Koch, W. F.,** Complication in determination of nitrite by ion chromatography, *Anal. Chem.,* 51, 1571, 1979.
67. **Stevens, T. S., Davis, J. C., and Small, H.,** Hollow fiber ion-exchange suppressor for ion chromatography, *Anal. Chem.,* 53, 1488, 1981.
68. **Hoover, T. B. and Yager, G. D.,** Comparison of collection procedures for the re-injection ion chromatography of water, *J. Chromatogr. Sci.,* 22(10), 435, 1984.
69. **Rokushika, S., Quin, Z. Y., and Hatano, H.,** Micro-column ion chromatography with hollow fiber suppresssor, *J. Chromatogr.,* 260(1), 81, 1983.
70. **Nagashima, H., Kuboyama, K., and Ono, K.,** Simultaneous multi-elementary analysis for fluorine, chlorine, bromine, and sulfur in organic compounds by ion chromatography, *Jpn. Analyst,* 34(7), 381, 1985.
71. **Mirna, A., Wagner, H., Kloetzer, E., and Fausel, E.,** Use of ion chromatography for the investigation of meat and meat products, *Lebensmittelchem. Gerichtl. Chem.,* 38(1), 18, 1984.
72. **Stevens, T. S., Jewett, G. L., and Bredeweg, R. A.,** Packed hollow-fiber suppressors for ion chromatography, *Anal. Chem.,* 54, 1206, 1982.
73. **Jones, V. K. and Tarter, J. G.,** Single-injection ion-chromatographic analysis of both anions and cations, *J. Chromatogr.,* 312, 456, 1984.
74. **Dasgupta, P. K.,** Annular helical suppressor for ion chromatography, *Anal. Chem.,* 56, 103, 1984.
75. **Dasgupta, P. K.,** Ion-chromatographic separation of anions with ion-interaction reagents and an annular helical suppressor, *Anal. Chem.,* 56, 769, 1984.
76. **Dasgupta, P. K., Bligh, R. Q., and Mercurio, M. A.,** Dual-membrane annular helical suppressors in ion chromatography, *Anal. Chem.,* 57, 484, 1985.
77. **Siemer, D. D. and Johnson, V. J.,** Silicone rubber tubing for elimination of background conductivity in anion chromatography, *Anal. Chem.,* 56, 1033, 1984.
78. **Sunden, T., Cedergren, A., and Siemer, D. D.,** Carbon dioxide-permeable tubing for post-suppression in ion chromatography, *Anal. Chem.,* 56, 1085, 1984.
79. **Faucher, M. and Bisson, M.,** New analytical method for the chemical characterization of acidic precipitations, *Eau Qúebec,* 15(4), 377, 1982.
80. **Watanabe, I., Tanaka, R., and Kashimoto, T.,** Determination of potassium bromate by ion chromatography, *J. Food Hyg. Soc. Jpn.,* 23(2), 135, 1982.
81. **Hoover, T. B. and Yager, G. D.,** Determination of trace anions in water by multidimensional ion chromatography, *Anal. Chem.,* 56, 221, 1984.
82. **Tyree, S. Y. and Bynum, M. A. O.,** A method for nitrate and phosphate in saline water, *Limnol. Oceanogr.,* 29(6), 1337, 1984.
83. **Cox, J. A. and Tanaka, N.,** Application of ion-exchange reactions between membranes and resins, *Talanta,* 32, 34, 1985.
84. **Lee, D. P.,** New anion-exchange phase for ion chromatography, *J. Chromatogr. Sci.,* 22(8), 327, 1984.
85. **Crowther, J. and McBride, J.,** Determination of anions in atmospheric precipitation by ion chromatography, *Analyst (London),* 106, 702, 1981.
86. **Marko-Varga, G., Csiky, I., and Jonsson, J. A.,** Ion chromatographic determination of nitrate and sulfate in natural waters containing humic substances, *Anal. Chem.,* 56, 2066, 1984.
87. **Jancar, J. C., Constant, M. D., and Herwig, W. C.,** Determination of inorganic anions in beer by ion chromatography, *J. Am. Soc. Brew. Chem.,* 42(2), 90, 1984.
88. **Evans, K. L., Tarter, J. G., and Moore, C. B.,** Pyrohydrolytic-ion chromatographic determination of fluorine, chlorine and sulfur in geological samples, *Anal. Chem.,* 53, 925, 1981.
89. **Wilson, S. A. and Gent, C. A.,** Determination of fluoride in geological samples by ion chromatography, *Anal. Lett.,* 15(A10), 851, 1982.
90. **Maketon, S. and Tarter, J. G.,** Ion-chromatographic determination of fluoride and chloride in soils, *LC Liq. Chromatogr. HPLC Mag.,* 2(2), 124, 1984.
91. **Evans, K. L. and Moore, C. B.,** Combustion-ion chromatographic determination of chlorine in silicate rocks, *Anal. Chem.,* 52, 1908, 1980.
92. **Nadkarni, R. A. and Pond, D. M.,** Applications of ion chromatography for determination of selected elements in coal and oil shale, *Anal. Chim. Acta,* 146, 261, 1983.
93. **Nakaoka, H., Umoto, F., Kasano, M., Ikeda, N., Ichimura, K., Ueda, E., and Itano, T.,** Determination of fluoride, chloride and bromide in mineral-spring water by ion chromatography, *Jpn. Analyst,* 30(10), T97, 1981.
94. **Matusiewicz, H. and Natusch, D. F. S.,** Ion-chromatographic determination of soluble anions present in coal fly-ash leachates, *Int. J. Environ. Anal. Chem.,* 8(3), 227, 1980.
95. **Hara, H., Nagara, K., Honda, K., and Goto, A.,** Particle size distribution of inorganic components in atmospheric aerosol. I. Eluent extraction of components for ion-chromatographic determination, *Taiki Osen Gakkaishi,* 15(9), 380, 1980.

96. **Fujii, T.,** Rapid determination of nitrogen oxides in exhaust gases by ion chromatography, *Jpn. Analyst,* 31(4), 214, 1982.
97. **Cole, D. E. C. and Scriver, C. R.,** Microassay of inorganic sulfate in biological fluids by controlled flow anion chromatography, *J. Chromatogr.,* 225, 359, 1981.
98. **De Jong, P. and Burggraaf, M.,** Ion chromatographic method for simultaneous determination of inorganic phosphate, bromide, nitrate, and sulfate in human serum, *Clin. Chim. Acta,* 132(1), 63, 1983.
99. **Forrest, J., Spandau, D. J., Tanner, R. L., and Newman, L.,** Determination of atmospheric nitrate and nitric acid employing a diffusion denuder with a filter pack, *Atmos. Environ.,* 16(6), 1473, 1982.
100. **Mizobuchi, M., Ohmae, H., Umoto, F., Tanaka, T., Ichimura, K., Ueda, E., and Itano, T.,** Phosphate ion and combined phosphorus in urine, *Clin. Chem. (Winston Salem, N.C.),* 28(8), 1823, 1982.
101. **Whittaker, J. W. and Lemke, P. R.,** Ion-chromatographic determination of the principal inorganic ion in bulk antibiotic salts, *J. Pharm. Sci.,* 71(3), 334, 1982.
102. **Ishibashi, W., Kikuchi, R., and Yamamoto, K.,** Simultaneous determination of chlorine and fluorine in tantalum metal by pyrolysis-ion chromatography, *Jpn. Analyst,* 30(9), 604, 1981.
103. **Ishibashi, W., Kikuchi, R., and Yamamoto, K.,** Ghost peaks in ion chromatography using a sodium carbonate-sodium hydrogen carbonate solution as an eluent, *Jpn. Analyst,* 31(4), 207, 1982.
104. **Wu, W. S., Arai, D. K., Nazar, M. A., and Leong, D. K.,** Determination of nitrites in metal cutting fluids by ion chromatography, *Am. Ind. Assoc. J.,* 43(12), 942, 1982.
105. **Dulski, T. R.,** Determination of acid concentrations in speciality alloy pickling baths by ion chromatography, *Anal. Chem.,* 51, 1439, 1979.
106. **Smith, R. E.,** Determination of chloride in sodium hydroxide and in sulfuric acid by ion chromatography, *Anal. Chem.,* 55, 1427, 1983.
107. **Smith, R. E. and Davis, W. R.,** Determination of chloride and sulfate in chromic acid by ion chromatography, *Plat. Surf. Finish,* 71(11), 60, 1984.
108. **Siemer, D. D.,** Separation of chloride and bromide from complex matrices prior to ion-chromatographic determination, *Anal. Chem.,* 52, 1874, 1980.
109. **Tarter, J. G.,** Determination of non-oxidizable species using electrochemical detection in ion chromatography, *J. Liq. Chromatogr.,* 7(8), 1559, 1984.
110. **Tanaka, T., Hiiro, K., Kawahara, A., and Wakida, S.,** Ion chromatographic determination of hypophosphate, phosphite and phosphate ions, *Jpn. Analyst,* 32(12), 771, 1983.
111. **Smith, F., McMurtrie, A., and Galbraith, H.,** Ion chromatographic determination of sulfur and chlorine using milligram and submilligram sample weights, *Microchim. J.,* 22(1), 45, 1977.
112. **Chen, L. and Mu, S.,** Determination of sulfate, nitrate and phosphate in soils by ion chromatography, *Fenxi Huaxue,* 11(2), 88, 1983.
113. **Wang, C. Y., Bunday, S. D., and Tartar, J. G.,** Ion chromatographic determination of fluorine, chlorine, bromine, and iodine with sequential electrochemical and conductometric detection, *Anal. Chem.,* 55, 1617, 1983.
114. **Vinjamoori, D. V. and Ling, C. S.,** Personal monitoring method for nitrogen dioxide and sulfur dioxide with solid-sorbent sampling and ion-chromatographic determination, *Anal. Chem.,* 53, 1689, 1981.
115. **Steiber, R. and Merrill, R.,** Determination of arsenic as the oxidate by ion chromatography, *Anal. Lett. Part A,* 12(3), 273, 1979.
116. **McCormick, M. J.,** Determination of total sulfur in fuel oils by ion chromatography, *Anal. Chim. Acta,* 121, 233, 1980.
117. **Trujillo, F. J., Miller, M. M., Skogerboe, R. K., Taylor, H. E., and Grant, C. L.,** Ion chromatographic determination of thiosulfate in oil shale leachates, *Anal. Chem.,* 53, 1944, 1981.
118. **Takamine, K., Tanaka, S., and Hashimoto, Y.,** Simultaneous determination of hydrogen chloride and sulfur dioxide in the atmosphere by ion chromatography after collection on alkali filters, *Jpn. Analyst,* 31(12), 692, 1982.
119. **Jenke, D. R., Mitchell, P. K., and Pagenkopf, G. K.,** Anion content of snow by suppressed and non-suppressed ion chromatography, *J. Chromatogr. Sci.,* 21(11), 487, 1983.
120. **Fratz, D. D.,** Automated determination of salts in water-soluble certifiable color additives by ion chromatography, *J. Assoc. Off. Anal. Chem.,* 63(4), 882, 1980.
121. **Itoh, H. and Shinbori, Y.,** Determination of anions in seawater by ion chromatography, *Jpn. Analyst,* 29(4), 239, 1980.
122. **Sayre, W. G. and Constable, D. C.,** Ion-chromatographic determination of selenite and selenate in aqueous solutions, *Acqua Aria,* No. 2, 113, 1983.
123. **Gustafson, F. J., Markwell, C. G., and Simpson, S. M.,** Determination of chloride, bromide and iodide in silver halides by ion exchange and ion interaction chromatography, *Anal. Chem.,* 57, 621, 1985.
124. **Lai, S. T., Nishina, M. M., and Sangermano, L.,** Multi-dimensional column ion-chromatographic determination of tetrafluoroborate and phosphate, *HRC CC J. High Resolut. Chromatogr. Commun.,* 7(6), 336, 1984.

125. **Mulik, J., Puckett, R., Williams, D., and Sawicki, E.,** Ion-chromatographic analysis of sulfate and nitrate in ambient aerosols, *Anal. Lett.*, 9(7), 653, 1976.
126. **Dick, W. A. and Tabatabai, M. A.,** Ion-chromatographic determination of sulfate and nitrate in soils, *Soil Sci. Soc. Am. J.*, 43, 899, 1979.
127. **Korth, W. and Ellis, J.,** Ion chromatographic determination of chloride and fluoride in electrolyte from the halogen tin-plating process, *Talanta*, 31(6), 467, 1984.
128. **Takamatsu, T., Kawashima, M., and Koyama, M.,** Ion chromatographic determination of arsenite and arsenate in sediment extract, *Jpn. Analyst*, 28(10), 596, 1979.
129. **Schwabe, R., Darimont, T., Möhlmann, I., Pabel, E., and Sonneborn, M.,** Determination of inorganic compounds and organic acids in different types of water by ion chromatography, *Int. J. Environ. Anal. Chem.*, 14(3), 196, 1983.
130. **Takami, K., Ohkawa, K., Kuge, Y., and Nakamoto, M.,** Determination of total nitrogen and phosphorus in river water and industrial waste water by ion chromatography, *Jpn. Analyst*, 31(7), 362, 1982.
131. **Lash, R. P. and Hill, C. J.,** Evaluation of ion chromatography for determination of selected ions in geothermal well-water, *Anal. Chim. Acta*, 108, 405, 1979.
132. **Darimont, T., Schulze, G., and Sonneborn, M.,** Determination of nitrate in drinking water by means of ion chromatography, *Fresenius' Z. Anal. Chem.*, 314, 383, 1983.
133. **Hill, C. J. and Lash, R. P.,** Ion-chromatographic determination of boron as tetrafluoroborate, *Can. Res.*, 13(1), 53, 1980.
134. **Hill, C. J. and Lash, R. P.,** Ion chromatographic determination of boron as tetrafluoroborate, *Anal. Chem.*, 52, 24, 1980.
135. **Hansen, L. D., Richter, B. E., Rollins, D. K., Lamb, J. D., and Eatough, D. J.,** Determination of arsenic and sulfur species in environmental samples by ion chromatography, *Anal. Chem.*, 51, 633, 1979.
136. **Smith, D. L., Kim, W. S., and Kupel, R. E.,** Determination of sulfur dioxide by adsorption on a solid sorbent followed by ion chromatography, *Am. Ind. Hyg. Assoc. J.*, 41(7), 485, 1980.
137. **Bouyoucos, S. A., Melcher, R. G., and Vaccaro, J. R.,** Collection and determination of sulfuryl fluoride in air by ion chromatography, *Am. Ind. Hyg. Assoc. J.*, 44(1), 57, 1983.
138. **Tanaka, S., Yoshimori, T., and Hashimoto, Y.,** Solvent separation and determination by ion-chromatography of sulfuric acid, ammonium hydrogen sulfate and ammonium sulfate in the atmosphere, *Jpn. Analyst*, 32(12), 735, 1983.
139. **Itoh, H. and Shinbori, Y.,** Ion chromatography of anions on a "fast run" column, *Jpn. Analyst*, 31(4), T39, 1982.
140. **Koch, W. F. and Stolz, J. W.,** Analysis of chloride doped cadmium sulfide by ion chromatography, *Anal. Chem.*, 54, 340, 1982.
141. **Latimer, J. N., Bush, W. E., Higgins, L. J., and Shay, R. S.,** in Handbook of Analytical Procedures, Report RMO-3008, U.S. Atomic Energy Commission, February 1984.
142. **Green, L. W. and Woods, J. R.,** Ion-chromatographic determination of anions in waste water precipitate, *Anal. Chem.*, 53, 2187, 1981.
143. **Stevens, T. S., Turkelson, V. T., and Albe, W. R.,** Determination of anions in boiler blow-down water with ion chromatography, *Anal. Chem.*, 49, 1176, 1977.
144. **Bynum, M. A. O., Tyree, S. Y., and Weiser, W. E.,** Effects of major ions on the determination of trace ions by ion chromatography, *Anal. Chem.*, 53, 1935, 1981.
145. **Bodek, I. and Smith, R. H.,** Determination of ammonium sulfamate in air using ion chromatography, *Am. Ind. Hyg. Assoc. J.*, 41(8), 603, 1980.
146. **Zolotov, Yu. A., Shpigun, O. A., and Bubchikova, L. A.,** Determination of arsenic, molybdenum, tungsten, and chromium by ion chromatography, *Dokl. Akad. Nauk SSSR*, 266(5), 1144, 1982.
147. **Zolotov, Yu.A., Shpigun, O. A., Bubchikova, L. A., and Sedel'nikova, E. A.,** Ion-chromatography — method for automatic determination of ions. Determination of selenium, *Dokl. Akad. Nauk SSSR*, 263(4), 889, 1982.
148. **Zolotov, Yu.A., Shpigun, O. A., and Bubchikova, L. A.,** Ion-chromatographic separation and determination of selenium, arsenic, molybdenum, tungsten, and chromium as their oxyanions, *Fresenius' Z. Anal. Chem.*, 316, 8, 1983.
149. **Kim, W. S., McGlothlin, J. D., and Kupel, R. E.,** Sampling and analysis of iodine in the industrial atmosphere, *Am. Ind. Hyg. Assoc. J.*, 42(3), 187, 1981.
150. **Kan, M., Ohnishi, K., and Shintani, M.,** Application of ion-exchange chromatography to the determination of halogens and sulfur in organic compounds, *Yakugaku Zasshi*, 104(7), 763, 1984.
151. **Pabon, V. M. and Bodnya, V. A.,** Separation of mercury(II) and selenium(IV) by isopropyl(methyl)thiocarbamate from nitrate and halide solutions, *Vestn. Mosk. Univ. Ser. 2 Khim.*, 24(6), 575, 1983.
152. **Ficklin, W. H.,** Separation of tungstate and molybdate by ion chromatography and its application to natural waters, *Anal. Lett.*, 15(A10), 865, 1982.

153. **Chen, S. G., Cheng, K. L., and Vogt, C. R.,** Ion exchange separation of some aminopolycarboxylic acids and inorganic anions, *Mikrochim. Acta,* 1(5-6), 473, 1983.
154. **Chakraborti, D., Hillman, D. C. J., Irgolic, K. J., and Zingaro, R. A.,** Hitachi Zeeman graphite-furnace atomic-absorption spectrometer as a selenium-specific detector for ion chromatography. Separation and determination of selenite and selenate, *J. Chromatogr.,* 249(1), 81, 1982.
155. **Pettersen, J. M.,** Determination of free chromate in ligninsulfonate dispersants by ion chromatography with atomic absorption spectrometric detection, *Anal. Chim. Acta,* 160, 263, 1984.
156. **Nakamura, K. and Morikawa, Y.,** Studies on high-speed liquid chromatography in cosmetic analysis. XI. Determination of sulfur in cosmetic products by ion chromatography, *Jpn. Analyst,* 32(4), 224, 1983.
157. **Oehme, M. and Stray, H.,** Comparison of ion-chromatographic and potentiometric techniques for determination of gaseous and particulate fluoride in air, *Fresenius' Z. Anal. Chem.,* 306, 356, 1981.
158. **Cheney, J. L. and Duke, D. L.,** Effects of gaseous hydrochloric acid and source particulate characterization, *Anal. Lett.,* 16(A4), 309, 1983.
159. **Wilson, S. A. and Gent, C. A.,** Determination of chloride in geological samples by ion chromatography, *Anal. Chim. Acta,* 148, 299, 1983.
160. **Oikawa, K. and Saito, H.,** Ion chromatographic determination of anions in rainwater, *Chemosphere,* 11(9), 933, 1982.
161. **Dolzine, T. W., Esposito, G. G., and Rinehart, D. S.,** Determination of hydrogen cyanide in air by ion chromatography, *Anal. Chem.,* 54, 470, 1982.
162. **Watanabe, I., Tanake, R., and Kashimoto, T.,** Determination of potassium bromate by ion chromatography, *Shokuhin Eiseigaku Zasshi,* 22(3), 246, 1981.
163. **Oikawa, K., Saito, H., Sakazume, S., and Fujii, M.,** Ion chromatographic determination of bromate in bread, *Jpn. Analyst,* 31(8), E251, 1982.
164. **Saitoh, H. and Oikawa, K.,** Simultaneous determination of nitrite and nitrate ions in foods by ion-chromatography with an ultra-violet spectrophotometric detector, *Jpn. Analyst,* 33, E441, 1984.
165. **Hansen, L. D., Richter, B. E., Rollins, D. K., Lamb, J. D., and Eatough, D. J.,** Determination of arsenic and sulfur species in environmental samples by ion chromatography, *Anal. Chem.,* 51, 633, 1979.
166. **Gjerde, D. T. and Fritz, J. S.,** Sodium and potassium benzoate and benzoic acid as eluents for ion chromatography, *Anal. Chem.,* 53, 2324, 1981.
167. **Pohlandt, C.,** Determination of anions by ion chromatography, *S. Afr. J. Chem.,* 33(3), 87, 1980.
168. **Katoh, K.,** Determination of bromide ion in seawater by solvent extraction and ion chromatography, *Jpn. Analyst,* 32(10), 567, 1983.
169. **Mueller, K. P.,** Ion chromatography in precipitation water, *Fresenius' Z. Anal. Chem.,* 317, 345, 1984.
170. **Yan, D. R., Roessner, B., and Schwedt, G.,** Comparison of titrimetric, spectrophotometric and ion-chromatographic methods for determination of hydrogen carbonate in drinking and mineral waters, *Anal. Chim. Acta,* 162, 451, 1984.
171. **Legrand, M., De Angelis, M., and Delmas, R. J.,** Ion chromatographic determination of common ions at ultra trace levels in Antarctic snow and ice, *Anal. Chim. Acta,* 156, 181, 1984.
172. **Smee, B. W., Hall, G. E. M., and Koop, D. J.,** Analysis for fluoride, chloride, nitrate, and sulfate in natural waters, using ion chromatography, *J. Geochem. Explor.,* 10(3), 245, 1978.
173. **Tabatabai, M. A. and Dick, W. A.,** Simultaneous determination of nitrate, chloride, sulfate, and phosphate in natural waters by ion chromatography, *J. Environ. Qual.,* 12(2), 209, 1983.
174. **Morrow, C. M. and Minear, R. A.,** Determination of bromide in natural waters by ion chromatography using a concentrator column, *Water Res.,* 18(9), 1165, 1984.
175. **Forrest, J., Spandau, D. J., Tanner, R. J., and Newman, L.,** Determination of atmospheric nitrate and nitric acid employing a diffusion denuder with a filter pack, *Atmos. Environ.,* 16(6), 1473, 1982.
176. **Alkezweeny, A. J., Laulainen, N. S., and Thorp, J. M.,** Physical, chemical and optical characteristics of a clean air mass over northern Michigan, *Atmos. Environ.,* 16(10), 2421, 1982.
177. **Slanina, J., Lamoen-Doornenbal, L. V., Lingerak, W. A., Meilof, W., Klocklow, D., and Niessner, R.,** Application of a thermo-denuder analyser to determination of sulfuric acid, nitric acid and ammonia in air, *Inst. J. Environ. Anal. Chem.,* 9(1), 59, 1981.
178. **Stevens, R. K., Dzubay, T. G., Russwurm, G., and Rickel, D.,** Sampling and analysis of atmospheric sulfates and related species, *Atmos. Environ.,* 12(1 to 3), 55, 1978.
179. **Solomon, P., Derrick, M., Moyers, J., and Hyde, P.,** Performance comparison of three samplers of suspended air-borne particulate matter, *J. Air Pollut. Control. Assoc.,* 32(4), 373, 1982.
180. **Esposito, G.,** Determination of hydrogen cyanide in air by ion chromatography, *Anal. Chem.,* 57, 1168, 1985.
181. **Miller, D. P.,** Ion-chromatographic analysis of Palmes tubes for nitrite, *Atmos. Environ,* 18(4), 891, 1984.
182. **Wagenknecht, J. H., Jansson, R. E. W., and Stover, F. S.,** Analysis of mixtures of chlorine and chlorine dioxide, *Anal. Lett. Part A,* 14(3), 197, 1981.
183. **Fortune, C. R. and Dellinger, B.,** Stabilization and analysis of sulfur (IV) aerosols in environmental samples, *Environ. Sci. Technol.,* 16(1), 62, 1982.

184. **Dasgupta, P. K.,** Ion-chromatographic determination of sulfur (IV), *Atmos. Environ.*, 16(5), 1265, 1982.
185. **Cole, D. E. C.,** Micro-assay of plasma sulfate, *Clin. Chem. (Winston Salem, N. C.)*, 30(8), 1421, 1984.
186. **Cole, D. E. C. and Landry, D. A.,** Determination of inorganic sulfate in human saliva and sweat by controlled-flow anion chromatography. Normal values in adult humans, *J. Chromatogr.*, 337, 267, 1985.
187. **Bartonek, G. and Werner, H.,** Simultaneous ion-chromatographic determination of chlorine and sulfur in plant matrix, *GIT Fachz. Lab.*, 27(12), 1075, 1983.
188. **Viswanadham, P., Smick, D. R., Pisney, J. J., and Dilworth, W. F.,** Comparison of ion chromatography and titrimetry for determination of sulfur in fuel oils, *Anal. Chem.*, 54, 2431, 1982.
189. **Pohlandt, C. and Cameron, A.,** The determination by ion chromatography of chlorine, bromine, phosphorus, and sulfur in organic materials, Report M153, Council for Mineral Technology, Randburg, South Africa, September 1984.
190. **McCormick, M. J. and Dixon, L.,** Determination of sulfite in fixers and photographic effluents by ion chromatography, *J. Chromatogr.*, 322(3), 478, 1985.
191. **Cox, J. A. and Tanaka, N.,** Determination of anionic impurities in selected concentrated electrolytes by ion chromatography, *Anal. Chem.*, 57, 383, 1985.
192. **Franklin, G. O. and Fitchett, A. W.,** Fast chemical characterisation of pulping and bleaching process liquors by ion chromatography, *Pulp. Pap. Can.*, 83(10), 40, 1982.
193. **Moses, C. O., Nordstrom, D. K., and Mills, A. L.,** Sampling and analysing mixtures of sulfate, sulfite, thiosulfate, and polythionate, *Talanta*, 31, 331, 1984.
194. **Nordmeyer, F. R., Hansen, L. D., Eatough, D. J., Rollins, D. K., and Lamb, J. D.,** Determination of alkaline earth and divalent transition metal cations by ion chromatography with sulfate-suppressed barium and lead eluents, *Anal. Chem.*, 52, 852, 1980.
195. **Lamb, J. D., Hansen, L. D., Patch, G. G., and Nordmeyer, F. R.,** Iodate-suppressed lead eluent for ion chromatographic determination of bivalent cations, *Anal. Chem.*, 53, 749, 1981.
196. **Wimberley, J. W.,** Zn^{2+}-nitric acid eluent for the ion chromatographic separation of alkaline earth metals, *Anal. Chem.*, 53, 2137, 1981.
197. **Kifune, I. and Oikawa, K.,** Determination of traces of ammonia and lower amines in the atmosphere by ion-chromatography, *Jpn. Analyst*, 28(10), 587, 1979.
198. **Tanaka, K., Ishihara, Y., and Nakajima, K.,** Determination of alkali- and alkaline earth metal cations by ion chromatography with coulometric detection, *Jpn. Analyst*, 32(10), 626, 1983.
199. **Sato, H.,** Ion exchange chromatography of alkali-metal and alkaline earth metal ions with conductivity detection, *Jpn. Analyst*, 32(10), 610, 1983.
200. **Wimberley, J. W.,** Ion-chromatographic separation of cations on an anion separator column, *Anal. Chem.*, 53, 1709, 1981.
201. **Basta, N. T. and Tabatabai, M. A.,** Determination of total potassium, sodium, calcium, and magnesium in plant materials by ion-chromatography, *Soil Sci. Soc. Am. J.*, 49(1), 76, 1985.
202. **Hajós, P. and Inczédy, J.,** Preparation and ion chromatographic application of surface-sulfonated cation exchangers, *J. Chromatogr.*, 201, 253, 1980.
203. **Mizobuchi, M., Ohmae, H., Umoto, F., Tanaka, T., Ichimura, K., Ueda, E., and Itano, T.,** Concentration of three types of nitrogen in human urine, *Clin. Chem. (Winston Salem, N.C.)*, 29(2), 408, 1983.
204. **Buechele, R. C. and Reutter, D. J.,** Effect of methanol on the mobile phase on the ion chromatographic determination of some monovalent cations, *J. Chromatogr.*, 240(2), 502, 1982.
205. **Bouyoucos, S. A. and Melcher, R. G.,** Collection and ion-chromatographic determination of ammonia and methyl amines in air, *Am. Ind. Hyg. Assoc. J.*, 44(2), 119, 1983.
206. **Duval, D. L. and Fritz, J. S.,** Coated anion-exchange resins for ion chromatography, *J. Chromatogr.*, 295(1), 89, 1984.
207. **Duval, D. L., Fritz, J. S., and Gjerde, D. T.,** Indirect determination of cyanide by single-column ion chromatography, *Anal. Chem.*, 54, 830, 1982.
208. **Barron, R. E. and Fritz, J. S.,** Effect of functional-group structure and exchange capacity on the selectivity of anion exchangers for bivalent anions, *J. Chromatogr.*, 316, 201, 1984.
209. **Fritz, J. S., Duval, D. L., and Barron, R. E.,** Organic acid eluents for single-column chromatography, *Anal. Chem.*, 56, 117, 1984.
210. **Gjerde, D. T., Schmuckler, G., and Fritz, J. S.,** Anion chromatography with low-conductivity eluents. II, *J. Chromatogr.*, 187, 35, 1980.
211. **Gjerde, D. T., Fritz, J. S., and Schmuckler, G.,** Anion-chromatography with low-conductivity eluents, *J. Chromatogr.*, 186, 509, 1979.
212. **Pohlandt, C.,** Separation and determination of anions by ion-chromatography, Report NIM 2044, National Institute of Metallurgy, Randburg, South Africa, February 1980.
213. **Roberts, K. M., Gjerde, D. T., and Fritz, J. S.,** Single-column ion chromatography for the determination of chloride and sulfate in steam condensate and boiler feed water, *Anal. Chem.*, 53, 1691, 1981.
214. **Jenke, D. R. and Pagenkopf, G. K.,** Models for prediction of retention in non-suppressed ion chromatography, *Anal. Chem.*, 56, 88, 1984.

215. **Okada, T. and Kuwamoto, T.,** Determination of anions in environmental water by non-suppressed ion chromatography, *Jpn. Analyst,* 32(10), 595, 1983.
216. **Girard, J. E. and Badio, D. Y.,** Capacity variation of silica-based single-column ion chromatography columns, *Anal. Chem.,* 56, 2992, 1984.
217. **Okada, T. and Kuwamoto, T.,** Ion chromatographic determination of silicic acid in natural water, *Anal. Chem.,* 57, 258, 1985.
218. **Chang, C. A. and Fong, K. L.,** Substitution-inert metal complex mobile phases in non-suppressed anion chromatography, *J. Chromatogr.,* 312, 99, 1984.
218a. **Heckenberg, A. L. and Haddad, P. R.,** Studies on sample preconcentration in ion chromatography. II. Automated, single-pump pre-concentration system for non-suppressed ion chromatography with conductivity detection, *J. Chromatogr.,* 330(1), 95, 1985.
219. **Hertz, J. and Baltensperger, U.,** Determination of nitrate and other inorganic anions in salad and vegetables by ion chromatography, *Fresenius' Z. Anal. Chem.,* 318, 121, 1984.
220. **Okada, T. and Kuwamoto, T.,** Sensitivity of non-suppressed ion chromatography using divalent organic acids as eluents, *J. Chromatogr.,* 284(1), 149, 1984.
221. **Hill, R. A.,** Clean-up technique for samples containing high-level interferences prior to ion chromatographic analysis, *HRC CC J. High Resolut. Chromatogr. Chromatogr. Commun.,* 6(5), 275, 1983.
222. **Mackie, H., Speciale, S. J., Throop, L. J., and Yang Taiyan,** Ion chromatographic determination of the azide ion in a prealbumin fraction from human serum, *J. Chromatogr.,* 242(1), 177, 1982.
223. **Hern, J. A., Rutherford, G. K., and VanLoon, G. W.,** Determination of chloride, nitrate, sulfate, and total sulfur in environmental samples by single-column ion chromatography, *Talanta,* 30, 677, 1983.
224. **Okada, T. and Kuwamoto, T.,** Non-suppressor ion chromatography of inorganic and organic anions with potassium hydroxide as eluent, *Anal. Chem.,* 55, 1001, 1983.
225. **Matsushita, S., Tada, Y., Komiya, K., and Ono, A.,** Separation characteristics of silica-based ion exchanger for ion chromatography, *Jpn. Analyst,* 32(10), 562, 1983.
226. **Ivey, J. P.,** Eluent conductivity suppression by thermal decarboxylation, *J. Chromatogr.,* 281, 314, 1983.
227. **Willison, M. J. and Clarke, A. G.,** Analysis of atmospheric aeresols by non-suppressed ion chromatography, *Anal. Chem.,* 56, 1037, 1984.
228. **Okada, T. and Kuwamoto, T.,** Determination of silicate by non-suppressed ion chromatography, *Anal. Lett.,* 17(A15), 1743, 1984.
229. **Streckert, H. H. and Epstein, B. D.,** Comparison of suppressed and non-suppressed ion chromatography for determination of chloride in boric acid, *Anal. Chem.,* 56, 21, 1984.
230. **Glatz, J. A. and Girard, J. E.,** Factors affecting the resolution and detectability of inorganic anions by non-suppressed ion chromatography, *J. Chromatogr. Sci.,* 20(6), 266, 1982.
231. **Buchholz, A. E., Verplough, C. I., and Smith, J. L.,** Method for the simultaneous measurement of less than a part per million of chloride, nitrate and sulfate in aqueous samples by non-suppressed ion chromatography, *J. Chromatogr. Sci.,* 20(11), 499, 1982.
232. **Jupille, T., Burge, D., and Togami, D.,** High-speed analysis of acid-rain anions by single-column ion chromatography, *Chromatographia,* 16, 312, 1982.
233. **Ivey, J. P.,** Novel eluent for u.v. and conductometric detection of anions in unsuppressed ion chromatography, *J. Chromatogr.,* 267(1), 218, 1983.
234. **Dogan, S. and Haerdi, W.,** Separation and measurement of inorganic anions in natural waters by ion-exchange chromatography and conductometric detection, *Chimia,* 35(9), 339, 1981.
235. **Jenke, D. R. and Pagenkopf, G. K.,** Behavior of cations in nonsuppressed anion chromatography, *Anal. Chem.,* 55, 1168, 1983.
236. **Yamamoto, M., Yamamoto, H., Yamamoto, Y., Matsushita, S., Baba, N., and Ikushige, T.,** Simultaneous determination of inorganic anions and cations by ion chromatography with ethylene-diamine-tetraacetic acid as eluent, *Anal. Chem.,* 56, 832, 1984.
237. **Igawa, M., Saito, K., Tsukamoto, J., and Tanaka, M.,** Ion chromatographic separation of anions on silica-coated polyamide crown resin, *Anal. Chem.,* 53, 1942, 1981.
238. **Nakajima, M., Kimura, K., Hayata, E., and Shono, T.,** Ion chromatography on poly (crown ether)-modified silica possessing high affinity for sodium, *J. Liq. Chromatogr.,* 7(11), 2115, 1984.
239. **Jenke, D. R. and Pagenkopf, G. K.,** Effect of analyte concentration on retention behavior in non-suppressed ion chromatography, *J. Chromatogr. Sci.,* 22(6), 231, 1984.
240. **Dasgupta, P, K., De Cesare, K., and Ullrey, J. C.,** Determination of atmospheric sulfur dioxide without tetrachloromercurate(II) and the mechanism of the Schiff reaction, *Anal. Chem.,* 52, 1912, 1980.
241. **Jenke, D. R. and Pagenkopf, G. K.,** Optimization of anion separation by non-suppressed ion chromatography, *Anal. Chem.,* 56, 85, 1984.
242. **Fritz, J. S., Gjerde, D. T., and Becker, R. M.,** Cation chromatography with a conductivity detector, *Anal. Chem.,* 52, 1519, 1980.
242a. **Benson, J. R. and MacBlane, D.,** Cation chromatography using polymeric columns, *LC Mag.,* 3(2), 137, 1985.

243. **Sevenich, G. J. and Fritz, J. S.,** Addition of complexing agents in ion chromatography for separation of polyvalent metal ions, *Anal. Chem.*, 55, 12, 1983.
243a. **Sato, H.,** Ion exchange chromatography of bivalent metal ions by conductivity detection, *Jpn. Analyst,* 34(10), 606, 1985.
243b. **Yan, D. and Schwedt, G.,** Ion chromatographic analysis of magnesium, calcium, manganese, and zinc together with application examples for calcium/magnesium determination, *Fresenius' Z. Anal. Chem.*, 320, 121, 1985.
243c. **Blasius, E., Janzen, K. P., Simon, H., and Zender, J.,** Silica gels coated or substituted with cyclic crown ethers for ion chromatography, *Fresenius' Z. Anal. Chem.*, 320, 435, 1985.
244. **Schmidt, G. J. and Scott, R. P. W.,** Simple and sensitive ion chromatograph for trace metal determination, *Analyst (London),* 109, 997, 1984.
245. **Cassidy, R. M. and Elchuk, S.,** Application of high performance liquid chromatography to the analysis of nuclear materials, *J. Liq. Chromatogr.*, 4(3), 379, 1981.
246. **Cassidy, R. M. and Elchuk, S.,** Trace-enrichment methods for determination of metal ions by high-performance liquid chromatography. II, *J. Chromatogr. Sci.*, 19(10), 503, 1981.
247. **Kudoh, M., Kusuyama, T., Yamaguchi, S., and Fudano, S.,** High-performance liquid chromatographic determination of sodium sulfate in anionic surfactants, *J. Am. Oil Chem. Soc.*, 61(1), 108, 1984.
248. **DiNunzio, J. E. and Jubara, M.,** Donnan dialysis preconcentration for ion chromatography, *Anal. Chem.*, 55, 1013, 1983.
248a. **Yan, D. and Schwedt, G.,** Multi-element analysis by ion chromatography, *Fresenius' Z. Anal. Chem.*, 320, 325, 1985.
248b. **Schwedt, G.,** Multi-ion chromatography, *GIT Fachz. Lab.*, 29(7), 697, 1985.
249. **Elchuk, S. and Cassidy, R. M.,** Separation of the lanthanides on high-efficiency bonded phases and conventional ion exchange resins, *Anal. Chem.*, 51, 1434, 1979.
249a. **Nagashima, H.,** Determination of calcium and magnesium ions in human plasma and electrolyte solution for transfusion by ion chromatography with post column color reaction, *Jpn. Analyst,* 35(1), 7, 1986.
249b. **Gan, W., Hu, W., Dong, C., and Dong, M.,** Determination of rare earth metals by high-performance liquid chromatography with post-column derivatization, *Fenxi Haxue,* 13(8), 569, 1985.
249c. **Yan, D. R. and Schwedt, G.,** Trace analysis of aluminum and iron by ion chromatographic derivatization, *Fresenius' Z. Anal. Chem.*, 320, 252, 1985.
249d. **Saitoh, H. and Oikawa, K.,** Simultaneous determination of iron(II) and -(III) by ion chromatography with post-column reaction, *J. Chromatogr.*, 329, 247, 1985.
249e. **Brunt, K.,** Sulfate determination in industrial waste water by liquid chromatography with post-column solid-phase reaction detection, *Anal. Chem.*, 57, 1338, 1985.
249f. **Mori, T., Nishioka, C., Ishikawa, H., and Kuroda, H.,** Determination of potassium bromate in foods by high performance liquid chromatography, *Shokuhin Eiseigaku Zasshi,* 26(3), 260, 1985.
250. **Rokushika, S., Qiu, Z. Y., Sun, Z. L., and Hatano, H.,** Microbore packed-column anion chromatography using a u.v. detector, *J. Chromatogr.*, 280, 69, 1983.
251. **Miller, M. E. and Cappon, C. J.,** Anion-exchange chromatographic determination of bromide in serum, *Clin. Chem. (Winston Salem, N.C.),* 30(5), 781, 1984.
252. **Wilson, S. A. and Yeung, E.,** Quantitative ion chromatography with an ultra-violet absorbance detector without standards, *Anal. Chim. Acta,* 157, 53, 1984.
253. **Hayakawa, K., Hiraki, H., and Miyazaki, M.,** Novel ion chromatography using a conventional ion exchange column and a photometric detector, *Jpn. Analyst,* 32(8), 504, 1983.
254. **Ayers, G. P. and Gillett, R. W.,** Sensitive detection of anions in ion chromatography using u.v. detection at wavelengths less than 200 nm, *J. Chromatogr.*, 284(2), 510, 1984.
255. **Cochrane, J. A. and Hillman, D. E.,** Analysis of anions by ion chromatography using u.v. detection, *J. Chromatogr.*, 241(2), 392, 1982.
256. **Haddad, P. R. and Heckenberg, A. L.,** Studies on sample preconcentration in ion chromatography. I. Design of an automated single-pump pre-concentration system with direct u.v. absorbance detection, *J. Chromatogr.*, 318(2), 279, 1985.
257. **MacMillan, W. D.,** Determination of sodium pyrophosphate and sodium triphosphate in detergent granules by non-suppressed ion chromatography, *HRC CC J. High Resolut. Chromatogr. Chromatogr. Commun.*, 7(2), 102, 1984.
258. **Cooke, M.,** ODS phases for ion chromatography, *J. High Resolut. Chromatogr. Chromatogr. Commun.*, 7(9), 515, 1984.
259. **Hayakawa, K., Ebina, R., Matsumoto, M., and Miyazaki, M.,** Determination of inorganic ions in some vegetables by ultraviolet photometric ion chromatography, *Jpn. Analyst,* 33(7), 390, 1984.
260. **Rocklin, R. D.,** Determination of gold, palladium and platinum at the parts-per-billion level by ion chromatography, *Anal. Chem.*, 56, 1959, 1984.
261. **Hordijk, C. A. and Cappenberg, T. E.,** Sulfate analysis in pore water by radio-ion chromatography employing 5-sulfoisophthalic acid as a novel eluent, *J. Microbiol. Methods,* 3(3-4), 205, 1985.

262. **Heisz, O.,** Ion chromatography by h.p.l.c. with indirect u.v. detection, *GIT Fachz. Lab.*, 27(7), 596, 1983.

263. **Domazetis, G.,** Determination of anions by non-suppressed ion chromatography using an amine column, *Chromatographia,* 18(7), 383, 1984.

264. **Jenke, D. R., Mitchell, P. K., and Pagenkopf, G. K.,** Use of multiple detectors and stepwise elution in ion chromatography without suppressor columns, *Anal. Chim. Acta,* 155, 279, 1983.

265. **Naish, P. J.,** Indirect ultraviolet detection for ion chromatography — its optimisation and application, *Analyst (London),* 109, 809, 1984.

266. **Bowman, S.,** Analysis of soil extracts for inorganic and organic tracer anions via high-performance liquid chromatography, *J. Chromatogr.*, 285(3), 467, 1984.

267. **Heckenberg, A. L. and Haddard, P. R.,** Determination of inorganic anions at parts-per-billion levels using single column ion chromatography without sample preconcentration, *J. Chromatogr.*, 299(1), 301, 1984.

268. **Laurent, A. and Bourdon, R.,** Determination of anions by ion exchange chromatography, *Ann. Pharm. Fr.*, 36(9 and 10), 453, 1978.

269. **Shintani, H., Tsuji, K., and Oba, T.,** Determination of cations in serum by ion chromatography with ultra-violet and conductometric detector, *Jpn. Analyst,* 34(2), 109, 1985.

270. **Jenke, D. R. and Raghavan, N.,** Application of indirect photometric chromatography to pharmaceutical analysis, *J. Chromatogr. Sci.*, 23(2), 75, 1985.

271. **Matsushita, S.,** Simultaneous determination of anions and metal cations by single-column ion chromatography with ethylenediaminetetraacetate as eluent and conductivity and ultraviolet detection, *J. Chromatogr.*, 312, 327, 1984.

271a. **Brandt, G., Matuschek, G., and Kettrup, A.,** Determination of carbonate traces in high-purity water by means of ion chromatography, *Fresenius' Z. Anal. Chem.*, 321, 653, 1985.

271b. **Pillon, D. J. and Meezan, E.,** Liquid chromatographic determination of chloride in sweat from cystic fibrosis patients and normal persons, *Clin. Chem. (Winston Salem, N.C.),* 31(7), 1155, 1985.

271c. **Miyazaki, M., Hayakawa, K., and Choi, S. G.,** Simultaneous determination of sodium, ammonium, potassium, rubidium, cesium, magnesium, and calcium ions by a conventional high-performance liquid-chromatographic system with a common ion exchange column, *J. Chromatogr.*, 323(2), 443, 1985.

271d. **Okada, T. and Kuwamoto, T.,** Elimination of the peak of a major ion in anion chromatography with u.v. detection, *J. Chromatogr.*, 325(1), 327, 1985.

271e. **Yamamoto, A., Matsunaga, A., Sekiguchi, H., Hayakawa, K., and Miyazaki, M.,** Determination of potassium bromate in bread and Kamaboko by photometric ion chromatography, *Eisei Kagaku,* 31(1), 47, 1985.

271f. **Shintani, H.,** Improvement of ion chromatography with ultra violet photometric detection and comparison with conductivity detection for the determination of serum cations, *J. Chromatogr. Biomed. Appl.*, 42(1), 53, 1985.

272. **Haddard, P. R. and Heckenberg, A. L.,** High performance liquid chromatography of inorganic and organic ions using low-capacity ion exchange columns with indirect refractive index detection, *J. Chromatogr.*, 252, 177, 1982.

273. **Buytenhuys, F. A.,** Ion chromatography of inorganic and organic species using refractive-index detection, *J. Chromatogr.*, 218, 57, 1981.

274. **Nguyen, T. T. and Sporns, P.,** Liquid-chromatographic determination of flavor enhancers and chloride in food, *J. Assoc. Off. Anal. Chem.*, 67(4), 747, 1984.

274a. **Schweizer, A. and Schedt, G.,** Anion chromatography of hydrogen carbonate, chloride and sulfate in mineral waters with comparisons of methods, *Fresenius' Z. Anal. Chem.*, 320, 480, 1985.

275. **Toida, T., Togawa, T., Tanabe, S., and Imanari, T.,** Determination of cyanide and thiocyanate in blood plasma and red cells by high-performance liquid chromatography with fluorimetric detection, *J. Chromatogr. Biomed. Appl.*, 33, 133, 1984.

276. **Sun Haing Lee, and Field, L. R.,** Post-column fluorescence detection of nitrite, nitrate, thiosulfate, and iodide anions in high-performance liquid chromatography, *Anal. Chem.*, 56, 2647, 1984.

277. **Gooijer, C., Markies, P. R., Donkerbroek, J. J., Velthorst, N. H., and Frei, R. W.,** Quenched phosphorescence as a detection method in ion chromatography: determination of nitrite and sulfite, *J. Chromatogr.*, 289, 347, 1984.

278. **Haddad, P. R., Alexander, P. W., and Trojanowicz, M.,** Ion chromatography of magnesium, calcium, strontium, and barium ions using a metallic copper electrode as a potentiometric detector, *J. Chromatogr.*, 294, 397, 1984.

279. **Girard, J. E.,** Ion chromatography with coulometric detection for determination of inorganic ions, *Anal. Chem.*, 51, 836, 1979.

280. **Suzuki, K., Aruga, H., Ishiwada, H., Oshima, T., Inoue, H., and Shirai, T.,** Determination of anions with a potentiometric detector for ion chromatography, *Jpn. Analyst,* 32(10), 585, 1983.

281. **Edwards, P. and Haak, K. K.,** Pulse amperometric detector for ion chromatography, *Int. Lab.*, 13(5), 38, 1983.
282. **Alexander, P. W., Haddad, P. R., and Trojanowicz, M.,** Potentiometric detection in ion chromatography using a metallic copper indicator electrode, *Chromatographia,* 20(3), 179, 1985.
283. **Pohlandt, C.,** Determination of cyanides in metallurgical process solutions by ion chromatography, *S. Afr. J. Chem.*, 37(3), 133, 1984.
284. **Butler, E. C. V. and Gershey, R. M.,** Application of ion-exchange chromatography with an ion-selective electrode detector to iodine determination in natural waters, *Anal. Chim. Acta,* 164, 153, 1984.
285. **Ostrovidov, E. A., Kuleshova, L. A., Mitina, S. I., Vinogradova, G., Vorontsov, A. M., Kanev, A. S., and Rys'ev, O. A.,** Electrochemical detection of ions in liquid chromatography, *Zh. Anal. Khim.*, 35(9), 1677, 1980.
286. **Wang, W. N., Chen, Y. J., and Wu, M. T.,** Complementary analytical methods for cyanide, sulfide, certain transition metals and lanthanides in ion chromatography, *Analyst (London),* 109, 281, 1984.
287. **Imanari, T., Ogata, K., and Tanabe, S.,** Determination of thiocyanate, thiosulfate, sulfite, and nitrite by high performance liquid chromatography coupled with electrochemical detection, *Chem. Pharm. Bull.*, 30(1), 374, 1982.
288. **Pyen, G. S. and Erdmann, D. E.,** Automated determination of bromide in waters by ion chromatography with an amperometric detector, *Anal. Chim. Acta,* 149, 355, 1983.
289. **Bond, A. M., Heritage, I. D., Wallace, G. G., and McCormick, M. J.,** Simultaneous determination of free sulfide and cyanide by ion chromatography with electrochemical detection, *Anal. Chem.*, 54, 582, 1982.
290. **Rocklin, R. D. and Johnson, E. L.,** Determination of cyanide, sulfide, iodide, and bromide by ion chromatography with electrochemical detection, *Anal. Chem.*, 55, 4, 1983.
291. **Hershcovitz, H., Yarnitzky, Ch., and Schmuckler, G.,** Ion chromatography with potentiometric detection, *J. Chromatogr.*, 252, 113, 1982.
292. **Koch, W. F.,** Determination of trace levels of cyanide by ion chromatography with electrochemical detection, *J. Res. Natl. Bur. Stand. (U.S.),* 88(3), 157, 1983.
293. **Girard, J. E.,** Ion chromatography with coulometric detection for the determination of inorganic ions, *Anal. Chem.*, 51, 836, 1979.
294. **Pohlandt, C.,** The determination of cyanide in hydrometallurgical process solutions and effluents by ion chromatography, Report MINTEK, M128, Council for Mineral Technology, Randburg, South Africa, June 1984.
294a. **Haddad, P. R., Alexander, P. W., and Trojanowicz, M.,** Application of indirect potentiometric detection with a metallic copper electrode to ion chromatography of transition metal ions, *J. Chromatogr.*, 324(2), 319, 1985.
294b. **Mu, S., Han, K., Luo, Y., and Hou, X.,** Simultaneous determination of cyanide and sulfide by ion chromatography with electrochemical detection, *Fenxi Huaxue,* 13(6), 457, 1985.
294c. **Jones, V. K. and Tarter, J. G.,** Simultaneous analysis of anions and cations in water samples using ion chromatography, *Int. Lab.*, 15(7), 36, 1985.
295. **Ricci, G. R., Shepard, L. S., Colovos, G., and Hester, N. E.,** Ion chromatography by atomic-absorption spectrometric detection for determination of organic and inorganic arsenic species, *Anal. Chem.*, 53, 610, 1981.
296. **Roden, D. R. and Tallman, D. E.,** Determination of inorganic selenium species in groundwaters containing organic interferences by ion chromatography and hydride generation/atomic absorption spectrometry, *Anal. Chem.*, 54, 307, 1982.
297. **Downey, S. W. and Hieftje, G. M.,** Replacement-ion chromatography with flame photometric detection, *Anal. Chim. Acta,* 153, 1, 1983.
298. **Huber, J. F. K. and van Urk-Schoen, A. M.,** Rapid separation of alkali metals by ion exchange column chromatography, *Anal. Chim. Acta,* 58, 395, 1972.
298a. **Banerjee, S. and Steimers, J. R.,** Indirect ion chromatography with radioactive eluents, *Anal. Chem.*, 57, 1476, 1985.
299. **Wheaton, R. M. and Bauman, W. C.,** Ion exclusion — a unit operation utilizing ion exchange materials, *Ind. Eng. Chem.*, 45(1), 228, 1953.
300. **Waki, H. and Tokunaga, Y.,** Donnan exclusion chromatography. I. Theory and application to the separation of phosphorus oxy-anions or metal cations, *J. Chromatogr.*, 201, 259, 1980.
301. **Waki, H. and Tokunaga, Y.,** Donnan exclusion-ion exclusion chromatography, *J. Liq. Chromatogr.*, 5(Suppl. 1), 105, 1982.
302. **Tokunaga, Y., Waki, H., and Ohashi, S.,** Donnan exclusion chromatography, *J. Liq. Chromatogr.*, 5(10), 1855, 1982.
303. **Waki, H., Tsuruta, K., and Tokunaga, Y.,** Donnan exclusion chromatography with nonionic adsorption, *J. Liq. Chromatogr.*, 8(11), 2105, 1985.

304. **Tokunaga, Y. and Waki, H.,** Donnan exclusion chromatography. Application to complex formation studies, *J. Liq. Chromatogr.*, 5(11), 2169, 1982.
305. **Tanaka, K. and Ishizuka, T.,** Determination of phosphate in waste-water by ion exclusion chromatography with flow coulometric detection, *Water Res.*, 16(5), 719, 1982.
306. **Tanaka, K. and Ishizuka, T.,** Ion exclusion chromatography of condensed phosphates on a cation exchange resin, *J. Chromatogr.*, 190, 77, 1980.
307. **Tanaka, K. and Ishizuka, T.,** Elution behavior of acids in ion exclusion chromatography using a cation exchange resin, *J. Chromatogr.*, 174, 153, 1979.
308. **Tanaka, K. and Sunahara, H.,** Ion exclusion chromatography of phosphate, phosphite and hypophosphite on a hydrogen-form cation-exchange resin with flow coulometric detection, *Jpn. Analyst,* 27(2), 95, 1978.
309. **Tanaka, K., Nakajima, K., and Sunahara, H.,** Ion-exclusion chromatography of phosphate ion by use of cation exchange resin, *Jpn. Analyst,* 26(2), 102, 1977.
310. **Tanaka, K.,** Simultaneous determination of nitrate, nitrite and ammonium ions in biological nitrification-denitrification process water by ion exclusion chromatography coupled with cation and anion exchange resin columns, *Jpn. Analyst,* 31(12), T106, 1982.
311. **Tanaka, K.,** Simultaneous determination of nitrate and nitrite ions in biological nitrification-denitrification process water by ion-exclusion chromatography with ultraviolet detection, *Jpn. Analyst,* 30(10), 661, 1981.
312. **Tanaka, K., Ishizuka, T., and Sunahara, H.,** Ion-exclusion chromatography of the ammonium ion on an anion exchange resin, *J. Chromatogr.*, 177(1), 21, 1979.
313. **Tanaka, K.,** Determination of fluoride in metal industry waste waters by ion-exclusion chromatography with coulometric detection, *Jpn. Analyst,* 32(7), 439, 1983.
314. **Pohlandt, C.,** The separation of some weak-acid anions by ion exclusion chromatography, Report NIM-2107, National Institute for Metallurgy, Randburg, South Africa, May, 1981.
315. **Tanaka, K.,** Determination of bicarbonate ion in biological nitrification process water by ion-exclusion chromatography with coulometric detection, *Jpn. Analyst,* 30(6), 358, 1981.
316. **Tanaka, K., Ishihara, Y., and Sunahara, H.,** Liquid chromatography using flow coulometry for anions with a hydrogen-type cation exchange resin, *Jpn. Analyst,* 24(4), 235, 1975.
317. **Korkisch, J.,** Combined ion exchange-solvent extraction (CIESE): a novel separation technique for inorganic ions, *Sep. Sci.*, 1(2-3), 159, 1966.
318. **Korkisch, J.,** Combined ion-exchange — solvent extraction: a new dimension in inorganic separation chemistry, *Nature,* 210, 626, 1966.
319. **Korkisch, J.,** Combined ion-exchange — solvent extraction (CIESE) — a new separation principle in analytical chemistry, *Österr. Chem. Ztg.*, 67(9), 309, 1966.
320. **Korkisch, J.,** Ion exchange in mixed aqueous-organic solvent systems, *Mikrochim. Acta,* No. 4 and 5, 634, 1966.

320a. **Korkisch, J.,** Seperation of metals by combined ion-exchange solvent extraction in *Recent Developments in Seperation Science,* Vol. 8, Li, N. N. and Navratil, J. D., CRC Press, Boca Raton, FL, 1986, 105.

321. **Kuroda, R. and Hasoi, N.,** Ion exchange thin-layer chromatography of metal ions in organic solvent-hydrochloric acid-complexing agent media, *Chromatographia,* 14(6), 359, 1981.
322. **Sherma, J. and Van Lenten, F. J.,** Ion exchange paper chromatography of metal ions in mixed aqueous organic solvents containing mineral acid and a selective extractant, *Sep. Sci.*, 6(2), 199, 1971.
323. **Panse, M. and Khopkar, S. M.,** Combined ion exchange solvent extraction, *J. Sci. Ind. Res. (India),* 34, 5, 1975.
324. **De, A. K. and Bhattacharyya, C. R.,** Combined ion exchange-solvent extraction (CIESE) studies of metal ions on ion exchange papers. Synergistic effects and chromatographic separations, *Anal. Chem.*, 44, 1686, 1972.
325. **Shanker, R. and Venkateswarlu, K. S.,** Radio-tracer application in the study of combined ion exchange solvent extraction (CIESE) of some transition elements, in Proc. Chemistry Symp., Vol. 1, Aligarh Muslim University, Aligarh, December 21 to 23, 1972, 107.
326. **Yoshimura, K. and Waki, H.,** Ion-exchanger phase absorptiometry for trace analysis, *Talanta,* 32, 345, 1985.
327. **Waki, H. and Yoshimura, K.,** Ion-exchanger phase spectrophotometry as an ultrasensitive analytical method, *Ionics,* 10, 217, 1985.
328. **Ohzeki, K., Sakuma, T., and Kambara, T.,** Ion exchanger colorimetry by means of a coagulated material composed of finely divided anion and cation exchange resins, *Jpn. Analyst,* 28(11), 713, 1979.
329. **Pinfold, T. A. and Karger, B. L.,** The coagulation of finely-divided ion exchange resins and the use of the coagulated material for rapid ion removal, *Sep. Sci.*, 5(3), 183, 1970.
330. **Janauer, G. E., Bernier, W. E., Zucconi, T. D., and Ramseyer, G. O.,** Reactive ion exchange separations — a new approach using old chemistry, in *The Theory and Practice of Ion Exchange, Intern. Conf., University of Cambridge,* No. 33.1 to 33.8, SCI, London, 1976.
331. **Janauer, G. E., Ramseyer, G. E., and Lin, J. W.,** Selective separations by reactive ion exchange with common polystyrene-type resins. I. General considerations, *Anal. Chim. Acta,* 73, 311, 1974.

332. **Lin, J. W. and Janauer, G. E.,** Selective separations by reactive ion exchange. III. Preconcentration and separation of oxo anions, *Anal. Chim. Acta,* 79, 219, 1975.
333. **Bernier, W. E. and Janauer, G. E.,** In situ reduction on ion exchange resins as a method of preconcentration of selenium and other heavy metals from aqueous solutions, in *Trace Substances in Environmental Health,* Vol. 10, Hemphill, D. D., Ed., University of Missouri, Columbia, 1976, 323.
334. **Janauer, G. E.,** Innovative uses of ion exchange resins: higher preconcentration ratios and improved selectivity for trace environmental species, in *Trace Substances in Environmental Health,* Vol. 8, Hemphill, D. D., Ed., University of Missouri, Columbia, 1974, 231.
335. **Delayette-Mills, M., Karm, L., Janauer, G. E., and Chan, P.-K.,** Selective separations by reactive ion exchange. IV. Preconcentration of cadmium and zinc by in situ precipitation as hexacyanoferrate(II) salts on gel and macroporous ion exchange resins, *Anal. Chim. Acta,* 124, 365, 1981.
336. **Sansoni, B. and Bauer-Schreiber, E.,** Redox exchangers and their applications. XVI. Separation of hydrogen peroxide from aqueous solution on redox and ion exchangers, *Talanta,* 17, 987, 1970.
337. **Sansoni, B. and Dorfner, K.,** Redox ion exchangers with inorganic, anionic redox groups, *Angew. Chem.,* 71(4), 160, 1959.
338. **Sansoni, B.,** Novel application of ion exchange resins as electron exchangers, *Naturwissenschaften,* 39(12), 281, 1952.
339. **Sansoni, B. and Sigmund, O.,** Hydride-anion exchangers, *Naturwissenschaften,* 48(18), 598, 1961.
340. **Inczédy, J.,** Redox reactions in ion exchangers and their analytical applications, *Z. Chem.,* 2(10), 302, 1962.
341. **Khristova, R. and Novkirishka, M.,** Sorption and separation of ions by precipitation with ion exchange resins, *God. Sofii. Univ. Khim. Fak.,* 62, 347, 1967/1968.
342. **Tera, F., Ruch, R. R., and Morrison, G. H.,** Pre-concentration of trace elements by precipitation ion-exchange, *Anal. Chem.,* 37, 358, 1965.
343. **Koblyanskii, A. G. and Kameneva, N. Kh.,** The influence of precipitate formation on the processes of ion exchange on a cationic ion-exchange column, *Tr. Krasnodar. Inst. Pishch. Promsti.,* No. 11, 73, 1955.
344. **Komlev, A. I. and Tsimbalista, L. I.,** Use of precipitation chromatography in qualitative semi-microanalysis, *Zh. Anal. Khim.,* 8(4), 217, 1953.
345. **Weisz, H., Pantel, S., and Moesta, B.,** Amplification reactions with ion exchange resins. Determination of cations and anions, *Fresenius' Z. Anal. Chem.,* 306, 106, 1981.
346. **Weisz, H. and Moesta, B.,** Amplification reactions by multiplication with an ion exchanger in an inversion tube, *Anal. Chim. Acta,* 130, 193, 1981.
347. **Bhatnagar, D. V.,** Estimation of free nitric acid in uranyl nitrate by ion exchange, *J. Sci. Ind. Res. Sect. B (India),* 16, 23, 1957.
348. **Dizdar, Z. I. and Obrenovic, I. D.,** Determination of free acid in uranium(VI) solutions by means of cation exchangers, *Anal. Chim. Acta,* 21(6), 560, 1959.
349. **Campbell, M. H. and Adams, J. F.,** An analysis for free nitric acid and total nitrate ion in uranyl nitrate and plutonium nitrate solutions using a cation exchanger resin, Report HW-76363, General Electric Company, Hanford Atomic Products Operation, Richland, Wash., February 1963.
350. **Gaddy, R. H. and Dorsett, R. S.,** Determination of free acid by automated ion exchange and colorimetry, *Anal. Chem.,* 40, 429, 1968.
351. **Pérez-Bustamante, J. A.,** Anomalous ion-exchange behavior exhibited by aged acid-free uranyl nitrate solutions, *Mikrochim. Acta,* No. 3, 455, 1971.
352. **Fisher, S. and Kunin, R.,** Analysis of iron-pickling liquor by means of ion exchange, *Anal. Chem.,* 27, 1649, 1955.
353. **Barrachina, M. and Villar, M. A.,** Determination of the free acidity in hydrochloric acid solutions of iron, *An. R. Soc. Esp. Fis. Quim. Ser. B,* 63(1), 107, 1967.
354. **Oshchapovskii, V. V.,** Determination of free acid in solutions of quadrivalent cerium salts, *Ukr. Khim. Zh.,* 21(3), 384, 1955.
355. **Korgaonkar, V. G.,** Determination of free acid in uranyl nitrate solutions by anion exchange, *Indian J. Chem.,* 1(5), 195, 1963.
355a. **Sepulveda Munita, C. and Atalla, L. T.,** Radiochemical separation of thorium by isotope ion exchange, *J. Radioanal. Nucl. Chem.,* 91(1), 135, 1985.
356. **Fujimoto, M. and Suga, T.,** Characteristic coloration of normally weakly-coloured strongly-basic-anion-exchange resins charged with the colourless anions hydroxide, iodide and sulphide, studied by diffuse reflectance spectral measurements, *Bull. Chem. Soc. Jpn.,* 44(5), 1425, 1971.
357. **Fujimoto, M. and Iwamoto, T.,** Micro-analysis with the aid of ion exchangers. XIII. The resin spot test method for detection of millimicrogram amounts of heavy metals with 1-(2-pyridylazo)-2-naphthol(PAN) or 4-(2-pyridylazo)resorcinol (PAR), *Mikrochim. Acta,* No. 4, 655, 1963.
358. **Fujimoto, M.,** Ion-exchange resins as reaction media for microdetection tests, *Chemist Analyst,* 49 (1), 4, 1960.

359. **Fujimoto, M.,** Recent advances in resin spot test: 1960 to 1964. II, *Chemist Analyst,* 54(3), 92, 1965.
360. **Fujimoto, M.,** Recent advances in ion-exchange-resin spot tests: 1960 to 1964. I, *Chemist Analyst,* 54(2), 58, 1965.
361. **Fujimoto, M.,** Adsorption by finely divided particles of strongly basic anion exchangers at the interface between two immiscible liquids and its application to microchemical analysis, *Naturwissenschaften,* 47(11), 252, 1960.
362. **Fujimoto, M.,** Increase of sensitivity of precipitation reactions by use of ion exchange resins, *Bull. Chem. Soc. Jpn.,* 29(5), 646, 1956.
363. **Takiyama, K. and Suito, E.,** Qualitative analysis by the use of anion exchange resin, *Jpn. Analyst,* 4(1), 8, 1955.
364. **Noe, M.,** Systematic study of cations on the ultra-micro scale on grains of ion exchange resin, *Mikrochim. Acta,* 1(5-6), 663, 1975.
365. **Mooney, J. B.,** General micro-qualitative technique. Combination of ring-oven, cation-exchange paper and emission spectrograph, *Anal. Chem.,* 34, 1506, 1962.
366. **Machiroux, R., Merciny, E., and Schreiber, A.,** Qualitative-analysis scheme for cations, based on ion-exchange properties, *Mikrochim. Acta,* No. 5, 829, 1973.
367. **Miller, W. E.,** Ion-exchange resins as indicators, *Anal. Chem.,* 30, 1462, 1958.
368. **Légrádi, L.,** Ion-exchange resin indicators, *Magy. Kem. Foly.,* 66(2), 76, 1960.
369. **Qureshi, M., Qureshi, S. Z., and Zehra, N.,** Organic and inorganic ion-exchange beads as indicators in redox and precipitation titrations, *Talanta,* 19, 377, 1972.
370. **Freeman, D. H. and Paulson, R. A.,** Chemical micro-standards from ion-exchange resin, *Nature,* 218, 563, 1968.
371. **Freeman, D. H., Currie, L. A., Kuehner, E. C., Dixon, H. D., and Paulson, R. A.,** Development and characterization of ion-exchange bead micro-standards, *Anal. Chem.,* 42, 203, 1970.
372. **Hahn, P. B. and Schleien, B.,** Ion-exchange beads as calibrated microscopic radioactivity sources, *Anal. Chem.,* 42, 1608, 1970.
373. **Kayasth, S. R., Iyer, R. K., and Das, M. S.,** Preparation of ion-exchange resin based elemental standards for use in activation analysis. I. Use of cation-exchange resin, *J. Radioanal. Chem.,* 59(2), 373, 1980.
374. **Kayasth, S. R. and Desai, H. B.,** Ion-exchange resin as base for the preparation of standards for heavy metal ions: application to neutron-activation analysis. II, *Radiochem. Radioanal. Lett.,* 44(6), 403, 1980.
375. **Alder, J. F. and Batoreu, M. C. C.,** Ion exchange resin beads as solid standards for the electrothermal atomic-absorption spectrometric determination of metals in hair, *Anal. Chim. Acta,* 135, 229, 1982.
376. **Wilshire, J. P. and Brown, W. A.,** Determination of boric oxide by ion chromatography and ion chromatography exclusion, *Anal. Chem.,* 54, 1647, 1982.
377. **Fujii, T.,** Rapid determination of nitrogen oxides in exhaust gases by ion chromatography, *Kogai To Taisaku,* 18(9), 901, 1982.
378. **Fujii, T.,** Rapid determination of nitrogen oxides, sulfur oxides and hydrogen chloride in exhaust gases by ion chromatography, *Jpn. Analyst,* 31(11), 677, 1982.
379. **Bouyoucos, S. A.,** Determination of ammonia and methyl-amines in aqueous solutions by ion chromatography, *Anal. Chem.,* 49, 401, 1977.
380. **Gent, C. A. and Wilson, S. A.,** The determination of sulfur and chlorine in coals and oil shales using ion chromatography, *Anal. Lett.,* 18(6), 729, 1985.
381. **Merrill, R. M.,** Analysis of anions in geological brines using ion chromatography, SAND-84-2297, Sandia Laboratories (Tech. Rep.)SAND, March 1985.
382. **Schnitzler, M.,** Ion chromatography in the determination of organic water polutants by group parameters: simultaneous determination of adsorbable organic chlorine, bromine and sulfur, *GIT Suppl.,* No. 4, 32, 1985.
383. **Wilken, R. D. and Kock, H. H.,** Determination of chloride, nitrate and sulfate in pore waters of sediments, *Fresenius' Z. Anal. Chem.,* 320, 477, 1985.
384. **Tretter, H., Paul, G., Blum, F., and Schreck, H.,** Determination of anions in high-purity water by ion chromatography, *Fresenius' Z. Anal. Chem.,* 321, 650, 1985.
385. **Wagner, F., Valenta, P., and Nuernberg, H. W.,** Ion-chromatographic determination of the anions chloride, nitrate and sulfate in rain water, *Fresenius' Z. Anal. Chem.,* 320, 470, 1985.
386. **Roessner, B. and Schwedt, G.,** Methods for analysis of inorganic anions. V. Comparisons of the efficiency of ion chromatographic analysis systems, *Fresenius' Z. Anal. Chem.,* 320, 566, 1985.
387. **Tan, L. K. and Dutrizac, J. E.,** Determination of arsenic(III) and arsenic(V) in ferric chloride-hydrochloric acid leaching media by ion chromatography, *Anal. Chem.,* 57, 1027, 1985.
388. **Watanabe, I., Tanabe, K., Furuya, K., and Matsushita, H.,** Determination of elements and anions in leaching solution of coal fly ashes, *Jpn. Analyst,* 34(5), T45, 1985.
389. **Dokiya, Y., Hirose, K., Yoshimura, E., and Toda, S.,** Determination of phosphorus by inductively coupled plasma emission spectrometry and ion chromatography in air-borne particles, *Jpn. Analyst,* 34, T153, 1985.

390. **Fuchs, G. R., Lisson, E., Schwarz, B., and Baechmann, K.**, Determination of anions in atmospheric aerosols, *Fresenius' Z. Anal. Chem.*, 320, 498, 1985.
391. **Pimminger, M., Puxbaum, H., Kossina, I., and Weber, M.**, Simultaneous determination of some inorganic and organic anions by means of an ion-chromatographic triple column system, *Fresenius' Z. Anal. Chem.*, 320, 445, 1985.
392. **Ban, Y., Watanabe, I., Furuya, K., Matsushita, H., and Gohshi, Y.**, State analysis of sulfur in coal fly ash and its comparison with quantity of sulfate ion in leaching solution, *Jpn. Analyst*, 34(5), 264, 1985.
393. **Kalbasi, M. and Tabatabai, M. A.**, Simultaneous determination of nitrate, chloride, sulfate, and phosphate in plant materials by ion chromatography, *Commun. Soil Sci. Plant Anal.*, 16(7), 787, 1985.
394. **Sullivan, D. M. and Smith, R. L.**, Determination of sulfite in foods by ion chromatography, *Food Technol. (Chicago)*, 39(7), 45, 1985.
395. **Basta, N. T. and Tabatabai, M. A.**, Determination of potassium, sodium, calcium, and magnesium in natural waters by ion chromatography, *J. Environ. Qual.*, 14(3), 450, 1985.
396. **Tanaka, T.**, Ion chromatographic determination of divalent metal ions using EDTA as complexing agent, *Fresenius' Z. Anal. Chem.*, 320, 125, 1985.
397. **Richardson, D. E. and Jewell, I. J.**, Analysis of anions in paper industry process waters by single-column ion chromatography, *Appita*, 38(2), 113, 1985.
398. **Baltensperger, U. and Hertz, J.**, Determination of anions and cations in atmospheric aerosols by single-column ion chromatography, *J. Chromatogr.*, 324(1), 153, 1985.
399. **Deiana, S., Dessi, A., Micera, G., Gessa, C., and Solinas, V.**, Determination of vanadate(V) by conductometric anion chromatography, *J. Chromatogr.*, 320(2), 450, 1985.
400. **Maketon, S. and Tarter, J. G.**, Ion chromatographic determination of phosphorus in eight USGS standard rocks, *Anal. Lett.*, 18(A2), 181, 1985.
401. **Brandt, G. and Kettrup, A.**, Ion-chromatographic behavior of carbonate species, *Fresenius' Z. Anal. Chem.*, 320(5), 485, 1985.
402. **Ikeda, S., Satake, H., and Segawa, H.**, Ion chromatography of sulfur species in polysulfide solution utilizing cyanolysis, *Nippon Kagaku Kaishi*, No. 9, 1704, 1985.
403. **Haddad, P. R. and Jackson, P. E.**, Determination of ascorbate, bromate and metabisulfite in bread improvers using high-performance ion-exchange chromatography, *Food Technol. Aust.*, 37(7), 305, 1985.
404. **Weiss, J. and Goebl, M.**, Analysis of inorganic sulfur compounds by means ion chromatography, *Fresenius' Z. Anal. Chem.*, 320, 439, 1985.
405. **Jay, P. C. and Judd, J. M.**, Determination of common anion concentrations in surface and ground water samples by eluent suppressed ion chromatography, *Int. J. Environ. Anal. Chem.*, 19(2), 99, 1985.
406. **Peura, P. and Koskenniemi, J.**, Simultaneous determination of nitrate and nitrite by ion chromatography in leaves of *Urtica dioica* L. (Urticaceae) in Finland, *Acta Pharm. Fenn.*, 94(2), 67, 1985.
407. **Shpigun, O. A., Choporova, O. D., and Zolotov, Yu. A.**, Ion chromatographic determination of alkaline earth metals and some heavy metals, *Anal. Chim. Acta*, 172, 341, 1985.
408. **Green, M. J.**, Ion chromatographic analysis of perchlorate in perchlorate/sugar explosive devices, *LC Mag.*, 3(10), 894, 1986.
409. **Chen, S. S., Lulla, H., Sena, F. J., and Reynoso, V.**, Simultaneous determination of potassium nitrate and sodium monofluorophosphate in dentifrices by single column ion chromatography, *J. Chromatogr. Sci.*, 23(8), 355, 1985.
410. **Okada, T. and Kuwamoto, T.**, Trace analysis of anions by use of a black flush method and large injection volumes in ion chromatography, *J. Chromatogr.*, 350(1), 317, 1985.
411. **Matsushita, S.**, Determination of protein free and protein bound calcium and magnesium in biological samples by use of ultra filtration and ion chromatography, *Anal. Chim. Acta*, 172, 249, 1985.
412. **Nieto, K. F. and Frankenberger, W. T.**, Single column ion chromatography. II. Analysis of ammonium, alkali metals and alkaline earth cations in soils, *Soil Sci. Soc. Am. J.*, 49(3), 592, 1985.
413. **Margeson, J. H., Knoll, J. E., Midgett, M. R., Oldaker, G. B., and Reynolds, W. E.**, Determination of sulfur dioxide, nitrogen oxides and carbon dioxide in emissions from electric utility plants by alkaline permanganate sampling and ion chromatography, *Anal. Chem.*, 57, 1586, 1985.
414. **Iskandarani, Z. and Miller, T. E.**, Simultaneous independent analysis of anions and cations using indirect photometric chromatography, *Anal. Chem.*, 57, 1591, 1985.
415. **Miller, M. E., Cosgriff, J. M., and Schwartz, R. H.**, Anion exchange chromatography to determine the concentration of chloride in sweat for diagnosis of cystic fibrosis, *Clin. Chem. (Winston Salem, N.C.)*, 31(10), 1715, 1985.
416. **Mho, S. I. and Yeung, E. S.**, Detection method for ion chromatography based on double beam laser excited indirect fluorimetry, *Anal. Chem.*, 57, 2253, 1985.
417. **Haddad, P. R., Alexander, P. W., and Trojanowicz, M.**, Ion chromatography of inorganic anions with potentiometric detection using a metallic copper electrode, *J. Chromatogr.*, 321(2), 363, 1985.

418. **Subramanian, K. S., Meranger, J. C., Wan, C. C., and Corsini, A.,** Preconcentration of cadmium, chromium, copper and lead in drinking water on the poly(acrylic ester) resin, XAD-7, *Int. J. Environ. Anal. Chem.*, 19(4), 261, 1985.
419. **Wu, J. C. and Robinson, J. W.,** Speciation studies of zinc and magnesium in body fluids using interfaced HPLC and ultrasonic nebulizer flame AAS, *Spectrosc. Lett.*, 19(1), 61, 1986.
420. **King, J. N. and Fritz, J. S.,** Concentration of metal ions by complexation with sodium bis-(2-hydroxyethyl) dithiocarbamate and sorption on XAD-4 resin, *Anal. Chem.*, 57, 1016, 1985.
421. **Tabani, I. and Kratochvil, B.,** System design strategy for microcomputer-controlled data acquisition/processing of an ion exchange atomic absorption system for ion speciation, *Anal. Instrum. (N.Y.)*, 14(2), 169, 1985.

Practical Applications

RARE EARTH ELEMENTS

INTRODUCTION

According to a proposition of Vickery the term ''rare earths'' should include all elements of the IIIA group whose electronic 5d shell is empty or incomplete, i.e., ''rare earths'' = lanthanides (La through Lu) + Y + Sc. In fact, the similarity of the chemical properties of these elements justifies this definition so that it is employed throughout this chapter.

Since the analysis of natural and also industrial materials for the rare earths (predominantly present in the form of trivalent cations) is in most cases aimed at determining as many of these elements as possible, the first separation step based on ion exchange usually involves the isolation of the rare earths as a group. This is a relatively easy task because of the above-mentioned similarity of chemical properties of the elements and is most effectively achieved in mineral acid media using strongly acidic cation exchange resins. Consequently, the isolation of the rare earths, as a group, by means of such exchangers will be discussed in the first section of this chapter. This will then be followed by a section in which the utilization of the same cation exchange resins will be described with respect to the mutual separation, i.e., fractionation of the rare earths into individual elements using suitable organic complexing agents. This is an important separation step, especially for the lanthanides and Y, because it is extremely difficult to separate their mixture on an ion exchanger using only the differences in their hydrated ionic radii (which increase with increasing atomic number from La to Lu; with Y between Ho and Dy), as it would be the case when employing, e.g., hydrochloric acid (HCl) as the eluent. Mineral acids can be used, however, to separate Ce(IV) and also Sc from other rare earth elements on both cation and anion exchange resins. In some of these acids tetravalent Ce and trivalent Sc show a behavior which deviates considerably from that of the trivalent lanthanides and Y.

In subsequent sections the mineral acid systems are discussed which can be utilized for the isolation of the rare earths, as a group, by their adsorption (e.g., from nitric acid media) or nonadsorption (e.g., from HCl media) on strongly basic anion exchange resins. In most cases, these separations serve the purpose to separate the rare earths from constituents of the matrix (e.g., geological, biological, or industrial material) which interfere either with their determination when present as a mixture or with their fractionation (if a mutual separation of the rare earths should be necessary before the quantitative assay of the individual rare earth elements).

With the nitric acid-methanol systems discussed in one of the sections it is also possible to achieve a certain degree of fractionation of the rare earth elements (e.g., into several subgroups), so that in some cases the mutual separation of the rare earths by subsequent cation exchange in media containing complexing organic ligands becomes unnecessary.

A brief discussion of the anion exchange behavior of the rare earths in alkaline media is also presented and the separability of the rare earths in systems containing organic complexing agents is demonstrated. These methods have, however, received only marginal attention as compared with the great popularity enjoyed by the methods based on cation exchange.

Also, relatively little information is available with respect to the behavior of rare earth elements towards chelating ion exchange resins, a topic which is briefly discussed.

Before the development of the ion exchange methods in 1941 to 1947, separation of adjacent rare earth elements (especially of the heavy lanthanides) was very difficult and involved fractional crystallization, most commonly of double sulfates, fractional precipitation, or fractional decomposition supplemented by procedures based on the removal of Ce(IV) and of Eu, Sm, and Yb in their divalent states. Elaborate large-scale column techniques now allow the cation exchange separation, in a very pure state, of macroquantities of the rare earths which are commercially available. For these separations, displacement chromatography is employed utilizing complexing ligands of the type discussed.

CATION EXCHANGE RESINS

Mineral Acid Media

General

HCl Systems

All rare earth elements, including Sc and Y, are strongly adsorbed on strong acid cation exchange resins from dilute HCl solutions.[1-6] From Table 1* it is seen that this adsorption is highest from 0.1 to 0.5 *M* acid (K_d >10^3 to 10^5), intermediate in 1.0 to 1.75 *M* HCl media, and relatively low at higher acid concentrations. A minimum of adsorption is reached in 4 to 8 *M* HCl which is followed by a slight rise of the distribution coefficient at higher molarities.[6] In the range of minimum adsorption (distribution coefficients of ≲10) the rare earths (Ce) are probably present in a complexed anionic form containing chloride ion as the ligand.[7] In 10 to 11 *M* HCl media, distribution coefficients for La of >10 were measured, while the heavy lanthanides (Yb and Lu) show adsorption values which are lower by about one order of magnitude. In other words, the adsorption of the light members is higher than that of the heavy rare earths and, as can be seen from Table 1, the distribution coefficients decrease regularly, for a given HCl medium, with increasing atomic number. However, HCl solutions canot be employed for the effective separation of adjacent rare earth elements because of the small difference in distribution coefficients. Nevertheless, such media are exceedingly suitable for separations of the rare earths, as a group, from other metal ions which are either less strongly retained (such as mono- and divalent ions) or much more strongly adsorbed (e.g., Th)[8-10] on strong acid cation exchange resins.

In most cases (see Tables 2 to 17) the rare earth elements are adsorbed on strongly acidic cation exchange resins from 0.1 *M*[8,11-17] or 0.5 *M* HCl[1,8,9,18-29] under which condition simultaneous elution and separation from anions such as phosphate take place. However, most other metal ions are coadsorbed with the rare earths so that for their separation it becomes necessary to apply the measures mentioned further below. Other HCl systems from which the rare earths were adsorbed on cation exchangers are (the elements which pass into the effluent are given in parentheses): dilute HCl of pH 1.3[14] or 0.05 *M* HCl (H_3PO_4),[30] 0.2 *M* HCl (H_3PO_4),[31] 0.3 *M* HCl (Rh),[32] <0.5 *M* HCl,[33] 0.6 *M* HCl,[34] 1 *M* HCl (Be, Ca, Mg, and U),[9,29,35-38] 1.75 *M* HCl (Al, Fe[III], Ti, Ca, Mg, Mn[II], K, and Na),[39] 2 *M* HCl (most accompanying elements),[29,40-43] 0.5 to 2 *M* HCl,[44-46] 2.5 *M* HCl (alkali metals),[47] and 3 *M* HCl (most accompanying elements).[48-51]

A simultaneous removal of other elements is not only achieved by adsorbing the rare earths from stronger acid solutions such as from 1 to 3 *M* HCl media (see above), but also by the use of organic complexing agents which are added to the sorption solution to complex certain metals and, hence, prevent their coadsorption with the rare earths. For this purpose, the following systems (the elements which are complexed and separated from the rare earths are given in parentheses) have been employed: 0.1 *M* HCl - 0.02 *M* oxalic acid (Ru, Sb, Zr, and Nb),[52] 0.1 *M* HCl - 0.2 *M* oxalic acid (Ru and Rh),[53,54] ~0.08 *M* HCl containing EDTA (Fe[III]),[55] and 5% citric acid solution of pH 2.2 (Ti, Zr, Hf, and Sb).[9]

The coadsorption of Fe(III) and other elements (given in parentheses) can also be prevented by adsorbing the rare earths from the following mixed aqueous-organic solvent systems containing HCl and high concentrations of CIESE-active components such as acetone,[56-60] tetrahydrofuran,[61,62] and trioctylphosphine oxide,[61,62] 4:2 or 3:1 mixtures of acetone with 12 or 3 *M* HCl, respectively (trivalent Fe),[56,57] 60% acetone - 40% 2 *M* HCl (Mn),[58] 90%

* Tables for this chapter appear at end of text.

acetone - 10% 1 *M* HCl (Pu),[60] and 95% tetrahydrofuran - 5% 6 *M* HCl which is 0.1 *M* in trioctylphosphine oxide (Sc and other elements) (see Tables 15 and 18).[61,62]

Neutral chloride media such as 1 or 2 *M* NaCl solutions have also been used for the adsorption of Sc and the simultaneous elution of divalent metals such as Co, Ni, Pb, and Mn (see Table 20).[9]

After adsorption of the rare earths from very dilute HCl solutions (0.1, 0.2, 0.5, or 0.6 *M* in this acid), the elution of coadsorbed elements such as the alkali and alkaline earth metals, Fe(III), Al, Co, Mn, Cu, Ni, Zn, Cd, Ti, Be, U, In, Tl, etc., is in many cases effected by using 1.75 *M*,[1,2,19,20,45] 1.85 *M*,[26,30] or 2 *M* HCl[21,22,24,28,36,59] as the eluent. Under these conditions the distribution coefficients of the rare earth elements are still sufficiently high to prevent their elution, provided that the experimental conditions have been properly selected. When using HCl solutions of higher molarity such as 2.5 *M*,[63] 3 *M*,[40,44-46] and 4 *M* HCl[10,12] for the removal of coadsorbed elements, it is possible (especially if small columns and/or large volumes of the eluent are used) that part of the adsorbed rare earths will be coeluted with the elements to be removed. Therefore, it has been suggested to collect the eluate in two separate fractions of which the first contains the coadsorbed elements such as Fe, Al, Mg, Ca, and Na, and the second the rare earth elements (see Table 17).[10]

Eluents which are 0.8 *M*,[15] 1 *M*,[14,35,64] or 1.6 *M*[65] in HCl as well as 1 *M* ammonium chloride followed by 0.3 *M* HCl[32] have also been used for the elution of coadsorbed mono- and divalent ions as well as for Fe(III) and Ti. The latter element is readily eluted with 1 *M* HCl - 0.5 *M* H_2O_2[66]

For certain coadsorbed elements (given in parentheses) the following eluents containing oxalic acid have been found to be useful: 0.5 *M* HCl - 0.1 *M* oxalic acid (Sc, Fe, and Co),[23] 0.5 *M* oxalic acid (Zr, Hf, Nb, and Th),[25,32] and 2% oxalic acid (Ti).[67]

Very effective and suitable eluents for coadsorbed elements proved to be the following mixed aqueous-organic solvent systems containing HCl (the eluted elements are given in parentheses): 50% ethanol - 3 *M* HCl (Al, Fe[III], Ti, Ga, In, Be, Mg, U[VI], Bi, Tl, Pb, Cd, Zn, Cu, Mn, and Ni), 25% ethanol - 3 *M* HCl (Na, K, Ca, Mg, Sr, Al, Fe, and Ti),[34,68] 20% ethanol - 3 *M* HCl (major rock-forming elements),[25] and a 7:3 mixture of acetone and 4 *M* HCl (trivalent iron).[56] Any coadsorbed organic substances as, e.g., contained in natural waters can be removed by washing the resin bed with 40 or 80% acetone.[65]

Other eluents which can be utilized for the removal of certain coadsorbed elements (listed in parentheses) include: 0.5 *M* HNO_3 (Cs),[52] 1 *M* HNO_3 (Cs and Sr),[52-54] 2 *M* HNO_3 (alkali and alkaline earth metals such as Sr and Ba),[47,53,54,63] 1 *N* H_2SO_4 (Zr),[59] 1.2 *N* H_2SO_4 followed by 1.8 *M* HCl (Fe and other elements),[33] and 0.45 *M* HSCN (Co).[52]

The most widely used eluting agents for rare earth elements, which were adsorbed on strongly acidic cation exchange resins, are 4 *M*[2,8-14,18-20,26,30,34,39,55,58] and 6 *M* HCl.[15,21-25,36,40,44-46,56,68,69] Coeluted elements include Sr, Ba, Zr, and Hf.[2,20,34,39,40,65] Any coadsorbed Th is not eluted with the rare earths. The high popularity of these two eluents is explained by the fact that the rare earth adsorption on cation exchange resins is a minimum in these systems (see preceding paragraphs) so that rapid and quantitative elutions can be achieved. By means of the 6 *M* HCl eluent it is possible to separate the adsorbed rare earths into two fractions, namely, with $\approx$Gd $\rightarrow$ Lu in the first, and La $\rightarrow$ Eu in the second.[21]

HCl solutions of the following molarities have also been used for the elution of the rare earth group: 3 *M* HCl,[1,64] 4.8 *M* HCl,[33] 5 *M* HCl,[57,59,66,70] 6.6 *M* HCl,[41,42] 8 *M* HCl,[28,31] 11 *M* HCl,[32] and 12 *M* HCl.[35,37,71]

Various other eluents that have been employed for this elution include: 2 *M* HNO_3,[16,52] 2.5 *M* HNO_3,[63] 2.58 *M* HNO_3,[47] 0.3 *M* ammonium sulfate - 0.025 *M* H_2SO_4,[48-51] 1 *M* ammonium sulfate,[25] 1.5 *M* H_2SO_4,[9] and 2 *M* H_2SO_4.[72] The sulfate systems have been used for the selective elution of Sc which, unlike the other rare earth elements, forms an anionic sulfate complex (see later section) so that it is readily eluted. If Th is present, some of it may be coeluted with the Sc.

The elution of the rare earths, as a group, can also be effected by use of 0.4 *M* α-hydroxyisobutyrate (α-HIBA) of pH 6.[43]

Following elution of the rare earths from a cation exchange resin using the eluents mentioned above, coeluted elements (or metal ions such as iron used to isolate the rare earths from the eluate by coprecipitation) can be separated, if necessary, by means of additional anion exchange separation procedures utilizing the nonadsorbability of rare earth elements on anion exchange resins from HCl or sulfuric acid media. Thus, by use of the following HCl systems (mentioned in parentheses), the rare earths have been separated from trivalent Fe (6 and 6.2 *M* HCl; see Tables 9, 10, and 14)[28,51,73-75] and 12 *M* HCl (see Table 10),[65] U and Pu (11 *M* HCl; see Table 17),[32] and Zr and Hf (12 *M* HCl; see Table 10).[65] Separations of the rare earths from the latter two elements as well as from Ti and other metal ions can also be effected by combined cation-anion exchange procedures of the type illustrated in Tables 11 and 16. For this purpose, both strongly acidic and strongly basic anion exchange resins are employed in a composite column (two separate layers of the resins),[37] in a mixed bed column,[76] and as a combination of cation and anion exchange resin filters.[71] Concentrated HCl is used as the eluent for the rare earths from the cation exchange resin and as the sorption medium from which some coeluted elements are adsorbed on the anion exchanger.

Very dilute sulfuric acid media (e.g., of pH 1.5) have variously been used for the purification of Sc[48-51] and to separate undesirable Sc activities[26] from the other rare earth elements by adorption of the Sc on strongly basic resins (see Table 9). On the other hand, 0.25 *M* sulfuric acid has been employed (see Table 19)[8] to adsorb Zr and Hf on such a resin to separate it from Sc, which passes into the effluent because of its very weak adsorbability at this acidity (see Table 95; in chapter on Actinides, Volume II).

In case that mutual separations of the rare earths are required following their elution, as a group, from the cation exchange resin, the chromatographic procedures outlined in Tables 7, 8, 10, and 17 can be employed. For this fractionation of individual rare earths, either conventional elution chromatography using α-HIBA,[32,65,77] α-hydroxybutyrate (α-HBA),[25] and ammonium citrate[65] or gradient elution (concentration or pH gradients) with α-HIBA (see Table 7)[43-46,64] can be utilized.

From HCl solutions containing both water and a water-miscible organic solvent, the rare earth elements are usually much more strongly adsorbed on cation exchange resins, e.g., Dowex® 50, than from pure aqueous solutions of comparable acidity.[78-82] Thus, for instance, the distribution coefficient of Ce(III) in 90% organic solvent - 10% 9 to 12 *M* HCl (overall acid concentration = 0.9 to 1.2 *M*) is $>10^3$ to 10^4 irrespective of the kind of organic solvent present (which may be methanol, ethanol, *n*-propanol, isopropanol, methyl glycol, acetic acid, tetrahydrofuran, and acetone), while in a pure aqueous solution of the same acidity, the coefficient has a value of 265 (see Table 1).[79] It has also been shown that the distribution coefficients of the rare earth elements (e.g., La, Nd, Sm, and Eu) in, e.g., methanol or ethanol strongly increase with increasing concentration of the organic solvent.[80,83] Similar observations were made with Sc[84,85] and Pr.[86]

As has already been mentioned in a preceding paragraph of this section, acetone-HCl media can be used for the effective separation of rare earth elements from Fe and other elements. Thus, for example, in 90% acetone - 0.5 *M* HCl medium, the distribution coefficients of La and Fe(III) have values of $>10^4$ and 1.3, respectively, so that Fe passes into the effluent practically unadsorbed, while the rare earth element is very strongly retained by, e.g., Dowex® 50.[87] A similar behavior as shown by trivalent Fe was observed for U (K_d = 14) and other elements readily forming anionic chloride complexes. Also, in the case of Ce adsorption from 90% acetone - 10% HCl media (see above), it was observed[79] that Fe(III) and other elements forming chloride complexes are not retained (distribution coefficients <1) by the cation exchanger from systems with 0.15 to 1.2 *M* overall acidity. Similar observations were made at comparable acidities and concentrations of organic solvents in systems containing tetrahydrofuran or methyl glycol.[79]

As has been mentioned previously in this section, Sc is not retained by Dowex® 50 from 95% tetrahydrofuran - 5% 6 or 12 *M* HCl media which are 0.1 *M* in trioctylphosphine oxide (TOPO).[88] Under these conditions the distribution coefficient of Sc is less than one, while the coefficients of Yb and Ce are 460 and 21,600, respectively. Consequently, Sc can be readily separated from the rare earth elements, but not from Zr, Hf, Mn, Ga, In, Cu, Fe, Co, Au, U, Zn, Cd, Hg, Bi, Sn, Pb, Ge, V, Mo, and the platinum metals which are coeluted with the Sc and thus separated from the adsorbed lanthanides and other elements (see Table 18). If under the same conditions tetrahydrofuran is replaced by other organic solvents such as hexone, acetone, methyl glycol, methanol, and acetic acid, the distribution coefficient of Sc has the values of 5.7, 8.3, 15, 105, and 251, respectively. Higher distribution coefficients for Sc have also been measured in comparable systems containing bis-(2-ethylhexyl)phosphate (HDEHP) or tributyl phosphate (TBP) in place of TOPO.

Also stated previously in this section was the fact that mixed aqueous-ethanol systems containing HCl can be used more effectively than pure aqueous systems for the elution of a large number of elements which were coadsorbed with the rare earths on cation exchange resins. This is because the distribution coefficients of the rare earths in these ethanol media are higher than in the pure aqueous solutions, while those of other elements such as Al, Fe, and Cu are lower in the ethanolic systems.[8] In other words, in the latter systems, the separation factors between the rare earths and the other elements are higher than in pure aqueous HCl solutions.

In Tables 2 to 17 numerous methods are presented of cation exchange separations of the rare earth elements in HCl media which have been applied to the determination of these elements in geological, biological, and industrial materials. In addition, several other procedures are also described in Tables 18 to 20 which were employed for the separation of rare earth elements from synthetic mixtures with other elements.

Nitric Acid Systems

From dilute, pure aqueous nitric acid solutions with molarities ranging up to 2 *M*, Sc, Y, and the lanthanides are adsorbed on strongly acidic cation exchange resins to about the same extent as from HCl systems of comparable acidity (see preceding section and Table 1).[2,89-91] At higher acid concentrations, their adsorption decreases and shows a minimum at Gd, after which the adsorption increases again to Yb and probably to Lu. Since this is not the case in HCl media, it seems to indicate different patterns of ion association of the rare earths in the two acids.

For their isolation from the various systems listed in Tables 21 to 23 the rare earths have been adsorbed on strongly acidic resins from the following nitric acid media: dilute HNO_3 of pH 5,[92] 0.1 *M* HNO_3,[93] 0.1 to 0.5 *M* HNO_3,[94] 1 *M* HNO_3,[95,96] and from a 1:1:3 mixture of HNO_3, acetic acid, and water.[97] Under these conditions essentially the same elements are coadsorbed with the rare earths as from the HCl media of comparable acidity (see preceding section).

To separate the rare earths from some of the coadsorbed metal ions, eluents which are 1.5[98] or 2 *M*[94-96] in nitric acid can be used. With these eluents mainly divalent ions such as the alkaline earths, Be, Mg, Mn(II), Ni, Zn, Pb, and U (as the uranyl ion) are removed. With 2 *M* nitric acid also Al and Bi can be eluted before the heavy rare earths and Sc pass into the eluate.[94] The light rare earths remaining adsorbed on the cation exchanger can then be eluted with the 4 *M* acid.[94]

Eluents with which the entire group of rare earths can be eluted effectively are 3 *M*,[98,99] 6 *M*,[95,96] and 8 *M* nitric acid.[96] In all cases, coadsorbed trivalent ions such as Fe and Al are coeluted. The Fe can be separated easily from the rare earths by anion exchange in 8 *M* HCl.[95]

Also, from mixed aqueous-organic solvent media containing nitric acid, the rare earths are adsorbed on strongly acidic resins to about the same extent as from comparable media containing HCl.[90,91,100,101] In some of the latter systems, however, elements forming anionic chloride complexes such as Fe can be readily separated from the rare earths (see previous section). Therefore, in these systems separations of greater selectivity can be achieved than in the nitric acid media, for which only comparatively few applications have been described in the literature. These methods were utilized in connection with the determination of the rare earths in geological and industrial materials (see Tables 21 and 22), as well as employed to separate them from synthetic mixtures with other elements (see Tables 20 and 23).

The fact that tetravalent Ce forms anionic nitrate complexes has been used to separate this element from Pr[102] and other trivalent lanthanides.[103] From very weakly acid nitrate solutions (e.g., pH 1.5) the trivalent lanthanides are adsorbed on a cation exchange resin (Amberlite® IR-120[103] or Cationite® KU-2[102]), while tetravalent Ce as the anionic complex passes into the effluent. At higher acid concentrations Ce(IV) is rapidly reduced to the trivalent state by the cation exchange resin, so that the anionic complex is not formed.

Sulfuric Acid Systems

In sulfuric acid solutions the rare earth elements, except Sc, show adsorption characteristics similar to those observed in HCl solutions, although the distribution coefficients are higher in sulfuric acid systems.[104] For Sc the case is reversed, i.e., this element can be readily eluted with dilute sulfuric acid solutions, e.g., 2 *N* sulfuric acid,[29,104] which do not remove the other rare earth elements adsorbed on a cation exchange resin. Similarly, other elements which, like Sc, form anionic sulfate complexes in dilute sulfuric acid media are also not coadsorbed with Y and the lanthanides. To these elements belong U, Th, and Zr which are not retained by strongly acid cation exchange resins when adsorbing the rare earths from 1 *N*[105-108] or 10% sulfuric acid[109] or from, e.g., 0.75 *M* ammonium sulfate solution.[10] Any residual, coadsorbed U can be removed (as anionic ascorbate complex; see in chapter on Actinides, Volume II) by passing 7 *M* ascorbic acid solution of pH 4 to 4.5 through the resin bed[105] before elution of the rare earths with 4 to 6 *M* HCl solutions.[104-106,110,111] Other eluents that have been employed for the elution of rare earths which were adsorbed from dilute sulfuric acid media are 6.5 *M* nitric acid,[108] 1:10 sulfuric acid,[112] and 1[29] or 2 *M* ammonium sulfate (e.g., of pH 1.2).[10,109] Ammonium sulfate solutions (3 and 1.5 *M*) have also been used for the separation of Ce from Y on a column containing a mixture of the Cationite® KU-2 and Pb sulfate.[113]

Coeluted Fe, or Fe which has been used to isolate the rare earths by coprecipitation with its hydroxide, can be removed by an anion exchange separation procedure based on the adsorption of trivalent Fe on Dowex® 1 from concentrated HCl.[105]

The fractionation of the rare earths into individual elements, following their group separation, can be effected by cation exchange chromatography utilizing ammonium α-HIBA as the eluent.[105]

Apart from a few studies conducted to elucidate the cation exchange behavior of Sc in mixed aqueous-organic solvent media containing sulfuric acid alone[85] or mixed with oxalic or tartaric acids,[114] little information is available on analytically useful data on the other rare earth elements in organic solvent systems.

Applications of cation exchange separations in sulfuric acid media to the analysis of industrial materials for the rare earth elements are presented in Table 24. To separate Sc and also Y from the lanthanides and other elements the methods illustrated in Tables 20 and 25 can be used.

Hydrofluoric Acid Systems

From dilute hydrofluoric acid (HF) solutions (e.g., 1 or 2.5 *M*) the rare earth elements (with the possible exception of Sc) are strongly retained by cation exchange resins such as Dowex® 50.[115-118] Thus, in the 1 *M* acid, distribution coefficients of $>10^3$, $>10^3$, and $\sim 10^3$

were measured for Y, La, and Eu, respectively.[64] These distribution values were found to decrease with increasing concentrations of HF, so that in the 10 *M* acid the corresponding distribution coefficients are ~100 (Y), >100 (La), and ~20 (Eu). It is possible, however, that the above-cited distribution coefficients of the rare earths are only apparent values, because Y and the lanthanides may precipitate as insoluble fluorides (except Sc which forms a soluble anionic fluoride complex). This assumption is corroborated by the fact that fission-rare earths had to be eluted with concentrated nitric acid in the presence of boric acid (this mixture dissolves insoluble fluorides such as those of the rare earths, and also of alkaline earth elements which are coeluted under these conditions; see Table 26). Nevertheless, the "adsorption" of the rare earth elements from HF media has found practical application not only for the isolation of rare earth fission products, but also with respect to the determination of rare earth elements in industrial materials consisting of matrix elements (i.e., U, Zr, Hf, Ta, and W) (see Table 26) which form soluble fluoride complexes that readily pass into the effluent unadsorbed. For the elution of "adsorbed" rare earth elements 6 *M* HCl has also been used.[116]

Thiocyanate Systems

Investigations of the cation exchange behavior of the lanthanides (La, Ce, Pr, Nd, Sm, Eu, Gd, Dy, Er, and Lu) on the resin Diaion® SK-1 in ammonium thiocyanate-HCl media have shown that the distribution coefficients decrease with increasing atomic number.[119] The complexes formed in these media are claimed to correspond to the formula $RE(SCN)_2Cl$ where RE is the rare earth metal. The distribution coefficients of the rare earth elements decrease with increasing concentration of ammonium thiocyanate. Thus, in 0.2 *M* thiocyanate - 0.5 *M* HCl the coefficients are in the range of ~400 to 1000, while at the same acidity but in presence of 2 *M* thiocyanate the coefficients are in the range of ~35 (Lu) to ~18 (La). Sc is less strongly adsorbed than the lanthanides, because it forms an anionic thiocyanate complex. This fact has been used to separate Sc from the lanthanides and other elements utilizing the separation procedure outlined in Table 27.[120]

It has been observed[121] that also in methanol media the adsorption of La on Dowex® 50 decreases with increasing concentration of thiocyanate.

Phosphoric Acid Systems

It has been shown[30] that La and Y are strongly adsorbed on the cation exchange resin Bio-Rad® AG50W-X8 from phosphoric acid media. The distribution coefficients of these two elements were found to be $>10^3$ at 0.1, 0.5, 1, 2, 4, and 6 *M* orthophosphoric acid. Since Sc is only weakly retained under these conditions (probably due to anionic complex formation), it can be separated from Y and other elements by using the procedure outlined in Table 28 in which a method is also described which is based on the nonadsorbability of Ce on Dowex® 50 when present as a polyphosphate.

Mutual separations of rare earth elements on cation exchange resins have also been effected by the use of organophosphorus compounds, i.e., organic derivatives of phosphoric acid such as TBP,[122] HDEHP,[123,124] and other compounds, e.g., ethylenediaminodi-(isopropyl-phosphonic acid), ethylenedinitrilo tetrakis-(methylphosphonic acid), and 1-hydroxyethylidenedi-(phosphonic acid).[125] For example, to separate radioactive isotopes of Tm, Eu, and Pm these elements are first adsorbed on a strongly acidic cation exchanger in the Na^+ form from 2 *M* ammonium thiocyanate - 0.1 *M* nitric acid or 2 *M* $NaNO_3$ - 0.1 *M* nitric acid and then their mutual separation is effected by elution with different concentrations of HDEHP in kerosine.[123]

Hydrobromic Acid Systems

In very dilute hydrobomic acid (HBr) solutions (0.1 to 0.5 *M*) all the rare earth elements show an adsorption behavior towards strongly acidic cation exchange resins which is very similar to that observed in HCl media of the same acidities (see Table 1).[87,126-128] At higher HBr concentrations the adsorbability of Sc decreases to a minimum (K_d = 4), near 4 *M* HBr and then increases sharply with increasing acid molarity to reach a $K_d > 5 \times 10^3$ in 12 *M* HBr. The adsorption function of Y resembles that of Sc, but at high HBr concentrations adsorbability is much lower. The adsorbability of Y increases from a minimum K_d = 1.6 near 6 *M* HBr to K_d = 23 in the 12 *M* acid. A behavior similar to that of Y is shown by the lanthanides (La, Eu, and Lu), i.e., the adsorbabilities decrease with increasing HBr concentration to minima near 6 *M* HBr and then increase with increasing HBr concentrations.[127]

As in the mixed aqueous-organic solvent mixtures containing HCl very high adsorption of the rare earths on strongly acidic cation exchange resins has also been observed in organic solvent media containing HBr,[87,126,128] and this fact was employed to separate Sc from other elements which are less strongly retained from acetone-HBr media (see Table 29).[126]

Perchloric Acid Systems

In dilute perchloric acid solutions the adsorption behavior of the rare earths towards strongly acidic resins is very similar to that observed in HCl and HBr media of comparable acidity. However, at high concentrations of perchloric acid the rare earths are much stronger retained than from HCl solutions of the same molarities.[6] Thus, the distribution coefficients of Sc, La, Y, and Lu were found to have the values of $\sim 10^6$ (10), $> 10^3$ (< 10), $> 10^2$ (> 1), and $\sim 10^2$ (< 1) in 9 *M* perchloric acid and 9 *M* HCl (values given in parentheses), respectively.[129] The adsorbability of Sc from concentrated perchloric acid is higher than that of any other trivalent ion.[6,130]

In the perchloric acid media the adsorption functions of the rare earths show minima near 4 to 5 *M* and rise steeply thereafter. In HCl the minima are very flat and occur near 6 *M*.[6]

The rare earth elements were also found to be adsorbed from 0.2 *M* $NaClO_4$ media containing various concentrations of aliphatic alcohols. The distribution coefficients decreased with increasing concentrations of alcohol.[131]

Applications

HCl Systems

Geological Materials A (Solid Samples)

In connection with the spectrographic determination of rare earths in terrestrial and lunar rocks and other geological materials, these elements were separated, as a group, using the one-column cation exchange procedure outlined in Table 2.

Similar separation methods have been employed for the mass spectrometric analysis of rare earth elements in rocks and minerals. These techniques are illustrated in Table 3.

Besides the X-ray spectrometric methods mentioned in Table 2, other techniques based on the X-ray fluorescence and polarographic determination of the rare earths in rocks, minerals, and ores[69] have been described. The cation exchange separation procedures which were used in connection with these analytical methods are presented in Table 4. Included into this Table 4 is an analytical scheme involving the separation of the rare earths, as a group, from matrix elements using cation exchange in ethanol-HCl medium prior to their determination by atomic absorption spectrophotometry.

Cation exchange separation of the rare earths in HCl media has also been used in conjunction with the determination of these elements in rocks using neutron activation analysis. The one-column procedures employed for this purpose are outlined in Table 5.

To determine rare earth elements in carbonatites the procedure outlined in Table 6 has been utilized.

Two-column procedures involving the successive application of cation exchange in HCl media to separate the rare earths, as a group, and to effect their fractionation into individual elements on a separate column of the same resin have been utilized for the analysis of minerals, rocks, lunar samples, and meteorites. For the mutual separations of the rare earths, eluents containing α-HIBA or α-HBA were used, and for the final determination of the separated elements radiometric, mass spectrometric, and spectrophotometric procedures were employed. The analytical separation schemes that have been utilized are presented in Tables 7 and 8. As can be seen from Table 8, a third column operation may serve to separate the rare earths from the complexing ligand.

Before the determination of rare earth elements in silicates, rocks, and meteorites, multicolumn methods combining cation exchange with anion exchange separation steps in sulfuric acid and HCl media for the removal of Sc and other elements have been used. These separation schemes are outlined in Table 9.

Geological Materials B (Liquid Samples)

To determine individual rare earth elements in mineral waters, the multicolumn procedures outlined in Table 10 can be used. In this analytical scheme, the rare earth elements are first separated as a group from all interfering elements by successive use of cation and anion exchange in HCl media. Then, the mutual separations of the rare earths are effected by cation exchange using eluents containing α-HIBA and citrate. Finally, the rare earths are determined spectrophotometrically following removal of α-HIBA by cation exchange in HCl media.

For the rapid isolation of radioactive Ce and radionuclides of other elements from natural waters, mixtures consisting of cation and anion exchange resins have been used. The principles underlying these separations are illustrated in Table 11.

Prior to its spectrophotometric determination, stable Ce can be isolated from seawater by using the separation procedure outlined in Table 12.

Biological Materials

In the analysis of biological matrices for Ce, Y, and Sc, methods based on one-column cation exchange procedures and one method making use of a combination of cation exchange with subsequent anion exchange removal of residual impurities have been employed. The separation schemes underlying these methods are illustrated in Tables 13 and 14.

Industrial Materials

Adsorption of Ce and other rare earth elements on cation exchange resins from HCl solutions in the presence or absence of acetone has variously been employed for their determination in steel and cast iron. For this purpose the separation schemes illustrated in Table 15 were used. From these it is seen that acetone-HCl eluents are employed in most cases for the removal of the Fe matrix. This is readily achieved owing to the CIESE effect predominating in these media.

In connection with the determination of some rare earths in nonferrous alloys and metals the separation methods outlined in Table 16 have been used. Included in Table 16 is a procedure in which a composite column consisting of a cation and an anion exchange resin bed is employed.

To isolate rare earth elements from matrices consisting predominantly of actinides such as irridiated U, U mill products, Pu, Th, and radioactive wastes, the separation procedures illustrated in Table 17 can be employed. In Table 17 a separation scheme is also presented in which a multicolumn procedure is utilized for the fractionation of rare earths and for the isolation of Rh.

Synthetic Mixtures

In Table 18 methods are illustrated which can be applied to the selective separation of Sc from numerous elements including the other rare earths, using organic solvent-HCl media as eluents. While the tetrahydrofuran-HCl eluent containing trioctylphosphine oxide prevents any adsorption of Sc on the cation exchanger (due to the strong CIESE effect shown by this system), the Sc is retained from the acetone HCl system.

For the effective separation of the rare earths, as a group, from a great number of accompanying elements, the separation schemes outlined in Table 19 can be used. Zr and Hf which are coeluted with the rare earths are removed by adsorption on an anion exchange resin from sulfuric acid solution.

To separate Sc and Y from numerous other elements on Dowex® 50, not only HCl but also a number of other eluents can be used. These separation procedures are outlined in Table 20, in which a method for the separation of rare earths from large quantities of phosphoric acid is also described.

Nitric Acid Systems

Geological Materials

In Table 21 two methods are shown, which have been applied to the determination of rare earth elements in rocks, minerals, soils, coal, and natural waters. The first of these techniques is based on a two-column procedure utilizing cation exchange in nitric acid media which is followed by an anion exchange separation step in strong HCl medium to separate the rare earths from Fe.

Industrial Materials

In connection with the determination of rare earth elements in materials of this type, the separation procedures outlined in Table 22 have been used.

Synthetic Mixtures

From Table 23 it is evident that separations of rare earth elements from bivalent metal ions can readily be effected by the use of eluents containing varying concentrations of nitric acid, while, however, no separation is achieved from trivalent metals.

Sulfuric Acid Systems

Industrial Materials

From the separation schemes illustrated in Table 24 it is seen that adsorption of the rare earths on strongly acidic cation exchange resins from dilute sulfuric acid and ammonium sulfate media has been used nearly exclusively for the determination of these elements in matrices consisting of elements which form stable anionic sulfate complexes (see Table 95 in chapter on Actinides, Volume II). This complex formation allows these elements (U, Th, and Zr) to be readily eluted, even if present in large excess over the rare earths. Included in Table 24 is a procedure which is based on the adsorption of a peroxide complex of Ce with H_2O_2.

Synthetic Mixtures

The separation methods outlined in Table 25 are based on the fact that Sc forms anionic sulfate complexes in dilute sulfuric acid solutions, so that its separation from other rare earth elements as well as from numerous other metal ions can be achieved by eluting Sc before desorption of the other elements with more concentrated eluents.

HF Systems

Adsorption of rare earth elements on strongly acidic cation exchange resins from dilute HF media has variously been employed for their separation from large amounts of U, Zr, and other metals forming stable anionic fluoride complexes so that these elements are not retained by the resins. Three methods of this type are illustrated in Table 26.

Phosphoric Acid Systems

To separate Sc from Y as well as from other elements and of Ce from some heavy metals, the procedures outlined in Table 27 can be used. In both cases, the elution of the rare earths is effected with phosphate-containing eluents.

Thiocyanate Systems

From the separation scheme shown in Table 28 it is seen that Sc can be eluted with 1 *M* NH_4SCN - 0.5 *M* HCl either before the rare earths or after other elements which are eluted ahead of the Sc using the same or similar eluents.

Media Containing Organic Complexing Agents

General

In the presence of a complexing ligand the distribution of trivalent rare earth cations between a solution and a cation exchange resin is the sum of two effects, the affinity of the cation for the resin and complexing agent, respectively. For a pair of rare earth elements the resin affinity terms are similar and, hence, the separation factor is essentially the ratio of the association constants of the two ions for the ligand.[132] Thus, the rare earth element with the highest affinity for the complexing agent will tend to pass into the solution phase and will therefore elute first from an ion exchange column. (This is also the element with the larger hydrated ionic radius and which consequently is also a little less strongly retained by the resin due to decreased hydrostatic attraction of the hydrated cation by the negatively charged fixed ion of the resin.) Similarly, the elution of mixtures of trivalent ions will generally follow the sequence of affinity of the complexing agent.

In a standard elution of the rare earths from a strongly acidic cation exchange resin of the polystyrene type, using a solution containing an organic complexing agent, the elements are eluted in the reverse order of their atomic numbers (i.e., Lu first and La last; with Y between Ho and Dy — if Sc is present it is eluted before the Lu) (for examples see Tables 31 to 60).

The order of elution of lanthanides varies slightly on using various chelating agents as eluting solutions. However, a different behavior was found on eluting Y with various organic ligands.[133] Its position in the elution order of lanthanides varies widely: using EDTA, Y is eluted between Dy and Tb. With DTPA the position of Y is near Nd, with HEDTA near Pr; near Eu with citrate at 10 to 20°C and is shifted to Dy-Ho with citrate at 87 to 100°C; near Dy-Ho with either lactate or thiocyanate.

On a polymethacrylic acid cation exchanger (Amberlite® IRC 50) the selectivity for lanthanides was found to increase with decreasing ionic radius. This is the reverse of the order given above for sulfonic acid resins.

On elution of the rare earths, the elements follow each other with increasing distance between the elution peaks, and when extremely small amounts of them are present (e.g., 10^{-12}-g quantities of single lanthanides), the elution peaks are sharp and show no tailing. This is, for instance, the case in studies involving nuclear reactions. In most analytical separations, however, the amounts of rare earths handled are much larger (usually in the microgram and milligram range), which results in a broadening and tailing of the elution curves, these effects being more pronounced the larger the amounts present. This, in particular, is a disadvantage when the heaviest rare earths, i.e., the trans-Eu elements, are to be separated. They appear so closely together in the elution curve that cross-contamination occurs and identification is made difficult. These difficulties can be overcome by the application of gradient elution techniques employed at normal or, still better, at elevated pressures[134] (i.e., utilizing pressurized ion exchange [PIX]),[135] e.g., by using the method of high performance liquid chromatography (HPLC) (see also section on trans-Pu elements in chapter on Actinides, Volume II).[136-139] This HPLC technique cannot only be performed on conventional ion exchange resins, but also on 5- to 10-μm bonded-phase strong-acid ion exchangers, e.g., Nucleosil®-SCX[140] or -SA.[141]

When gradient elution is utilized for the fractionation of the rare earth elements, the concentration of the complexing ligand is increased continuously (e.g., practically undissociated α-HIBA acid is increasingly dissociated with increase in pH value, i.e., more salt and, hence, more free ligands are formed at higher pH values than at lower values). This increase in the concentration of the ligand can be effected by an increase of the pH (this pH-gradient elution technique is based on mixing solutions with different pH values;[142,143] examples for this technique are presented throughout this chapter; see Tables 31,[144] 32,[145] 34,[146] 36,[147,148] 38,[144] 39,[149-152] 42,[153,154] 43,[155,156] 44,[157,158] 45,[97,159-167] 7,[43-46,64] 17,[32] 64,[168,169] 85,[170,171] 95,[172] and 99.[173] Similarly, the concentration of the ligand can be increased by an increase of the molarity of the organic complexing agent in the eluent solution (this concentration gradient elution procedure is carried out at constant pH) (examples for this technique are shown in Tables 32,[174] 33,[175] 36,[176] 37,[177,178] 40,[135,179] 44,[180] 91,[181] 93,[182-184] and 95[185,186]).

A third variant of elution chromatographic methods utilized for the continuous elution and fractionation of the rare earth elements is the technique of combined pH- and concentration gradient elution in which both parameters are varied during the elution process (several examples for this technique are presented in Tables 31,[187,188] 36,[189] 41,[190] 47,[191] 10,[65] 90,[192] and 95[185]).

As a consequence of the increase of the concentration of the complex-forming organic acid anion responsible for eluting the bands down the column, the slower bands still on the ion exchange bed are eluted more rapidly than would otherwise be the case. The net effect of this gradient elution is therefore to shorten the overall time required (~42 to 48 hr or only a few hours if HPLC is used) to complete the separation and to produce a series of peaks more uniform in spacing and in rare earth concentration.[193] These results can most effectively be obtained by application of the HPLC technique (see Tables 32,[174] 36,[189] 38,[194] 39,[149-152] 40,[135,179] 41,[190] and 46[195]).[135,142,196-198]

Less broadening and tailing of the elution curves is also achieved when using the slurry column technique which is illustrated in Tables 31,[144] 34,[146] 35,[199] 36,[176] 45,[159] 7,[64] 72,[200,201] and 90.[192,202]

Elutions of the rare earths at elevated temperatures are required with the higher cross-linked resins, e.g., Dowex® 50-X12 in order to attain resin-solution equilibrium quickly.[125,203,204] However, heated (jacketed) columns require specialized equipment and techniques and become cumbersome when many samples must be prepared. Therefore, those methods which allow the rare earth elements to be separated at room temperature on resins of lower cross-linking are of greater importance for the analytical chemist, particularly if his work is carried out on a routine basis.

With respect to the influence of the grain size of cation exchange resin, e.g., Dowex® 50-X12 on the mutual separability of rare earth elements, e.g., when using α-HIBA as eluent and a pH-gradient elution technique, it has been established[205] that the finer and more homogeneous (small scatter or dispersion) the resins are, the sharper are the peaks on the elution curves. Grain size in the range 0.074 to 0.015 mm is generally recommended for elution chromatography of the rare earths at normal pressure, while at elevated pressures, i.e., when using the HPLC technique, very fine resin particles (<30 μm) with a small dispersion in the particle grain size are employed.

As to the influence of pore size of the resin particles, it has to be mentioned here that slightly better separation of rare earths was found to be possible with macroreticular resins[206,207] than with the cation exchangers (standard resins) of normal pore sizes.

Besides the factors mentioned above, the separation efficiency of lanthanide pairs also depends on the flow rate of the solution passing through the resin bed. The optimum flow rate depends on the resin type, grain size, and temperature of the eluting solution. For every combination of these paramaters, there exists an optimum flow rate where the separation is most effective.

The specific column load is also of importance. On increasing the amount of the lanthanides mixture to be separated, a gradual increase in the peak width occurs, leading to a worsening of the separation. This effect becomes appreciably marked when the amount of the elements exceeds 5 to 10 mg related to 1 cm^2 of the column cross-section. A maximum column load of 30 to 40 mg of the elements related to 1 cm^2 of the column cross-section is recommended, provided that the ratio of the exchanger column height to its diameter is ≥ 50.

The separation efficiency is also affected by the ionic form of the resin and by radiation damage caused by high absorbed doses of ionizing emissions.[208,209]

Ion exchange based on elution chromatography (or elution development) utilizing organic complexing agents is now the most efficient technique available for the mutual separation of the rare earth elements for analytical purposes.

The chromatographic technique known as displacement chromatography (or displacement development) has also been used for some analytical separations of the rare earths (see Table 53),[210-218] but its main application is in preparative work.

A number of complexing agents can be used to effect rare earth separations on cation exchange columns. For the most part they are simply salts of α-hydroxycarboxylic acid or salts of aminopolyacetic acid, acetic acid, and others. Some of these reagents are mediocre and other are excellent. Some require very closely controlled conditions for success. For the majority of the eluting agents used at present, the distribution factor between two neighboring lanthanides is 1.3 to 1.4.

Hydroxycarboxylic Acids

The most important compounds of this group are α-HIBA, lactic, and citric acids (a comparison of separation factors of the rare earths and trans-Pu elements in solutions containing these acids is presented in Table 30). Among these, the first mentioned acid is by far the most frequently used reagent for the chromatographic separation of the rare earth elements on an analytical scale, so that its characteristics will be discussed in the first paragraph of this section.

α-HIBA

The ammonium salt of this complexing agent is one of the most suitable eluents for the analytical fractionation of the rare earth elements.[125,134,193,219-227] It is superior to other α-hydroxycarboxylic acids, e.g., lactic acid or citric acid, with respect to separation factors (the smallest of the separation factors for all pairs of elements is larger for α-HIBA than it is for any other complexing agent) and to EDTA with respect to flow rate and other properties. One of the greatest advantages is that it can be employed successfully at room temperature using a resin of lower cross-linking such as Dowex® 50-X4 or -X8. Thus, at 25°C and at pH 4 to 4.8 the separation factors of adjacent rare earths in the series Lu, Yb, Tm, Er, Ho, Y, Dy, Tb, Gd, Eu, Sm, Pm, Nd, Pr, Ce, and La are 1.44, 1.37, 1.53, 1.37, 1.77, 1.1, 1.84, 1.57, 1.54, 2.06, 1.57, 2.09, 2.36, 1.63, and 2.29, respectively. An important exception is the separation of Y from Dy, where citrate solutions seem to give more favorable results (see Table 10).[65]

It has been shown that the above values of the separation factors at room temperature are higher for most of the rare earths than when working at elevated temperatures, e.g., at 87°C, with a resin of higher cross-linking such as Dowex® 50-X12.

A decrease of the separation factor is also observed in the presence of indifferent electrolytes, e.g., ammonium perchlorate,[228] but the factor increases when α-HIBA solutions are employed, in which part of the aqueous phase is replaced by aliphatic alcohols or other organic solvents. In this case, the separation factor is found to increase with an increase of the alcohol concentration.[229-234] Furthermore, this effect is not limited to alcohols, since it has been shown that the separation of Eu and Pm is improved in the order pyridine > acetone > dioxan > methanol > water.[235,236] The effect of some polyhydric alcohols, e.g., glycol and glycerol, on the separation of rare earths with the aid of α-HIBA has also been investigated.[237] A better fractionation of certain rare earth pairs (e.g., Pm-Sm and Nd-Pm) is claimed to be possible when α-hydroxy-α-methylbutyric acid (αHαMBA)[220,238-240] is used in place of α-HIBA.

With α-HIBA as the complexing ligand, a large number of cation exchange separation procedures have been developed and utilized in connection with the determination of the rare earths in a great variety of different matrices. These procedures are illustrated in this section (see Tables 31,[144,187,188,241] 32,[145,174] 33,[175,242] 34,[146] 35,[199,243] 36,[147,148,176,189,244] 37,[177,178,245] 38,[144,246] 39,[149-152] 40,[135,179] 41,[190] 42,[153,154,247] and 49[248]) as well as in the section dealing with cation exchange separations in mineral acid media (see Tables 7,[43-46,64] 8,[24] 10,[65] 17,[32] and 24[105]), and in the section describing anion exchange separation procedures of the rare earths in nitric acid (see Tables 72,[200,201,249,250] 75,[251] and 79[252,253]), and HCl media (see Tables 87,[254] 90,[169,192,255] 93,[182-184] 94,[256] 95,[172,185,257] 99,[173] and 100[258]). Thus, in most cases the fractionation of the rare earths into individual elements is preceded by either a cation or anion exchange procedure (or both) in mineral acid media, to separate the rare earths, as a group, from matrix elements which, especially if present in large excess over the rare earths, may interfere with their chromatographic separation.

In many cases elution chromatographic separation of the rare earths using α-HIBA as the eluent is followed by an additional cation exchange step (which has to be performed for each eluate received) to isolate the separated rare earth element (or group) from the eluate. By means of this process not only appreciable concentration of the rare earths is achieved (especially if large volumes of α-HIBA eluent were required for their fractionation), but also α-HIBA is eliminated (passes into the effluent when adsorbing the rare earths on the cation exchanger from the acidified eluate). Examples for this technique are shown in this section (see Tables 34,[146] 36,[148] and 49[248]) and in the sections dealing with the cation exchange separation of the rare earths in HCl media (see Tables 8[24] and 10[65]) and in those in which their anion exchange behavior in nitric acid (see Table 72)[200,201,249,250] and HCl (see Table 100)[258] is described.

The α-HIBA may also be removed from the eluate fraction by extracting it into organic solvents, but in this case no concentration of the rare earths is obtained. As with the cation exchange method, a concentration of the rare earths, as well as the removal of the α-HIBA, can be achieved by evaporation of the eluate fraction followed by wet ashing (and/or dry ashing) of the organic complexing agent. However, this process is much more time consuming and less reliable than is the additional separation of the rare earths by means of cation exchange, which in some cases has also been used in connection with the treatment of eluate fractions obtained by the use of the complexing ligands discussed in the following paragraphs of this section (see also Table 49).[248,259]

For the fractionation of rare earth elements α-HBA can be used (see Tables 8[25] and 42[139]).

Lactate

Solutions containing lactate as complexing agent have frequently been employed[198,204,220,228,260-267] for the mutual separation of lanthanides (see Tables 45[97,159-167] and 46[268,269]) and also from trivalent trans-Pu elements[275] (see in chapter on Actinides, Volume II). Analytical separations of Y, Tb, Eu, and Sm are more easily performed in these media than in citrate solutions (the separation can be improved by the addition of a small amount of phenol to the eluting solution). However, with lactate solutions, the reproducibility of the elution volume (but not of separation factors) is jeopardized by the fact that lactic acid may contain a variable amount of dimer. α-HIBA and also glycollic acid behave like lactic acid with respect to pH changes, but no dimerization seems to occur in these acids. The dimer of lactic acid can be hydrolyzed by boiling its dilute solution for several hours.[204] Another disadvantage of lactic acid (and also citric acid) as an eluting agent is that the solutions prepared are often invaded by microorganisms or molds which change their composition.

To remove lactic acid from an eluate fraction, either an additional cation exchange procedure may be used (see section on α-HIBA) or this complexing agent is first eliminated by extracting it into diethyl ether (see Table 95[270,271]).

Methods based on the application of lactate as eluent for the rare earth elements are presented in Tables 33,[242] 43,[155,156,272,273] 44,[131,157,158,180,274] 45,[97,159-167] 46,[195,268] 47,[191,276] 48,[277] and 64.[168,169] Lactate elutions have also been applied quite frequently after group separation of the rare earth elements using anion exchange in HCl media. These methods are presented in Tables 85,[170,171] 88,[278] 90,[202] 91,[181] and 95.[185,186,270,271]

Citrate

The cation exchange behavior and separations of lanthanides involving the use of citrate solutions have been described by many investigators.[29,65,169,177,245,248,258,259,273,277,279-297] At present, however, this method is not very frequently used for analytical purposes, because better eluents, such as media containing α-HIBA, have been found with which more efficient fractionations of the rare earth elements can be achieved.

The comparatively few analytical procedures in which citrate is utilized are illustrated in Tables 37,[177,245] 43,[273] 48,[277,279,280] 49,[248,259,281,282] 50,[283,284] and 100.[258] Other applications of citrate eluents are outlined in Tables 10[65] and 20.[29]

It must, however, be pointed out here that by use of this complexing agent, ion exchange technology and rare earth chemistry have received great impetus, since it allowed the first successful separation of the rare earths in the World War II Manhattan Project and later macroscale separations of these elements, developed by Spedding and collaborators.[298]

Aminopolyacetic Acids

Very effective cation exchange separations of the rare earth elements are achieved by the use of eluents containing soluble salts of aminopolyacetic acids.[299]

The following compounds have been recommended: EDTA (ethylenediaminetetraacetic acid),[136,137,214,300-327] NTA (nitrilotriacetic acid),[216,256,300,328-338] HEDTA (*N*-[2-hydroxyethyl]ethylene-diamine-*NN'N* triacetic acid),[215,304,312,339-350] HIDA (*N*-[2-hydroxyethyl]-iminodiacetic acid),[217,218] DTPA (diethylenetriamine-*NNN'NN* penta-acetic acid),[311,332,347,351-353] IDA (iminodiacetic acid),[354] ethylenediamine-*NN'*-disuccinic acid,[355] DCTA (1,2-diaminocyclohexane-*NNN'N'*-tetraacetic acid), ethyleneglycol-bis(2-aminoethyl)ether *NNN'N'*-tetraacetic acid, and bis(2-aminoethyl)ether-*NNN' N'*-tetraacetic acid.

Among these, EDTA is the complexing agent which is most frequently employed for the micro- and macroscale separation of the rare earth elements. The stability constants of the rare earth-EDTA complexes vary from $10^{14.72}$ for La to $10^{19.65}$ for Lu, which is a difference ($10^{4.9}$) higher than that observed in citrate media. Consequently, EDTA is a better eluent than citrate, however, the disadvantage is that H_4EDTA (the free acid) precipitates in acid media; hence, no part of the resin bed through which EDTA passes can be in the acid form. The pH in the columns has therefore to be kept low enough (e.g., pH $\geqslant$ 4) to ensure that the separation can be performed at a reasonable speed, but care has to be taken that at these low pH values H_4EDTA does not precipitate.

Before the determination of separated rare earth elements EDTA may have to be removed from the eluates. This can be accomplished by wet ashing with nitric and perchloric acids[356] or nitric and sulfuric acids.[326]

In Tables 51,[357,358] 52,[359-361] 53,[210,211] and 54[356,362-365] methods are outlined in which EDTA solutions have been used for analytical separations involving the rare earth elements. An additional example for an application for this complexing ligand is illustrated in Table 85.[366]

The application of other aminopolyacetic acids is shown in Tables 53[215-218] and 94.[256]

HEDTA is more soluble than EDTA so that ion exchangers in the H^+ form can be used. No precipitation of the free acid in the resin bed occurs.

Other Organic Complexing Agents

Besides the complexing ligands already treated in the preceding sections, the adsorption characteristics and separations of the rare earth elements on cation exchange resins have also been investigated by use of the following compounds and systems: sulfosalicylic acid (see Table 55),[367-370] ammonium acetate-acetic acid (see Table 56),[133,371-375] ammonium acetate-dimethyl formamide,[376] ammonium acetate-nitric acid,[377] 2 *M* acetic acid (see Table 38),[246] monochloroacetic acid (see Table 25),[378] ammonium formate-formic acid (see Tables 48[277] and 59),[372,379,380] oxalic acid and ammonium oxalate (see Table 57),[281,381-383] and oxalic acid in the presence of acetone,[384] glycollic acid (see Tables 58[77,385] and 43),[386] tartaric acid in the absence or presence of acetone,[297] thenoyltrifluoroacetone (TTA) in dioxane-water,[226] ascorbic and malic acids (see Table 60),[387,388] organophosphorous compounds (see Table 61),[389,390] glycine,[391] and pyruvic acid.[392]

The order of elution efficiency of aliphatic monocarboxylic acids is acetate $\simeq$ propionate > isobutyrate > butyrate > formate for Dy and propionate > acetate > isobutyrate > butyrate > formate in the case of Y.[372]

A considerable amount of data concerning distribution coefficients, separation factors, etc. in solutions containing these complexing agents have been accumulated.

Comparatively little information is available, however, on investigations which clearly demonstrate the overlapping of the elution bands or make it possible to estimate the influence of different eluents upon the height of a theoretical plate. Therefore, it is not attempted here to make any generally valid statements as to the merits of the different eluents.

From Tables 38, 48, and 54 to 61 it is seen that many of the complexing ligands mentioned above have been applied for analytical separations involving the rare earth elements.

Applications

α-HIBA Systems

Geological Materials A (Solid Samples)

The separation procedures outlined in Table 31 have been used in connection with the determination of rare earth elements in rocks and marine sediments. All of these methods are based on one-column techniques in which α-HIBA solutions are used to fractionate the rare earths on columns of the strongly acidic cation exchange resin Dowex® 50.

Elution chromatographic procedures have also been applied with respect to the analysis for rare earth elements of monazite and monazite sand as well as to determine rare earth impurities in high-purity rare earth oxides. These techniques are illustrated in Table 32.

Geological Materials B (Liquid Samples)

To determine rare earths in seawater[393] these elements, as a group, are first separated from the sample (pH 5) by extraction with HDEHP-heptane and then the individual elements are obtained by cation exchange chromatography on a column of Dowex® 50 employing pH-gradient elution with α-HIBA as the eluent. The recovery for ^{144}Ce averaged 60%. The decontamination factor is better than 10^4 for neighboring mixed fission products.

In connection with the determination of stable Ce and other rare earth elements in seawater, individual separations of the rare earths on Dowex® 50 utilizing α-HIBA have also been described by other investigators.[394,395]

Industrial Materials

Ferrous materials such as iron and steel can be analyzed for the rare earths as well as for other elements by using the separating schemes outlined in Table 33 which are applied after neutron irradiation of the samples.

To determine the concentrations and isotopic distribution of microgram levels of fission product rare earths in nuclear fuels, the separation procedures illustrated in Table 34 have been employed.

For the separation of Pm and Y from neutron-irradiated Sm oxide and Nb and Zr metals, respectively, the methods illustrated in Table 35 have been used.

From Table 36 it is seen that numerous methods have been developed with respect to the determination of the rare earths in fission product mixtures utilizing chromatography on cation exchange resins with α-HIBA solutions as eluents. Included in Table 36 is a method based on HPLC in which concentration and pH-gradient elution are employed simultaneously to fractionate the rare earths.

Rare earth impurities contained in rare earth oxides can be determined radiometrically or coulometrically after application of the separation methods presented in Table 37 which include one procedure based on HPLC.

Prior to the determination of rare earth elements in garnets and glasses the separation procedures shown in Table 38 may be utilized. Included in Table 38 is a method based on the HPLC technique which can be used for the determination of the rare earths in Co-based alloys.

Synthetic Mixtures

From Tables 39 to 41 it is evident that HPLC can most effectively be employed for mutual separations of the rare earth elements using α-HIBA solutions as eluents. This fractionation of the rare earths into individual elements is achieved by elutions utilizing a pH gradient, a concentration gradient, or a combined pH-concentration gradient.

Cation exchange chromatography at normal or somewhat elevated pressure has also been employed for mutual separations of the lanthanides using α-HIBA eluents. Procedures of this type are illustrated in Table 42. In this table a method has been included which is based on the elution of Tb with 2-hydroxybutyric acid using HPLC.

In an ion chromatographic system a mixture of 14 lanthanoids (20 μℓ of a solution containing ≃10 μg/mℓ of each element) can be separated well on a column (30 × 0.4 cm) of Nucleosil® SA (10 μm) with a linear gradient elution program from 0.01 to 0.04 *M* α-HIBA (pH 4.6) over 30 min at 1 mℓ/min.[396] Detection is at 600 nm after postcolumn reaction with arsenazo I.

Lactate Systems

Geological Materials

For the determination of rare earth elements in monazite and other rare earth minerals as well as of radioactive Ce in natural waters, the one-column separation procedures outlined in Table 43 have been used. In all these methods, lactate eluents were employed either for the mutual separation of the rare earths or for separations of Ce from other radioactive species contained in the water samples.

Industrial Materials

Elution chromatography, including concentration and pH-gradient elution techniques, utilizing lactate eluents has been employed to fractionate rare earth elements before their determination in industrial materials of the type listed in Table 44.

Synthetic Mixtures

In Table 45 numerous methods are presented which can be used for the mutual separation of rare earth elements employing lactate eluents and the technique of pH-gradient elution. Included in Table 45 is a procedure which is based on the use of a mixed-bed ion exchange column to separate 14 rare earth elements, as well as a technique using a mixed lactate-EDTA eluent.

Elution chromatography at elevated pressures has been employed to separate the light lanthanides and Y using the procedures outlined in Table 46.

To separate Y from other rare earth elements and Sr, and to achieve a separation of Sc from Ca, the procedures illustrated in Table 47 have been used.

Citrate Systems

Geological Materials

In connection with the determination of radioactive Ce and other radionuclides in environmental materials such as natural waters, marine sediments, and fish samples, the separation schemes outlined in Table 48 have been employed. In the first method listed not only is citrate used as an eluent for the rare earths, but also media containing formate and lactate as complexing ligands.

Industrial Materials

In Table 49 three methods are described which are based on separations of the rare earths and accompanying elements in citrate and other media utilizing one- and multicolumn procedures for the isolation from U and "pure" Y.

Synthetic Mixtures

A scheme for the tracer-level separation of rare earths and numerous other radioelements is presented in Table 50 utilizing citrate and other complexing ligands. Included in Table 50 is a method which can be used for the mutual separation of the light rare earths (Ce group) employing a mixed citrate-acetate eluent.

Polyamino Polycarboxylic Acid Systems

Geological Materials

From the procedures outlined in Table 51 it is seen that EDTA solutions can be used for cation exchange separations of the rare earths before their radiometric or gravimetric determination. These methods were applied to the analysis of rocks and monazite.

Industrial Materials

In connection with the separation of rare earth impurities from actinide and rare earth matrices, the methods illustrated in Table 52 have been employed. To isolate the rare earths from the actinide materials, EDTA is used to form anionic complexes with the rare earths and Th which pass into the effluent unadsorbed. In the analysis of La oxide, an EDTA solution is employed for the chromatographic separation of the light lanthanides.

Synthetic Mixtures

For the mutual separation of rare earth elements by displacement chromatography using EDTA and other polyamino polyacetate ligands as eluents, the procedures outlined in Table 53 have been employed in which the analytical use of the retaining ions ("ion barriers") Cu^{2+}, H^{+}, Fe^{3+}, VO^{2+}, and Ni^{2+} is illustrated.

Elution chromatography on cation exchange resins using EDTA as the eluent has also variously been used to separate the rare earths from each other or from other elements. These methods are presented in Table 54.

Miscellaneous Organic Complexing Agents

The application of sulfosalicylic acid, acetate buffer solutions, oxalic acid, glycollic acid, formic acid, ascorbic acid, malic acid, and organophosphorous compounds for separations of the rare earths from each other and accompanying elements on strongly acidic cation exchange resins is illustrated by the procedures outlined in Tables 55 to 61.

ANION EXCHANGE RESINS

Mineral Acid Media

General

Nitric Acid Systems

Sc, Y, and the trivalent lanthanides from Pm to Lu are not adsorbed on strongly basic anion exchange resins from pure aqueous nitric acid solutions with acid concentrations ranging from 0.1 to 14 *M*.[397,398] Trivalent La, Ce, Pr, and Nd are weakly adsorbed and the retention decreases in this order. In about 5 to 6 *M* nitric acid, the distribution coefficients of La, Ce, and Pr are ~7, ~6 to 8, and ~5, respectively, while at lower or higher acidities these coefficients are even smaller.[397-399] The adsorption of Nd is even lower. The small difference between the values of the distribution coefficients does not, therefore, allow efficient separations of adjacent rare earths, but these can be separated, as a group, from the elements strongly retained from nitric acid media (see Table 49 in chapter on Actinides, Volume II) such as Pu (see Tables 64 and 68) and Th (for examples see Tables 68, 69, and 76).

Ce in the tetravalent state exhibits a behavior similar to that of Th (see Table 49 in chapter on Actinides, Volume II),[400] i.e., it is appreciably adsorbed on strongly basic resins from strong nitric acid solutions (most probably as the species $Ce[NO_3]_6^{2-}$).[401] Therefore, cerium can be readily separated from the other rare earths as well as from numerous other elements. In the absence of an oxidizing agent, however, it is rapidly reduced by the anion exchange resin and eluted as Ce(III). To prevent this reduction, the adsorption of Ce(IV) and the separations are carried out in the presence of holding oxidants such as Na or K bromate[401-403] or Pb dioxide.[400] The latter has the advantage that it rapidly oxidizes cerium in nitric acid. It thus eliminates the need for a preoxidation step, and being insoluble in nitric acid, it can be mixed with the resin, e.g., Dowex® 1, to produce an oxidizing anion exchange resin bed. For the elution of Ce thus separated from the rare earth elements, and other metal ions showing little adsorption in strong nitric acid media, dilute HCl or nitric acid solutions,[400] or nitric acid eluents containing reducing agents (to convert Ce to the virtual nonadsorbed trivalent state) such as hydrazine sulfate[402,403] or hydroxylamine hydroxychloride[401] can be used. Examples for applications involving the adsorption of Ce(IV) on anion exchange resins from strong nitric acid media are presented in Tables 64, 69, and 76.

As in the case of the adsorption of actinides, e.g., U (see chapter on Actinides, Volume II), the rare earth elements are much more strongly retained on strong base anion exchange resins from solutions containing large amounts of soluble inorganic nitrates[17,397,399,404-406] than from pure aqueous nitric acid solutions. This behavior of the rare earths in strong nitrate solutions is explained in terms of anionic nitrate complex formation. The complex in the anion exchanger is probably $(R_s)_2M(NO_3)_5$ (R_s = resin matrix).[404] This effect of increased adsorption has been observed in weak nitric acid media containing large amounts of the nitrates of Li,[404,405] ammonium,[397,399] Mg,[17,406] Al, and Ca. The extent of adsorption of the rare earths from such media is determined by the same three factors as observed in the case of U adsorption (see in chapter on Actinides, Volume II). Thus, the distribution coefficients of the rare earths are highest in nitrate solutions of Li and decrease in magnitude for solutions of Al, Ca, and ammonium. An increase in the free nitric acid concentration causes a decrease of the distribution coefficients, while the reverse is true when the concentration of nitrate salt is increased. Nitrate complexing, and hence adsorption of the rare earth elements, increases with decreasing atomic number of these elements, i.e., Lu is eluted first and La last.

Applications of weak acid media containing large quantities of Mg nitrate are presented in Table 77.[17,406] Other nitrate systems have also been employed for the chromatographic separation of adjacent rare earths, and of individual rare earths from other metal ions, using as eluents 3 to 4, 6, or 7 *M* Li nitrate[405] or 10 *M* ammonium nitrate.[397] The mutual separation of the rare earth elements can also be performed at elevated temperatures or by the use of gradient elution techniques.

Although anion exchange in these systems can be of advantage in purification or concentration processes, in radiochemical separations, particularly in the presence of high levels of radiation, application of these systems to the analysis for general trace impurities is not simple. Further separation and concentration may be necessary to attain the necessary sensitivity or analytical precision, and preparation of reagents of acceptable purity can be time consuming. On the other hand, the use of water-miscible organic solvents (see below) offers a major advantage in that most solvents can be readily purified and subsequently volatilized to concentrate the separated rare earth element.

If the greater part of the aqueous phase of a dilute nitric acid solution is replaced by a water-miscible organic solvent such as an aliphatic alcohol or acetic acid, the adsorption of rare earth elements on strong base anion exchange resins, e.g., Dowex® 1, is increased in a similar manner as when soluble inorganic nitrates are present.[398,407-414] From Tables 62 and 63 it is seen that the adsorption of the rare earths (i.e., of Ce, Yb, Y, and Sc taken as representatives for these elements) increases with acid concentration and decreases with a decrease of the percentage of the nonaqueous components. An increase of the adsorption of the rare earth elements is also observed in systems containing various aliphatic alcohols, i.e., their retention by Dowex® 1 increases rather regularly with increasing chain length of the alcohol present,[407,408,411] although observations[415] have shown that just the reverse may be true in some cases.

As a general rule, the adsorption of the nitrate complexes of the rare earth elements in a given organic solvent-nitric acid mixture decreases regularly with increasing atomic number until about Tb. Then the adsorption of the heavy rare earths is practically constant. The least strongly adsorbed element of the rare earth group is Sc (see Tables 62 and 63).

These mixed aqueous-organic solvent systems containing nitric acid can be used for separations of individual rare earths (including their mutual separation) (see Tables 65, 66, 70 to 74, 78, and 80) and of the heavy lanthanides from the light lanthanides (see Tables 65, 67, 78, and 81), as well as to effect a group separation of all rare earth elements from accompanying elements (see Tables 65 to 67, 73, 74, and 81).

Many of the methods outlined in these tables have been applied in connection with the determination of rare earth elements in geological and industrial materials, mainly using methanol-nitric acid systems in which nitrate salts are more readily soluble than in other organic solvent media of comparable usefulness, such as systems containing acetic acid or other alcohols or mixtures containing both species of organic solvents (see Tables 66, 67, 73, 80, and 81). Mutual separations of rare earth elements can also be achieved in mixtures of ethanol with formic acid and nitric acid (see Table 80).

Strong adsorption of the rare earth elements is also obtained in mixed aqueous-organic solvent media which contain, besides nitric acid, also nitrates of ammonium[251-253,416,417] or Li.[251-253,418] This fact has been utilized with respect to the determination of rare earth impurities in industrial materials (see Tables 74 and 75) and for the mutual separation of rare earth elements (see Table 79).

Coadsorbed with the rare earths from mixed aqueous-organic solvents containing nitric acid are not only the elements which are strongly adsorbed from pure aqueous nitric systems, but also a few other metal ions such as U, Pb, and Bi. Nevertheless, these systems show a much higher selectivity of separation of the rare earths, as a group, from other metal ions, especially from iron and other common elements than, for instance, comparable media containing HCl. Besides, these systems allow the mutual separation of the rare earths, once adsorbed on the anion exchange resin from a suitable organic solvent-nitric acid system, by simply decreasing the concentration of the solvent (e.g., by methanol concentration-gradient elution) and/or the molarity of nitric acid. These two facts as well as other advantages mentioned previously explain the high popularity of these systems. An additional advantage of these media is that the presence of phosphoric acid (up to 0.3 *M*) was found to have no effect on the separation or recovery of the rare earths.[419]

For the elution of rare earth elements which were adsorbed from mixed aqueous-organic solvent media containing nitric acid (and also nitrate salts) the eluents mentioned in Tables 65 to 67, 70 to 75, and 78 to 81 were employed.

HCl Systems

From Table 1 (see in chapter on Actinides, Volume II) it is evident that the lanthanides and Y are among the elements which show negligible or no adsorption on strongly basic resins from pure aqueous HCl solutions with concentrations ranging up to 12 *M*.[3,4,348,420,421] (In the 9 to 12 *M* acid the predominant chloride complex species are $LaCl_3$ and $LuCl_2^+$.)[422] Consequently, separations of the rare earth elements in such media can only be effected from elements which are strongly retained under these conditions such as U, Pu, other actinides, Fe(III), Co, Mo(VI), some Pt metals, etc. A behavior similar to that of the lanthanides and Y is shown by the alkali and alkaline earth metals, Al, Th, and a number of other elements so that these cannot be separated from the group of the rare earths.

It has been shown that Sc is adsorbable to a very slight extent (distribution coefficients of $\simeq$1) on strongly basic resins from media which are 9 [423] or 12 to 13.3 *M*[371,382,420,424] in HCl (in 9 to 12 *M* HCl the chloride complex species $ScCl_2^+$ is predominant).[422] Consequently, Sc can be separated from the lanthanides and Y, e.g., by using the procedures outlined in Table 101. For the same purpose 90% acetic acid - 10% 4 *M* HCl may be used, from which Sc was also found to be slightly adsorbed on such resins as, for instance, Dowex® 1 (see Table 101).[425] Adsorption of Sc on Dowex® AG1 from 9 *M* HCl has been used in connection with its determination in rocks and meteorites (see Table 83).[423] Strongly basic resins were also found to retain Sc from very dilute HCl solutions,[4,426] e.g., 0.1 *M* in the absence of chloride salts or heavily salted with Li chloride. Thus, in the activation analysis of corals the Sc-activity is separated from the lanthanides by adsorbing it from 11.9 *M* lithium chloride - 0.4 *M* HCl medium (see Table 91).[427] Also, on the weakly basic resin Amberlite® CG4B Sc is retained from 8.9 and 11.3 HCl media with distribution coefficients of 3.3 and 10, respectively. In the concentrated acid distribution values of 1.8, 4.0 and 4.7 were measured for Yb, Sm, and La, respectively.[428] However, from all these media only small amounts of Sc can be retained and early breakthrough occurs so that separation methods based on this adsorption are of limited applicability.

On the other hand, the nonadsorbability of the rare earth elements, as a group, from 6 to 12 *M* HCl media has been used very frequently in connection with the determination of rare earth elements in geological (see Tables 82 to 90), biological (see Table 91), and industrial materials (see Tables 92 to 100). Following the application of this separation principle, which is often used as a first purification step, the individual rare earth elements are obtained by utilizing ion exchange chromatography mainly on cation exchange resin columns employing solutions of organic complexing agents such as α-HIBA or lactic acid as the eluents. Subsequently, the individual rare earths can be determined separately and free from interferences.

When using mixed aqueous-organic solvent systems containing HCl, all the rare earths can be adsorbed on strongly basic anion exchange resins.[78,413,429-432] This adsorption has been found to be virtually independent of the rare earth element present, but increased retention of the rare earths was observed with increasing organic solvent and HCl concentrations. Furthermore, the extent of adsorption is dependent on the kind of organic solvent present in the mixtures. Thus, in 90% organic solvent - 10% 6 *M* HCl media, the distribution coefficient is <1 with methanol, methyl glycol, formic acid, or acetic acid, 10 with ethanol, 60 with tetrahydrofuran, about 100 with *n*-propanol or acetone, about 150 with isopropanol, about 200 with *n*-butanol, or about 500 with isobutanol, irrespective of the rare earth cation present.[431] In other words, fractionation of the individual elements of the rare earth group in any of these media is impossible. Because, under these conditions, Th, alkaline earth elements, and practically all those metal ions which are not retained on anion exchange resins from pure aqueous HCl solutions of any molarity are also adsorbed to about the same extent as the rare earth elements, separations carried out in these mixed aqueous-organic solvent systems are of very low selectivity. The nonadsorbability of rare earth elements from methanol-HCl media has been utilized in connection with their determination in silicate rocks (see Table 87)[254] and uranyl chloride (see Table 93).[433]

Sulfuric Acid Systems

Sc is adsorbed from dilute aqueous sulfuric acid solutions on strong base anion exchange resins, e.g., Dowex® 1 in the sulfate form (see Table 95 in chapter on Actinides, Volume II).[434] Y and the lanthanides from La through Lu, as well as many other metal ions, are not retained. A similar preferential adsorption of Sc on anion exchangers is observed from very dilute sulfuric acid solutions containing ammonium or K sulfate[435] and on the weakly basic anion exchange resin Amberlite® CG-4B.[436] In the presence of sulfate salts the other rare earths are also weakly adsorbed on strongly basic resins, most probably as the species $(RE[SO_4]_3)^{3-}$ (RE = Ce[III] or any other trivalent lanthanide element).[399]

The adsorption of Sc and the nonadsorbability of the other rare earth elements on strongly basic resins from such systems have variously been employed for analytical separations involving these elements (see Tables 103 and 104). When adsorbed Sc is eluted with 0.5 *N* sulfuric acid, U(VI) is retained by the anion exchange resin, while some of the other coadsorbed elements (see Table 95 in chapter on Actinides, Volume II) pass into the eluate together with the Sc.

As from HCl or nitric acid media containing organic solvents, the rare earth elements are much more strongly adsorbed from mixed aqueous-organic systems containing sulfuric acid than from pure aqueous solutions of this acid. These systems have not as yet been used for analytical purposes.

Thiocyanate Systems

Sc as anionic thiocyanate complex is strongly adsorbed on Dowex® 1 in the thiocyanate form from 2 *M* ammonium thiocyanate - 0.5 *M* HCl. The distribution coefficient is greater than 10^4.[437] If the concentration of ammonium thiocyanate is reduced to 1.0, 0.8, 0.6, 0.4, or 0.2 *M* the distribution coefficients of Sc are 2818, 5040, 905, 250, and 56, respectively. Under these conditions the coefficients of La, Sm, and Lu are <1. Ca and Al are not adsorbed at all. Cu, Zn, Cd, and Hg show distribution coefficients of $>10^4$ at all thiocyanate concentrations, so that these elements, and also U, Th, and In, are coadsorbed with the Sc. This does not interfere with the separations of Sc, because on its elution with 3 *M* HCl most of these elements are retained by the resin as anionic chloride complexes. Fe(III) is also retained as an anionic thiocyanate complex. Similar observations concerning the preferential adsorption of Sc on strong and weak base anion exchange resins have been made when using solutions of K thiocyanate.[438,439]

On elution from Dowex® 1 with ammonium thiocyanate the rare earth elements and Y show no simple elution sequence of the members.[440] These elements elute sooner than the trivalent actinides under all conditions, which means that the latter form more stable anionic thiocyanate complexes. The adsorption of rare earths from thiocyanate media has also been investigated by several other research workers.[441]

From mixed aqueous-organic solvent media containing ammonium thiocyanate the heavy rare earths Tb and Y were found to be increasingly adsorbed on Dowex® 1-X8 with increasing concentration of the organic component in the mixtures.[442] Thus, in 80% methanol (or acetone or tetrahydrofuran) - 1.0 *M* ammonium thiocyanate media distribution coefficients of about 20 to 40 were measured for these two elements. In the pure aqueous 1.0 *M* thiocyanate solution the adsorption is lower by about one order of magnitude.

Thiocyanate systems have been used for the separations illustrated in Table 105 and for the removal of Sc in an analytical scheme for the determination of rare earth elements in silicate rocks by neutron activation analysis.[443]

HF Systems

Sc forms an anionic fluoride complex which is strongly adsorbed on Dowex® 1 from dilute HF media.[426,444,445] Y and the lanthanides are precipitated as insoluble fluorides. The retention of Sc on the resin decreases with an increase in the concentration of this acid.[426,445] Thus, the distribution coefficients of Sc in 0.5, 1.0, 2.5, 5, 10, and 15 *M* acid are 3000, 2000, 720, 150, 30, and 4, respectively.[444] Because the distribution coefficient of Ti in 15 *M* acid is 10, a separation of this element from Sc can be achieved, with Sc passing into the eluate first when 15 *M* acid is used as the eluent. Because V(IV) is practically not retained from 0.5 to 2.5 *M* HF, it can be removed from a column of Dowex® 1, while Sc and Ti are still strongly retained under these conditions.[444] This separation principle is presented in Table 106.

Phosphoric Acid Systems

The rare earth elements can be adsorbed on strongly basic resins, e.g., Dowex® 1 from dilute solutions of orthophosphoric acid with adsorbabilities which differ quite distinctly for the individual members of this group.[446] From the point of view of distribution coefficients in 0.1 *M* phosphoric acid, three classes can be distinguished: (1) elements showing negligible or very small adsorbability. These are Y, La, and light to middle lanthanides from Ce to Er; (2) elements with distribution coefficients from tens to a few hundred. These are the heaviest lanthanides: Tm, Yb, and Lu; (3) elements showing very high adsorbability. In this class there is only Sc with a distribution coefficient of 10^3.

This adsorption behavior of the rare earths has been utilized for mutual separations of some elements of this group (see Table 107).

HBr and Hydroiodic Acid (HI) Systems

Sc, Y, and the lanthanides are not adsorbed on strongly basic resins from pure aqueous solutions containing HBr or HI.[447,448] However, from mixed aqueous-organic solvent systems containing HBr the rare earths are retained to some extent by, e.g., Dowex® 1, especially from media containing high concentrations of the organic component.[448,449] Thus, in 90% organic solvent - 0.6 *M* HBr media, Ce(III) shows distribution coefficients of <1, 7, 38, 58, <1, 8, 24, and 35 in the presence of methanol, ethanol, *n*-propanol, isopropanol, methyl glycol, acetic acid, tetrahydrofuran, and acetone, respectively.[448] At lower concentrations (e.g., 60% or less) of the organic components, no or only negligible adsorption is observed. If the concentration of HBr is increased (e.g., in the 90% organic solvent mixtures), the adsorption of Ce was found to increase somewhat in the systems containing aliphatic alcohols, while the reverse was true in the other organic solvents.

On the weakly basic resin Amberlite® IRA-68 (bromide form) Tm is adsorbed from $\sim$1.0 to 12 *M* HBr, with distribution coefficients increasing from 1 in the 1 *M* to 24 in the 12 *M* acid.[450] Under similar conditions U(VI) is much more strongly adsorbed, a fact which has been utilized for its separation from Eu (see in chapter on Actinides, Volume II).

Measurable adsorption on the weakly basic anion exchange resin AG3-X4A (bromide form) (styrene-divinylbenzene matrix with functional groups of $-NR_2$ type) has also been reported for some rare earth elements in 4 to 12 *M* HBr solutions.[451] In the 12 *M* acid, distribution coefficients of 15, 12, and 6 were measured for Eu, Tm, and Sc.

Alkaline Systems

The anion exchange behavior of the rare earth elements in carbonate solutions has variously been investigated,[452-461] but no actual separations of these elements from one another have been reported. It has been found that the distribution coefficients of the rare earths increase with increasing atomic number from Nd to Dy or Ho, whereas the values decrease with increasing atomic number from Er to Lu. The adsorption increases with a decrease of the concentration of carbonate. Thus, in 1, 0.67, 0.34, 0.1, or 0.01 *M* solutions of Na carbonate the distribution coefficients of Y have values of about 6, 20, 100, 3000, or 10^4, respectively.[457]

Adsorption of Y and lanthanides on Dowex® 1 from carbonate and bicarbonate solutions has been utilized for their separation from Sr,[459,460] Ba,[459] Th,[461] and U.[461] These methods are illustrated in Table 108.

With the same resin in the hydroxide form a precipitation separation of La from Ba can be effected using the procedure outlined in Table 109.[462] A method based on the same principle has been utilized in connection with the determination of ^{141}Ce and ^{143}Ce in a heavy water moderator.[463] The aqueous phase (adjusted to pH 2 to 6) is passed through Dowex® 2-X10 (OH^- form) causing precipitation of the Ce (together with ^{56}Mn and ^{51}Cr) which is then determined radiometrically.

Systems Containing Organic Complexing Agents

Polyamino Polycarboxylic Acid Systems

It has been observed[464-467] that the distribution coefficients of the rare earth elements on the strongly basic anion exchange resin Amberlite® IRA-400 (EDTA form) in EDTA media increase regularly from La to Eu (maximum) and then decrease to Lu. Therefore, the order of elution from this resin using EDTA systems as eluents is Lu, Yb, Tm (+Sc), Er, Y, Ho, La, Dy, Ce, Tb, Pr, Nd, Gd, Pm, Sm, and Eu.[468] In other words, there is no monotonical relation between the distribution coefficient and the atomic number or the ionic radius, respectively, which appears at the separation of lanthanides on cation exchange resins.

On the strongly basic resin Dowex® 1 of different cross-linking, the separation factors of the rare earths were found to increase with the degree of cross-linking, but the best separations in EDTA and DCTA media were obtained with the -X4 resin.[469,470] An overall increase in separation efficiency of various pairs of lanthanides (e.g., Tb-Pm and Tb-Gd) was also observed with increase in temperature, which additionally not only influences the time of separation, but sometimes also the order of elution.[471,472]

Examples for practical application of EDTA and DCTA systems for separations of various rare earth elements are presented in Tables 110 and 111.

Oxalic Acid Systems

Adsorption of the rare earth elements on strongly basic resins is very high at low concentrations of oxalic acid and decreases when increasing the concentration of this acid[473] or that of a mineral acid, e.g., HCl,[474] which may be present simultaneously. For example, at an oxalic acid concentration of 0.0192 *M,* Ce, Lu, and Sc show the distribution values of 1840, 52,500, and 28,600, respectively, while the same elements have coefficients of 36, 558, and 679 in 0.2 *M* oxalic acid.[473] In 0.1 *M* oxalic acid a distribution coefficient for Sc of 8100 was measured, which decreased to 400, 60, and 8 when increasing the concentration of added HCl from 0 to 0.04, 0.1, and 0.4 *M,* respectively.[474]

Analytical applications of separations involving Sc and other rare earth elements utilizing oxalic acid media are presented in Tables 112 and 113.

Miscellaneous Organic Complexing Agents

Sc is adsorbed on strongly basic resins from systems containing malonic, ascorbic, and tartaric acids and this fact has been utilized for its separation from a large number of other elements (see Table 114).[475,476] For the elution of the Sc, dilute HCl or nitric acid solutions (e.g., 1 to 4 *M* HCl or 0.5 *M* nitric acid) can be used.

Mutual separations of rare earth elements can be effected on anion exchange resins using citric acid solutions (see examples listed in Table 115)[477,478] and also in media containing xylenol orange (see Table 116)[479] and 2-hydroxyisobutyric (α-HIBA)[480] or lactic acid[481] dissolved in organic solvent-water mixtures. Applications of the latter two complexing agents to the determination of rare earth impurities in industrial materials are presented in Table 117.

Adsorption of rare earth elements (Sc, Y, and Eu) on strongly basic resins from acetic acid solutions was observed only at very high concentrations of this acid (~14 to 17.5 *M*).[482,483] While in glacial acetic acid (17.5 *M*) the distribution coefficients are greater than 10^3, the retention of the rare earths by Dowex® 1-X8 decreased rapidly with decreasing concentration of the acid. Thus, for example, in 16, 16.5, and 17 *M* acetic acid the distribution coefficients of Sc have the values of 1, 3.6, and 20, respectively. A similar decrease of adsorption is shown by Y, but Eu is more strongly adsorbed under these conditions, so that separations of the light rare earths from the Y-group elements seem to be possible in such media.

Examples for separations of the rare earths, as a group, from U in acetic acid media containing organic solvents are outlined in Table 116 (see in chapter on Actinides, Volume II).

On the weakly basic anion exchange resin Dowex® 3W-X8 the complexes of Ce(III), Fe(III), and Al with 4-hydroxybenzoic acid were separated using 30% methanol as mobile phase.[484]

Applications

Nitric Acid Systems

Geological Materials

In connection with the determination of stable Ce and of radioactive isotopes of this and other rare earth elements in marine environmental samples, the procedures outlined in Table 64 have been used. These methods are based on anion exchange in pure aqueous nitric acid solutions and individual rare earths are separated by cation exchange using ammonium lactate as eluent.

For the separation of the rare earths from geological matrices, nitric acid systems containing high percentages of methanol have much more frequently been employed than pure aqueous media, because from methanolic solutions the lanthanides are appreciably retained on strongly basic resins. Furthermore, a fractionation of the adsorbed rare earths into smaller groups can be achieved on the same resin by proper selection of the concentrations of methanol and nitric acid in the eluent used. This separation principle is illustrated by the methods outlined in Table 65.

Also utilized for the analysis of rocks was the fact that the rare earths are not only adsorbed from methanol-nitric acid media, but also from similar systems containing high percentages of ethanol (see Table 66) or acetic acid (see Table 67). The latter solvent serves to separate the rare earths, as a group, and further fractionation into subgroups is then accomplished by use of a second anion exchange separation procedure utilizing methanol-nitric acid systems.

Industrial Materials

From matrices predominantly consisting of the actinide elements Th, Pu, and U, the rare earths (usually present as traces) can be isolated by adsorption of the first two actinides on strongly basic resins from 8 *M* acid and by preliminary removal of the U using TBP extraction. These separation principles are outlined in Table 68 and have been employed prior to the spectrographic determination of rare earth impurities contained in industrial materials of this type. A similar technique has also been used in connection with radioactivation analysis of rare earth elements in metallic Th and Th oxide.[485]

Similar methods also using anion exchange in pure aqueous nitric acid systems have been applied to the determination of Y and other rare earths in steels and of Ce in alloys and ^{90}Sr. In the latter two applications, the separations of Ce from the matrix elements are effected by adsorption of tetravalent Ce from strong nitric acid media in the presence of holding oxidants. These separation techniques are illustrated in Table 69.

A two-column separation scheme is presented in Table 70 which is based on the preliminary removal of Pu by anion exchange in pure aqueous nitric acid medium and subsequent isolation of the Ce by means of anion exchange in a methanol-nitric acid system.

Adsorption of Nd on strongly basic anion exchange resins from methanol-nitric acid media has variously been employed in connection with the determination of the burn-up of nuclear fuels. For this purpose, both one- and multicolumn procedures have been employed which are illustrated in Tables 71 and 72.

From Table 72 it is seen that after the anion exchange separation step cation exchange is invariably used to further purify the Nd employing elution with α-HIBA.

Methanol-nitric acid media, which in most cases also contain acetic acid, have also been used for the separation of Ce and other rare earths from large amounts of Fe, Mg, and other constituents of alloys. In Table 73 procedures are outlined which were used to determine Ce and other rare earth elements in ferrous alloys and cast iron.

The use of one-column procedures utilizing methanol-nitric acid media has been described for the analysis of rare earth compounds, Am samples, glass, and Al foils for certain rare earth elements present as impurities. The separation principles underlying these methods are illustrated in Table 74, which shows that separations of the rare earths are also achieved in methanol systems containing ammonium nitrate.

A method for the preirradiation separation of the rare earths using methanol-nitrate systems has been described in connection with the determination of rare earth impurities in a Y oxide matrix. After neutron activation, the rare earths are separated by cation exchange chromatography using α-HIBA. The analytical scheme which was used for this purpose is outlined in Table 75.

Synthetic Mixtures

From Tables 76 and 77 it is seen that anion exchange in pure aqueous nitrate solutions can be used for the mutual separation of the rare earth elements as well as to separate Sc from Th and U, and Eu from Th. These separations can be effected in strong nitric acid solutions, e.g., in the presence of a holding oxidant for Ce (see Table 76) or by using Mg nitrate solutions or dilute nitric acid solutions heavily salted with this Mg salt (see Table 77).

The systems most frequently used to effect mutual separations of the rare earth elements on strongly basic resins are those utilizing methanol-nitric acid media. Some of these separations are illustrated in Table 78.

As can be seen from Table 79, similar possibilities for the mutual separation of the rare earths exist in dilute nitric acid-methanol media containing the nitrates of ammonium and Li. Also, ethanol-nitric systems which may, in addition, contain formic acid have been utilized for the same purpose. These systems are illustrated in Table 80.

In nitric acid media containing high percentages of propanol or acetic acid, the separations outlined in Table 81 can be effected.

HCl Systems

Geological Materials A (Solid Samples)

Applications utilizing one-column techniques which are based on the virtual nonadsorbability of the rare earth elements, as a group, on strongly basic resins from 6, 9, 10, and

12 *M* HCl solutions are listed in Table 82. These methods have been used in connection with the determination of the rare earths in rocks using neutron activation analysis. The same analytical technique has also been employed with respect to the determination of Sc in both terrestrial rocks and meteorites. In Table 83 the ion exchange procedures are outlined which have been used for the purification of Sc after neutron irradiation of the samples. In one of the methods described in Table 83 the very weak adsorbability of Sc from strong HCl medium is utilized to coadsorb Sc with Co and Zn, while in the other case experimental conditions have been selected which prevent its retention by the resin from 9 *M* HCl. The latter result is easily achieved by, e.g., using a larger volume of this acid.

In connection with the determination of rare earth elements in rocks, two-column procedures have been used in which anion exchange removal of impurities in 6 and 9 *M* HCl is combined with a second ion exchange separation step utilizing adsorption of the rare earths on either an anion exchange resin from a methanol-nitric acid medium or on a cation exchanger from HCl solution. In the former case, a fractionation of the rare earth group into light and heavy elements is achieved, while with the cation exchange procedure the separation of the rare earths, as a group, from accompanying elements is accomplished. These two methods are outlined in Table 84.

Methods based on two successive column operations utilizing anion exchange in HCl media and cation exchange employing the organic complexing agents lactate and EDTA have been applied to the analysis of both Fe and stony meteorites following the neutron irradiation of the samples. The principles underlying these methods are illustrated in Table 85.

An analytical scheme that has been suggested for the spectrographic determination of the rare earths and a number of other elements in silicates involves the repeated application of anion exchange in HCl solutions and subsequent separation of the rare earths from accompanying elements on a column of a cation exchange resin. This method is outlined in Table 86.

Anion exchange in methanol-HCl media has been utilized for the analysis of silicates for some rare earth elements and numerous other metal ions. For this purpose, the scheme illustrated in Table 87 can be used. It consists of one anion exchange separation procedure and two cation exchange steps of which the first is performed in HCl media of varying concentrations, while the second separation is used to achieve fractionation of the lanthanides.

HCl or nitric acid solutions containing 70 to 90% methanol have also been suggested for the analysis of soils for rare earth elements and actinides using Dowex® 1-X8.[413]

The nonadsorbability of the rare earth elements on strongly basic resins from concentrated HCl media has also been utilized in connection with the determination of rare earths and other metal ions in environmental materials such as fallout samples, deposits, and airborne particulate matter. The methods used for this purpose are outlined in Table 88, from which it is seen that the analysis of long-range fallout debris also requires the application of a cation exchange separation step if some fractionation of the radioactive rare earth nuclides is required.

Geological Materials B (Liquid Samples)

In connection with the analysis of seawater for ^{144}Ce and Sc, and of rainwater for ^{147}Pm, the one-column procedures illustrated in Table 89 have been employed. In these methods the anion exchange separation step performed in 8, 10, and 12 *M* HCl solutions is mainly used to separate the rare earths from Fe which, as the hydroxide, was used as a coprecipitant.

Other techniques used for the determination of rare earth elements in seawater are based on the application of two- or three-column methods in which the final ion exchange separation step is performed on a column of a cation exchange resin, while the first is always carried out using the anion exchange resin-HCl system. On the cation exchanger, the separation of the individual rare earth elements is effected using either α-HIBA or lactate eluents. These separation schemes are outlined in Table 90.

Biological Materials

The estimation of some rare earth elements, in particular, of Eu in marine organisms, is of great importance for reasons related to the marine ecology. Radionuclides of Eu with long lives originating from waste discharge, are entering the marine environment where they are rapidly taken up and retained by the marine biosphere.

In Table 91 several methods are shown in which the nonadsorbability of rare earth elements on anion exchange resins from HCl media is utilized for the separation from elements interfering with their determination in biological samples. To these elements belong Fe, Sc, and P. Since most of the ion exchange separation schemes shown in Table 91 are applied after neutron irradiation of the ashed biological samples, a number of radiochemical purification steps are necessary, including the use of two-column procedures of which the second separation step is performed on a cation exchange column, e.g., to obtain the individual rare earths. Ion exchange procedures for the separation of rare earths and other elements have also been used in connection with neutron activation analysis of biological material with high radiation levels.[486] The fully automatic separations, based on selective sorption, are carried out simultaneously with a peristaltic pump in direct connection with the anion exchange system. The entire separation takes 40 min.

Industrial Materials

Several rare earths have large neutron capture cross-sections and, hence, act as reactor poisons if they are present in nuclear fuels (maximum permissible concentrations of these elements in U used as fuel is in the range of 0.04 to 0.1 ppm). Therefore, the determination of rare earth elements in nuclear materials, mainly U samples, has received broad attention, since the content of lanthanides besides that of Cd and B belongs to the most important criteria of the purity of U used for nuclear purposes. For this assay a number of different analytical methods have been used such as mass spectrometric analysis, emission spectrography, and radiometric methods. These techniques are employed in connection with one- or two-column procedures as outlined in Tables 92 and 93, from which it is evident that in all cases best separation of U from the rare earths is achieved by anion exchange in HCl media.[487] Included in Table 92 are two methods that have been used to separate Ce from irradiated U targets because of the similarity of the ion exchange separation steps involved. In Table 92 a procedure is also described in which the adsorption of U from very dilute HCl media is used to separate it from Y.

In Table 94 a four-column separation scheme is shown that has been used in connection with the isolation of rare earths from U targets that were subjected to neutron irradiation. Activation analysis in conjunction with anion exchange in concentrated HCl has also been applied to the radiometric microdetermination of rare earth metals in Th oxide.[488]

To separate rare earth elements from fission product mixtures the radioanalytical schemes presented in Table 95 have been used. These separations involve the preliminary use of anion exchange in 8, 10, and 12 *M* HCl media to isolate the rare earths as a group, which is then fractionated into individual elements by application of cation exchange procedures. For the latter purpose, eluent systems containing either α-HIBA or lactate are utilized.

In connection with the determination of Sc in metals such as Be, Ga, and Al, the separation procedures outlines in Table 96 have been used.

To determine Y in steel and to separate it from an irradiated Sr carbonate target, the separation schemes illustrated in Table 97 can be employed.

In Table 98 separation methods are illustrated which have been used in connection with the determination of Ce in Fe and alloys. In all cases trivalent Fe is removed by adsorption on a strongly basic anion exchange resin from HCl media. Other elements such as Ni, Al, and Mn are separated from the Ce by adsorption of this rare earth element on the same resin in the nitrate form from an acetic acid-nitric acid system (see two-column procedure in Table 98).

The strongly basic anion exchange resin OAL (chloride form) has been utilized to separate La, Nd, Pr, Sm, and Y from Fe which is adsorbed on this exchanger from 8 *M* HCl.[489] This method was employed to separate milligram amounts of the elements.

Similar anion exchange methods utilizing HCl systems have been used to determine rare earth elements in steels, organic filter media, and high-purity quartz following neutron activation of the samples. These procedures are illustrated in Table 99 in which a two-column method is also outlined, which makes it possible to separate individual rare earth elements by cation exchange in α-HIBA media of varying pH values before the radiometric measurements. Determinations of the rare earth elements in steels are important, because these elements are often used as steel additives to improve mechanical and oxidation properties of steels while acting as strong desulfurizing agents. Usually, a total of 0.1 to 0.3% rare earths is added in the form of misch metal, an alloy containing La, Ce, Pr, and Nd in various proportions.

The multicolumn procedures outlined in Table 100 have been employed in connection with the determination by activation analysis of Gd impurity in high-purity Eu oxide samples.

Synthetic Mixtures

In Table 101 several methods are listed that have been used to separate Sc from rare earths and other accompanying elements, utilizing the fact that Sc is adsorbable on anion exchangers from strong HCl media and from acetic acid-HCl systems. Included in Table 101 is a two-column procedure in which a second column operation, anion exchange in a 8 *M* nitric acid medium, is used to separate Sc from elements which were coadsorbed with it from 13 *M* HCl.

The nonadsorbability of Sc on strongly basic resins from various dilute HCl systems has been employed to separate this element from Cu and U as well as from Y. The systems which were used for these separations are listed in Table 102.

Sulfuric Acid Systems

Geological and Industrial Materials

The adsorption of Sc on Dowex® 1 from very dilute sulfuric acid solution has been used in connection with the determination by neutron activation of this element in rocks, ores, and meteorites. For this purpose, the separation scheme outlined in Table 103 was utilized. Included in Table 103 is a method which is based on the nonadsorbability of rare earth elements from a sulfuric acid medium containing ammonium thiocyanate.

Synthetic Mixtures

In Table 104 three methods are illustrated which have been used for the separation of Sc from Y, lanthanides, and numerous other elements employing anion exchange in dilute sulfuric acid media. From all these systems the Sc is first adsorbed on the anion exchanger, together with other elements, and then selectively eluted with stronger acid.

Thiocyanate Systems

The retention of Sc by anion exchange resins from dilute HCl-thiocyanate media has been applied to the separation of this element from lanthanides and a number of other elements. From the separation methods outlined in Table 105, it is seen that for this purpose both strongly and weakly basic resins can be utilized. Included in Table 105 are procedures for the separation of lanthanides from trivalent Fe and In which are adsorbed as thiocyanate complexes, while the rare earths pass into the effluent unadsorbed.

HF Systems

The fact that Sc is more strongly adsorbed on Dowex® 1 from dilute HF solution than from the concentrated acid has been employed for its separation from V and Ti by using the procedure outlined in Table 106.

Phosphoric Acid Systems

Except for a method which is outlined in Table 107, phosphoric acid solutions have not been used for any other separations of rare earth elements on anion exchange resins.

Alkaline Systems

Adsorption of rare earth elements on strongly basic resins from carbonate and bicarbonate solutions has been applied for the separation illustrated in Table 108. The elements separated from the rare earths are ^{90}Sr, ^{140}Ba, and U.

A procedure for the separation of La from Ba which is based on the precipitation of insoluble La hydroxide on Dowex® 1 (OH^- form) (acting as the precipitating agent) is presented in Table 109. Precipitations of this type are also applicable to several other carrier-free separations. Thus, separations of ^{90}Y from ^{90}Sr on OH^--form anion exchangers and ^{89}Sr from ^{140}Ba on resins in the CrO_4^{2-} form can be obtained. Mutual separations of the rare earth metals are also possible by fractional precipitations on anion exchange resins in the hydroxide, carbonate, and oxalate forms.[490]

Polyamino Polycarboxylic Acid Systems

Industrial Materials

The separation schemes illustrated in Table 110 have been utilized in connection with the determination of rare earth impurities in high-purity Y and Er oxides. These separations are based on the use of EDTA solutions which are employed for the selective elution of the rare earth elements.

Synthetic Mixtures

Both EDTA and DCTA solutions have been applied to the mutual separation of rare earth elements employing the methods presented in Table 111.

Oxalic Acid Systems

From the methods illustrated in Table 112, it is seen that Sc can be separated from V and Ti by using anion exchange in oxalic acid media.

To separate rare earth metals (Y, Ce, and Pr) from a number of radionuclides such as ^{90}Sr, $^{95}Zr/^{95}Nb$, etc., the procedures outlined in Table 113 have been employed. Included in Table 113 is a method in which oxalic acid is used for the mutual separation of two lanthanide elements.

Malonic, Ascorbic, and Tartaric Acid Systems

From the separation schemes outlined in Table 114 it is seen that Sc can be separated on strongly basic resins from numerous other elements following its adsorption from media containing malonic, ascorbic, and tartaric acids. From such systems adjusted to the pH values indicated in Table 114, the Sc is adsorbed as anionic complex with these organic compounds.

Citric Acid Systems

In Table 115 two methods are presented which have been used for the mutual separation of pairs of lanthanides on strongly basic resins employing citric acid solutions as eluents.

With an anion exchange resin in the citrate form it is also possible to separate milligram amounts of rare earths from microgram quantities of Th by controlling the concentration of citric acid to $\geq$1% and the pH value of the eluent to $\geq$2.5.[491] Under these conditions the amount of citrate ion required to form anionic complexes with Th is just sufficient, so that this element is retained by the resin while the rare earths pass into the effluent. This separation can be completed within 2 to 3 hr using a small resin bed (3.0 $\times$ 0.44 cm ID).

Xylenol Orange Systems

Adsorption of Sc as a negatively charged complex with xylenol orange has been applied to the separation of this element from Ce and Y using the technique outlined in Table 116.

CHELATING RESINS

The adsorption behavior of the rare earth elements towards iminodiacetic acid resins has been studied by several investigators,[392,492-496] and it was shown that the selectivity of the Dowex® A-1 chelating resin for the trivalent lanthanides follows the order La $<$ Pr $<$ Nd $>$ Sm $>$ Gd $>$ Tb $\sim$ Dy $<$ Er $>$ Yb at 25°C, pH 2.0, and μ = 0.100 (NaCl).[492] At this pH value the distribution coefficients of the rare earth elements (including Sc and Y) are $\lesssim 10^2$ and increase to $>10^3$ at lower acidities, to remain virtually constant in the pH range from $\sim$4 to 6.[392] For the elution of the rare earth elements which were adsorbed, as a group, on this chelating resin, the use of 1 *M* Na carbonate at elevated temperatures is recommended.[497]

In the pH range from 2 to 6 the rare earths are less strongly adsorbed on the carboxylic resin Zeo-Karb® 226 from media containing 1,10-phenanthroline, which makes this resin behave as a chelating one (see in chapter on Actinides, Volume II).[498] Also, in the absence of this complexing agent the adsorption of the rare earths on this resin is considerably lower than on either iminodiacetic or sulfonic acid resins.

The adsorption and/or separation of rare earth elements on resins containing P in their functional groups[493,499,500] and resins (of the M type)[501] containing PO7,[502] i.e., 2-ethylhexyl(2-ethylhexyl)phosphonic acid (e.g., 50%), has been studied by several investigators.[493,499-502] For the elution of the rare earth elements from these adsorbents usually nitric acid solutions are employed. Thus, from resin I or II (see Table 118) all rare earths, except Sc, are eluted with 1 *M* nitric acid,[499] while from the M-type resin Er is eluted with the 2.5 *M* acid, leaving Tm further retained. The latter can be eluted with 6 *M* nitric acid.[501] On the other hand, sulfuric acid was found to be the most effective mineral acid for eluting the lanthanides from the phosphonic acid resin Duolite® C-63.[493]

Studies of the complexation properties of a poly(dithiocarbamate) resin for 14 rare earths and other elements have shown that maximum uptake of the lanthanides ($\sim$85 to 100%) is obtained from solutions of pH 4 to 5.[503] This chelating resin was found to be applicable in ICP-emission spectrographic analysis, e.g., for the rare earths, because it is readily destroyed by digestion with concentrated nitric acid so that the retained elements need not be eluted before their quantitative determination. Another advantage of this resin in comparison to, e.g., Dowex® A-1, is that it does not retain the alkali and alkaline earth elements, for the removal of which an additional separation step may be necessary when using the iminodiacetic acid resin (see Table 118).[497]

Analytical applications of chelating resins which include a method for the detection of Ce are illustrated in Table 118.

The synthesis and application of a chelating resin based on arsenazo I for separations involving microamounts of rare earths have also been described.[504]

MISCELLANEOUS

In addition to the procedures described in the Tables of this *Handbook*, separations on ion exchange resins have also been employed in connection with the determination of Y in monazite sand.[601]

Table 1
DISTRIBUTION COEFFICIENTS OF RARE EARTH ELEMENTS ON DOWEX® 50 IN HCl MEDIA[1,2,6]

Rare earth cation	Molarity of HCl (*M*) 0.1	0.2	0.5	1.0	1.75	2.0	3.0	4.0
La(III)	$>10^5$	$>10^4$	$>10^3$	$<10^3$	70	~60	>10	~10
Ce(III)	$>10^5$	$>10^4$	$>10^3$	265	67	48	—	10.5
Sm(III)	$>10^4$	$>10^4$	1330	217	56	39	15.4	8.6
Gd(III)	$>10^4$	$>10^4$	1220	183	48.2	30.2	11.4	6.1
Er(III)	$>10^4$	$>10^4$	990	165	41.8	27.2	10.7	6.0
Y(III)	$>10^4$	$>10^4$	$<10^3$	$>10^2$	40.1	>10	~10	<10
Yb(III)	$>10^4$	$>10^4$	960	153	39.9	27.4	12.2	7.4
Sc(III)	$>10^4$	3200	500	120	38.8	28.8	14.9	11.7

Table 2
EMISSION-SPECTROGRAPHIC DETERMINATION OF RARE EARTH ELEMENTS IN GEOLOGICAL MATERIALS AFTER SEPARATION BY ONE-COLUMN CATION EXCHANGE PROCEDURES[a]

Material	Ion exchange resin, separation conditions, and remarks	Ref.
Rocks	Bio-Rad® AG50W-X8 (200-400 mesh; H^+ form) (pretreated with 4 and 1 *M* HCl) Column: 25 × 2 cm ID containing 20 g of the resin to a height of 10 cm and operated at a flow rate corresponding to the back pressure of the resin bed a) Sample solution with an acid strength of <10% (e.g., 0.5 *M* HCl) (adsorption of rare earths, Ba, Sr, Zr, Hf, and other accompanying elements) b) 1.7 *M* HCl (400 mℓ) (elution of all major constituents and most of the trace elements) c) 4 *M* HCl (500 mℓ) (elution of the rare earths) Inductively coupled plasma source spectrometry was used for the determination of the rare earth elements Before the separation, the sample (0.5 g) is decomposed by acid treatment as well as by fusion with NaOH or KHF_2 For the ranges of concentrations most likely to be encountered the precision is ~1—2% The technique is well suited for the determination of all the rare earth metals except Tb and Tm A similar method has been used in connection with the determination of the rare earths in silicates;[68] the sample (0.3 g) is fused with Na_2O_2 (3 g), the melt is dissolved in hot 0.6 *M* HCl, and the solution is passed through a column (18 × 1.5 cm) of Dowex® 50W-X8 (200-400 mesh; H^+ form) pretreated with 0.6 *M* HCl (30 mℓ) at ≃3.0 mℓ/min; the matrix elements are eluted with 3 *M* HCl in 25% ethanol (300 mℓ) and the rare earths are then eluted with 6 *M* HCl (200—250 mℓ); ICP AES is used for their determination	18, 19
Rocks	Bio-Rad® AG50W-X8 (200-400 mesh; H^+ form) Column: 19 × 2.1 cm ID containing 20 g (60 mℓ) of the resin; operated at a flow rate of 3.0± 0.5 mℓ/min a) 1.75 *M* HCl (up to ~550 mℓ) (adsorption of the rare earths, Sr, Ba, Zr, and Hf; elution of Al, Fe[III], Ti, Ca, Mg, Mn[II], K, and Na)	39

Table 2 (continued)
EMISSION-SPECTROGRAPHIC DETERMINATION OF RARE EARTH ELEMENTS IN GEOLOGICAL MATERIALS AFTER SEPARATION BY ONE-COLUMN CATION EXCHANGE PROCEDURES[a]

Material	Ion exchange resin, separation conditions, and remarks	Ref.
	b) 4 M HCl (>200 mℓ) (elution of the rare earths together with Sr, Ba, Zr, and Hf) The rare earths were determined by emission spectrography; the method has been used for the analysis of standard rock samples for very small traces (<1 ppm) of rare earths	
Terrestrial and lunar rocks	Dowex® 50W-X8 (100-200 mesh; H^+ form) (preconditioned with 4 and 0.1 M HCl), or cationite KU-2 (H^+ form) Column: 12 × 1.6cm ID operated at a flow rate of 1 mℓ/min a) 0.1 M HCl (5 mℓ) (adsorption of rare earths and other elements) b) 4 M HCl (~100 mℓ) (elution of rare earths and coadsorbed elements) From this eluate, the rare earths are isolated by coprecipitation with Y oxalate (Y is added as collector and internal standard) and determined spectrographically; before the ion exchange separation, the sample (100 mg) was fused with sodium carbonate, silica was removed by conventional means, and the rare earths were isolated by a hydroxide precipitation A similar cation exchange technique has also been used in connection with the spectrographic or X-ray spectral determination of ≈0.1—1 ppm of rare earth metals in natural materials, e.g., rocks;[12] in this method the 4 M HCl eluent (~45 mℓ) is used to elute Fe, Al, Mg, Ca, and Na, before the rare earths are collected in subsequent eluate fractions of the same eluent; these rare earth-containing fractions are located radiometrically by measuring the ^{90}Y activity which was added to the sample solution before the separation The cation exchange procedure outlined above has also been used for the determination of Eu and Dy in rocks by neutron activation and high resolution X-ray spectrometry;[13] for the cation exchange separation, the resin Dowex® 50W-X8 was employed	11—13
Silicate rocks	Dowex 50-X8® (200-400 mesh; H^+ form) Column: 38 × 1.7 cm; the resin was pretreated with 3 M HCl a) 2 M HCl (4—5 mℓ) (adsorption of rare earths and elution of accompanying elements) (flow rate ≯0.25 mℓ/min) b) 3 M HCl (330 mℓ) (elution of more abundant elements) (flow rate: 15 mℓ/hr) c) 6 M HCl (700 mℓ) (elution of Sc, Y, Ce, Nd, La, Sr, and Ba with small amounts of Ca, Mg, and Al) (flow rate: 20 mℓ/hr) The rare earths were determined spectrographically	40

Table 2 (continued)
EMISSION-SPECTROGRAPHIC DETERMINATION OF RARE EARTH ELEMENTS IN GEOLOGICAL MATERIALS AFTER SEPARATION BY ONE-COLUMN CATION EXCHANGE PROCEDURES[a]

Material	Ion exchange resin, separation conditions, and remarks	Ref.
Fluorspar	Dowex® 50W-X8 (100-200 mesh; H^+ form) Column: 40 × 2.5 cm filled with the resin to a height of 20 cm and operated at a flow rate of 3—5 mℓ/min a) 1 *M* HCl (adsorption of the rare earths and elution of Ca) b) 1 *M* HCl (1000—1200 mℓ) (elution of the residual Ca) c) 12 *M* HCl (800—1000 mℓ) (elution of the rare earths) The rare earths were determined by emission spectrography; prior to the separation, the sample (20 g) is decomposed with H_2SO_4 and before dissolution in HCl the sulfates are converted into carbonates by treatment with ammonium carbonate solution	35
Rocks	Strongly acidic cation exchange resin (100-200 mesh; H^+ form) Column: 14 × 1.5 cm conditioned with 1 *M* HCl and operated at a flow rate of 0.5 mℓ/min a) 1 *M* HCl sample solution (150 mℓ) (adsorption of rare earths together with some matrix elements; Mg and Ca pass into the effluent) b) 2 *M* HCl (80 mℓ) (elution of remaining Ca) c) 2 *M* HNO_3 (200 mℓ) (elution of Ba and other impurities) d) 6 *M* HCl (200 mℓ) (elution of the rare earths) The eluate(d) is evaporated to dryness and the aqueous solution (25 mℓ) of the residue is passed 30 times through a disk of cation exchange resin paper to adsorb the rare earths; then the resin paper is ashed and the rare earths are determined by emission spectrography Before the separation, the sample (1 g) is decomposed by fusion with Na_2O_2	38

[a] See also References 19a and 19b.

Table 3
MASS-SPECTROMETRIC DETERMINATION OF RARE EARTH ELEMENTS IN GEOLOGICAL MATERIALS AFTER SEPARATION BY ONE-COLUMN CATION EXCHANGE PROCEDURES

Material	Ion exchange resin, separation conditions, and remarks	Ref.
Lunar and terrestrial rocks	Bio-Rad® AG50W-X8 (200-400 mesh; H^+ form) Column of 2 cm ID containing 60 mℓ (20 g) of the resin and operated at a flow rate of 3.0 ± 0.5 mℓ/min a) 0.5 *M* HCl (adsorption of rare earths and other elements) b) 1.75 *M* HCl (550 mℓ) (elution of major rock-forming elements) c) 4 *M* HCl (500 mℓ) (elution of lanthanides, Y, Sc, Zr, Hf, and Ba) Spark-source mass spectrometry was used for the determination of the rare earth elements which, in most cases, were present in ultratrace amounts (submicrogram quantities); before the separation, the sample (1 g) is decomposed with HF-HCl-$HClO_4$; the precision is $\simeq \pm 5\%$ (down to 0.1 μg) and $\simeq \pm 10\%$ for 10 ng As little as 0.001 ppm can be determined; the detection limit is ≃0.5 ng for all the lanthanoids; recoveries of 10 ng of La, Ce, and Lu were 99 ± 7%	20
Rocks and minerals	Dowex® 50W-X8 (200-400 mesh; H^+ form) Column of 1.8 cm ID filled with the resin to a height of 18 cm and pretreated with 2 *M* HCl; flow rate = back pressure of resin column a) 0.5 *M* HCl (~20 mℓ) (adsorption of the rare earths [together with a composite rare earth spike] and other elements) b) 2 *M* HCl (350 mℓ) (elution of coadsorbed elements) c) 6 *M* HCl (100 mℓ) (elution of the rare earths in two fractions of about equal size with ≃Gd through Lu in the first fraction and ≃La through Eu in the second) Each fraction, after concentration, was submitted to mass spectrometry; with this stable isotope-dilution technique a precision of $\simeq \pm 1\%$ is attainable Before the separation, the sample (0.5 to 1 g) is decomposed with HCl or $HClO_4$-HF; the monoisotopic rare earths Pr, Tb, Ho, and Tm cannot be determined by this method	21
Basalts	Dowex® 50W-X8 (100-200 mesh; H^+ form) in a 15-mℓ column a) 0.5 *M* HCl (5—10 mℓ) (adsorption of the rare earths and matrix elements) b) 2 *M* HCl (130 mℓ) (elution of most coadsorbed elements) (flow rate: 1.5—2.5 mℓ/min) c) 6 *M* HCl (100 mℓ) (elution of the lanthanides) For additional purification of the rare earths, this separation is repeated (with eluents [a], [b], and [c] having volumes of 1, 5, and 4 mℓ, respectively; using a flow rate of ≃1 drop/30 sec) on a very small column (4 × 0.4 cm) (0.5-mℓ column) of the same resin (200-400 mesh) and then the lanthanides (La, Ce, Nd, Sm, Eu, Gd, Dy, Er, and Yb) are determined by mass spectrometry Before the ion exchange separation, the sample (100—500 mg) is decomposed with HF-HCl in the presence of the enriched isotopes ^{138}La, ^{142}Ce, ^{143}Nd, ^{147}Sm, ^{151}Eu, ^{157}Gd, ^{164}Dy, ^{170}Er, and ^{171}Yb which are used as spikes	22

Table 4
DETERMINATION OF RARE EARTHS IN GEOLOGICAL MATERIALS BY X-RAY FLUORESCENCE, ATOMIC ABSORPTION SPECTROSCOPY, POLAROGRAPHY, AND SPECTROPHOTOMETRY AFTER SEPARATION BY ONE-COLUMN CATION EXCHANGE PROCEDURES

Material	Ion exchange resin, separation conditions, and remarks	Ref.
Rocks and minerals	Dowex® 50W-X8 (200-400 mesh; H^+ form) Column of 2.1 cm ID filled with the resin to a height of 15 cm and operated at a flow rate corresponding to the back pressure of the resin bed a) 2 *M* HCl (450 mℓ) (adsorption of the lanthanides, Y, Sc, and other elements; into the effluent pass the main constituents of the matrix) b) 6.6 *M* HCl (280 mℓ) (elution of the rare earths and Ba) From the eluate, Ba is removed by precipitation as $BaSO_4$ and after filtration, the rare earths are collected on cation exchange resin paper (Amberlite® IR-120) Finally, the individual elements of the rare earths adsorbed on this resin were determined by X-ray fluorescence; the method yields a precision of ±10—20% for all elements down to concentration of 1 to 2 μg/g; detection limits normally range from 0.5 to several micrograms per gram A very similar separation procedure has been described by other investigators[42] and utilized for the determination of the rare earths in geological materials (standard rock samples) using inductively coupled plasma optical emission spectrometry; the rare earths were separated on a column (1.5 cm ID contained 18 g of Amberlite® cation exchange resin [100-200 mesh] employing eluents (a) (140 mℓ) and (b) (160 mℓ) and finally the rare earths were determined spectrometrically	41
Natural Ca-Ce-phosphate	Dowex® 50-X8 (H^+ form) a) Sorption solution of pH 1.0 (adsorption of rare earths and matrix elements; elution of phosphate) b) Dilute HCl of pH 1.3 (500 mℓ) (elution of phosphate) c) 1 *M* HCl (2000 mℓ) (elution of Mg, Ca, Fe, Ti, and other elements) d) 4 *M* HCl (500 mℓ) (elution of rare earths) e) 3.6 *N* H_2SO_4 (800 mℓ) (elution of Th) Following precipitation as sebacates, the rare earths were determined by X-ray fluorescence; in the various eluates, the other elements were determined using gravimetric (phosphate and Al), titrimetric (Ca and Mg), and spectrophotometric (Fe, Ti, and Th) procedures; before the ion exchange separation, the sample (1 g) is decomposed with H_2SO_4-HF	14
Silicate rocks and minerals	Dowex® 50-X8 (200-400 mesh; H^+ form) Column: 18 × 2 cm ID operated at a flow rate corresponding to the back pressure of the resin column a) 0.6 *M* HCl (100 mℓ) (adsorption of the rare earths and matrix elements) b) 25% Ethanol - 3 *M* HCl (300 mℓ) (elution of the matrix elements Na, K, Al, Fe, Ti, Ca, Mg, and part of the Sr) c) 4 *M* HCl (200—250 mℓ) (elution of Sc, Y, lanthanides, Ba, and residual Sr) The rare earths (Dy, Eu, Tm, Ho, and Yb) were determined by atomic absorption spectrophotometry based on electrothermal atomization; before the separation the sample (500 mg) is decomposed by HF-$HClO_4$	34
Ores	Cation exchange resin a) Dilute HCl - sample solution (adsorption of the rare earths)	69

Table 4 (continued)
DETERMINATION OF RARE EARTHS IN GEOLOGICAL MATERIALS BY X-RAY FLUORESCENCE, ATOMIC ABSORPTION SPECTROSCOPY, POLAROGRAPHY, AND SPECTROPHOTOMETRY AFTER SEPARATION BY ONE-COLUMN CATION EXCHANGE PROCEDURES

Material	Ion exchange resin, separation conditions, and remarks	Ref.
	b) 12.5% HCl (elution of the coadsorbed elements) c) 6 *M* HCl (150 mℓ) (elution of the rare earths) Sm is determined polarographically Before the separation, the sample (0.5 g) is fused with Na_2O_2 (5—7 g) and the melt is treated with triethanolamine - 5% NaOH solution (50 mℓ) and 0.1 *M* 1,2-bis-(2-amino-ethoxy)ethane-NNN'N'-tetraacetic acid and the mixture is filtered; the residue (containing the rare earths) is washed with 2% NaOH solution and then dissolved in 0.1 *M* ascorbic acid in 10% HCl (50 mℓ) at 70°C and 30 mℓ each of water and 8% HCl are added to obtain the sample solution (a)	
Y-containing rocks	Strongly acidic cation exchange resin in a microcolumn a) 0.2 *M* HCl (elution of Si) b) 1 *M* HCl - 0.5 *M* H_2O_2 (elution of Ti) c) 5 *M* HCl (elution of rare earths, Al, Fe, Ca, and Mg) The rare earths and the other elements are determined spectrophotometrically, e.g., with arsenazo III (Ce-group metals)	66

[a] See also Tables 127 and 128 in the chapter on Actinides, Volume II.

Table 5
RADIOANALYTICAL DETERMINATION OF RARE EARTH ELEMENTS IN ROCKS AFTER SEPARATION BY ONE-COLUMN CATION EXCHANGE PROCEDURES[a]

Ion exchange resin	Separation conditions and remarks	Ref.
Dowex® 50W-X8 (100-200 mesh; H^+ form)	Column: 25 × 0.5 cm ID a) 0.5 *M* HCl (containing carriers for all the rare earth metals, except Pr, Nd, Dy, and Er) (adsorption of the rare earths and accompanying elements) b) 0.5 *M* HCl - 0.1 *M* oxalic acid (elution of Sc, Fe, and Co) c) 2 *M* HNO_3 (elution of alkali and alkaline earth elements) d) 6 *M* HCl (elution of the rare earths) The rare earth elements were determined radiometrically Before the ion exchange separation, the sample is irradiated with neutrons and dissolved in HCl-HF The chemical yield of the method is 98%	23
Amberlite® IRA-120 or Dowex® 50W-X8 (100-200 mesh; H^+ form)	Column: 30 × 1.0 cm ID containing the resin to a height of 20 cm (the resin was conditioned with 6 and 2 *M* HCl [100 mℓ each]) a) 1 *M* HCl (10 mℓ) (adsorption of the rare earths [in the presence of ^{144}Ce tracer] and other elements such as Ba) b) 2 *M* HCl (100 mℓ) (elution of coadsorbed elements) c) 6 *M* HCl (150 mℓ) (elution of the rare earths) (coeluted are Sr and Ba) The chlorides of the eluted rare earths are treated with HNO_3 and the nitrates irradiated with neutrons; finally, the rare earths were determined radiometrically Before the ion exchange separation, the sample (international standard rock) is decomposed with HF-$HClO_4$	36

[a] See also Table 40 in the chapter on Copper, Volume III.

Table 6
GRAVIMETRIC DETERMINATION OF RARE EARTH ELEMENTS[a] IN CARBONATITES AFTER SEPARATION BY A ONE-COLUMN CATION EXCHANGE PROCEDURE

Ion exchange resin	Separation conditions and remarks	Ref.
Dowex® 50W-X8 (100-200 mesh; H^+ form)	Column: 18 × 2 cm ID containing 20 g of the resin; a flow rate of 3.0 ± 0.2 mℓ/min is maintained throughout a) ~0.5 *M* HCl (adsorption of the rare earths and accompanying elements) b) 1.75 *M* HCl (600 mℓ) (elution of Fe[III], Al, Ti, U[VI], Be, Ca, etc.) c) 3 *M* HCl (800 mℓ) (elution of the rare earths) The rare earths were precipitated as oxalates and finally weighed as the oxides Before the ion exchange separation, the sample (5 g) is completely dissolved by treatment with acids and fusion with K-bisulfate and the rare earths are preconcentrated by a hydroxide precipitation The method was used to determine the sum of the rare earths contained in several South African carbonatites	1

[a] See Table 163 in the chapter on Actinides, Volume II.

Table 7
RADIOANALYTICAL DETERMINATION OF RARE EARTH ELEMENTS IN GEOLOGICAL MATERIALS AFTER SEPARATION BY TWO-COLUMN CATION EXCHANGE PROCEDURES

Material	Ion exchange resin, separation conditions, and remarks	Ref.
Minerals	Dowex® 50W-X8 (100-200 mesh; H^+ and NH_4^+ forms) A) First column operation (on H^+-form resin) (group separation of the rare earths in the presence of ^{153}Gd as a chemical yield monitor) Column: 20 cm × 2 cm²) a) 0.5—2 *M* HCl (adsorption of rare earths and other elements) b) 3 *M* HCl (60 mℓ) (elution of Fe, Be, Ti, Al, Cr, Mn, and most other coadsorbed elements) c) 6 *M* HCl (250 mℓ) (elution of the rare earths) In the case of tantalocolumbite, eluent (b) (70 mℓ) is also 1% in H_2O_2; to remove the oxidizing agent from the column, this is washed with 3 *M* HCl (20 mℓ) Subsequently, the rare earths are subjected to neutron irradiation and then the individual elements are separated by using the following procedure B) Second column operation (on NH_4^+ form resin) Column: 14 cm × 0.4 cm² a) 0.1 *M* HNO_3 (adsorption of the rare earths) b) 0.5 *M* α-HIBA (gradient elution of the rare earths; ammonia solution is added to obtain the desired ligand concentration [0.05—0.075 *M*]; the eluate is collected in 2-mℓ fractions) In the various fractions the rare earths were determined radiometrically Before the first cation exchange separation step, the sample (50 mg) of gadolinite (after addition of ^{153}Gd) is decomposed with HCl-HF-H_2SO_4 and $KHSO_4$ fusion and then a hydroxide precipitation is performed as a first preconcentration step; tantalocolumbite (500 mg + Gd tracer) is decomposed with $KHSO_4$ and treatment with HNO_3, H_2O_2, HF, and H_2SO_4 The method has been used for the determination of 11—14 rare earth elements in these minerals	44—46

Table 7 (continued)
RADIOANALYTICAL DETERMINATION OF RARE EARTH ELEMENTS IN GEOLOGICAL MATERIALS AFTER SEPARATION BY TWO-COLUMN CATION EXCHANGE PROCEDURES

Material	Ion exchange resin, separation conditions, and remarks	Ref.
Monazite	Dowex® 50W-X8(H^+ and NH_4^+ forms) A) First column operation (on H^+-form resin; 200-400 mesh) (group separation of the rare earths) Column: 6 × 0.5 cm ID a) Dilute sulfuric acid solution of the sample (50 mℓ) (adsorption of the rare earths and Th) b) 1 *M* HCl (150 mℓ) (elution of accompanying elements) c) 3 *M* HCl (60 mℓ) (elution of rare earths) Subsequently, the rare earths are irradiated with neutrons and the individual elements are separated using the following procedure B) Second column operation (on NH_4^+ form of the resin; −400 mesh) Column: 13 × 0.5 cm To separate the rare earths, pH-gradient elution using 0.25 *M* α-HIBA at a flow rate of 0.3 mℓ/min is employed; the pH is varied from 3.2—4.7 and in the fractions obtained, the 14 rare earths (La to Lu) are determined radiometrically Prior to the first column operation, the sample (~0.1 g) is digested with conc H_2SO_4 (1 mℓ) and the solution is diluted to 100 mℓ with water; from this solution, Tm can be isolated directly (following neutron irradiation), i.e., by using only separation step (B) for which the slurry column technique is utilized	64
Rocks and lunar samples	Dowex® 50-X8 (H^+ form) A) First column operation (on 50-100 mesh resin); column containing 1.5 g of the resin a) 2 *M* HCl (adsorption of the rare earths; elution of Na, Mn, and other elements) b) Water (removal of HCl) c) 0.4 *M* α-HIBA of pH 6 (15 mℓ) (elution of the rare earths as a group) B) Second column operation (200-400 mesh resin) Column: 7.2 × 0.2 cm ID Separation of individual rare earth elements using pH-gradient elution with 0.3 *M* α-HIBA (with increasing pH from 3.5—5.0) In the various fractions obtained, the rare earths (Eu, Sm, Ce, and La) were determined radiometrically Prior to the first cation exchange separation step, the sample (10 mg) is irradiated with neutrons and decomposed with HF-$HClO_4$ in the presence of La-carrier and ^{160}Tb as a yield tracer The chemical yield of the method is >90%; estimated standard deviations ranged from ±0.1 ppm for 1.2 ppm of Eu to ±10 ppm for 167 ppm of Ce	43

Table 8
DETERMINATION OF THE RARE EARTHS IN GEOLOGICAL MATERIALS BY MASS SPECTROMETRY AND SPECTROPHOTOMETRY AFTER SEPARATION BY TWO- AND THREE-COLUMN CATION EXCHANGE PROCEDURES

Material	Ion exchange resin, separation conditions, and remarks	Ref.
Meteorites	Diaion® SK 1 cation exchange resin (100-200 mesh; H^+ and NH_4^+ form) Column: 45 × 1.6 cm ID containing the resin to a height of 43 cm; the resin is preconditioned with 6 *M* HCl (1500 mℓ) and water A) First column operation (H^+ form of the resin) a) 0.5 *M* HCl (400 mℓ) (adsorption of the rare earths and other elements) b) 2 *M* HCl (elution of major constituents) c) 6 *M* HCl (150 mℓ) (elution of the rare earths) B) Second column operation (NH_4^+ form of the resin) Column: 25 × 0.4 cm ID; height of resin bed after preconditioning: 22 cm) a) Aqueous solution of the rare earth nitrates (adsorption of the rare earths) b) 0.5 *M* α-HIBA solution of pH 3.20 (elution of the rare earths and collection of three fractions containing Gd, Eu, and Sm) The α-HIBA is removed from the three fractions in turn by a column operation (on the resin in the H^+ form) using the following eluents Column: 15 × 0.4 cm a) 0.5 *M* α-HIBA fraction (adsorption of Gd, Eu, or Sm) b) 2 *M* HCl (removal of α-HIBA) c) 6 *M* HCl (elution of Gd, Eu, or Sm) In the eluates the rare earth elements (Gd, Eu, and Sm) were determined by mass spectrometry This three-column technique has been utilized in connection with the determination of the isotopic ratios of Gd, Sm, and Eu in the "Abee" enstatite chondrite	24
Silicate rocks and minerals	Dowex® 50W-X8 (200-400 mesh; H^+ and NH_4^+ forms) A) First column operation (H^+ form of the resin) Column of 2.2 cm ID containing resin to a height of 18 cm; the resin is pretreated with 3 *M* HCl (300 mℓ) and then with water followed by 6 *M* HCl - H_2O_2 (0.15%) (30 mℓ); the column is operated at a flow rate of 3.0 ± 0.5 mℓ/min a) Dilute HCl solution of the sample (adsorption of rare earths and other elements) b) 3 *M* HCl in 20% ethanol (250 mℓ) (elution of major elements) c) Water (to remove acid) d) 0.5 *M* oxalic acid (150 mℓ) (elution of Zr, Hf, and Th) e) 6 *M* HCl (300 mℓ) (elution of Y and other rare earths) B) Second column operation (NH_4^+ form of the resin) Column: 0.6 cm ID containing the resin to a height of 11 cm; the resin is pretreated with 0.24 *M* α-HBA at pH 5 (one-column volume) a) 0.24 *M* α-HBA (≃70 mℓ) (elution of Lu, Yb, Tm, Er, and Ho) (this fraction is discarded) b) 0.36 *M* α-HBA (15 mℓ) (elution of Y) The Y was determined spectrophotometrically using the arsenazo III method Before the first column operation, the sample (≃0.3 g) is fused with Na_2O_2 and the melt dissolved in 0.15% H_2O_2 - 6 *M* HCl	25

Table 9
DETERMINATION OF RARE EARTH ELEMENTS IN SILICATES AFTER SEPARATION BY MULTICOLUMN METHODS USING CATION AND ANION EXCHANGE RESINS

Ion exchange resin	Separation conditions and remarks	Ref.
Bio-Rad® AG 50W-X8 (200-400 mesh; H^+ form) and Bio-Rad® AGl-X8 (200-400 mesh; sulfate form)	A) First column operation (cation exchange resin) Column: 30 × 2 cm ID containing 45 mℓ of the resin a) 0.5 *M* HCl (100 mℓ) (adsorption of rare earths and base metals) b) 1.85 *M* HCl (300 mℓ) (elution of the base metals) c) 4 *M* HCl (400 mℓ) (elution of the rare earths) To separate the lanthanides from the coeluted Sc the following procedure is used B) Second column operation (anion exchange resin) Column of the same dimensions as under (A) containing 45 mℓ of the resin From dilute sulfuric acid solution (≈25 mℓ) of pH 1.5, the Sc is retained on this resin while the lanthanides pass into the effluent (for the complete elution of the lanthanides the column is washed with 250 mℓ of 0.02 *N* H_2SO_4), from which they are isolated by 8-hydroxyquinoline precipitation (with Al_2O_3 as carrier) and ignited to the oxides before spectrographic analysis Prior to the first column operation the sample (1 g) is fused with KHF_2, the insoluble Ca and rare earth metal fluorides are filtered off and converted into chlorides The removal of Sc from the sample by means of the second column operation is necessary, because Sc is used as an internal standard in the spectrographic determination of the lanthanides The cation exchange separation procedure outlined above has also been utilized in connection with the application of an Ar-N inductively coupled radiofrequency plasma (ICP) to the analysis of geological samples (silicate and phosphate reference materials) and related materials for their rare earth contents[27]	26
Zeo-Karb® 225 (<200 mesh; H^+ form) and Deacidite® FF (chloride form)	A) First column operation (cation exchange) Column containing 16 g of the resin, pretreated with 2.5 *M* HCl (~200 mℓ) and operated at a flow rate of ~1 mℓ/min a) 2.5 *M* HCl (~8 mℓ) (adsorption of rare earths and other elements; Li as well as K + Rb + Cs is collected in two separate fractions during the subsequent passage of ~100 mℓ 2.5 *M* HCl) (the first 20 mℓ are discarded, Li is collected in the next 20 mℓ, and after discarding the subsequent 20 mℓ, the K, Rb, and Cs fraction [50 mℓ] is obtained) b) Water (30 mℓ) (removal of HCl) c) 2.07 *M* HNO_3 (160 mℓ) (elution of Sr and Ba; these two elements are collected in one fraction [140 mℓ]) d) 2.58 *M* HNO_3 (200 mℓ) (elution of Eu, Sm, Nd, and Ce) B) Second column operation (column containing 2 g of the anion exchange resin) From 6.2 *M* HCl (1 mℓ) coeluted elements are adsorbed on this resin while Eu, Sm, Nd, and Ce pass into the effluent (following a rinse with 6.2 *M* HCl [10 mℓ]) in which they are determined by mass spectrometry; the same anion exchange procedure is used to purify the eluate containing Sr and Ba which are also determined mass spectrometrically	47

Table 9 (continued)
DETERMINATION OF RARE EARTH ELEMENTS IN SILICATES AFTER SEPARATION BY MULTICOLUMN METHODS USING CATION AND ANION EXCHANGE RESINS

Ion exchange resin	Separation conditions and remarks	Ref.
	The Li fraction (after evaporation to dryness and dissolution of the residue in water [1 mℓ]) is also passed through a column of this anion exchange resin in the oxalate form to remove impurities and in the effluent (obtained after washing with water [10 mℓ]), Li is determined by mass analysis; the latter is also used to determine K, Rb, and Cs in effluent (a) following application of this anion exchange technique Before the first ion exchange separation step, the sample ($\simeq$100 mg) is decomposed with HF-$HClO_4$ in the presence of isotopically enriched tracers; the method was used to determine the rare earths in rocks and meteorites	
Dowex® 50-X8 (100-200 mesh; H^+ form) and Dowex® 1-X8 (100-200 mesh; chloride and sulfate forms)	A) First column operation Column of 1 cm ID containing the cation exchange resin to a height of 10 cm a) 3 *M* HCl (20 mℓ) (adsorption of Sc and elution of some accompanying elements) b) Water (10 mℓ) (as a rinse) c) 0.3 *M* ammonium sulfate - 0.025 *M* H_2SO_4 (35 mℓ) (elution at a flow rate of 0.7 mℓ/min of Sc and coadsorbed elements in two fractions of which the second [15 mℓ] contains the Sc) To the Sc fraction 0.025 *M* H_2SO_4 is added (to make the solution 0.1 *M* in ammonium sulfate and 0.025 *M* in H_2SO_4) and foreign ions are separated from Sc using the following separation step B) Second column operation Column of 1 cm ID containing the anion exchange resin (10 g) (sulfate form) to a height of 18 cm a) Sc-containing fraction (adsorption of Sc and other elements, e.g., Th) b) 0.1 *M* ammonium-sulfate - 0.025 *M* H_2SO_4 (elution of coadsorbed elements which are eluted with 45 mℓ of this eluent before the Sc which is collected in a subsequent separate fraction (additional 130 mℓ of eluent) From this eluate fraction (130 mℓ) the Sc is isolated by coprecipitation with ferric hydroxide and the precipitate is dissolved in 6 *M* HCl C) Third column operation (column of 1 cm ID containing the anion exchange resin [2 g] [chloride form] to a height of 3 cm; separation of Sc from Fe) From 6 *M* HCl (10 mℓ), the Fe(III) is adsorbed while Sc passes into the effluent in which it is determined spectrophotometrically using the bromopyrogallol red method; also, arsenazo I or 4-(2-thiazolylazo)resorcinol can be used[51] Before the first ion exchange separation step, the sample (~0.5 g) is decomposed with $HClO_4$-HF and Sc is preconcentrated by a hydroxide precipitation; the above technique has been utilized for the determination of the Sc content of numerous samples of igneous rocks and marine sediments[50]	48—51

Table 10
DETERMINATION OF RARE EARTH ELEMENTS IN MINERAL WATERS AFTER SEPARATION BY MULTICOLUMN ION EXCHANGE PROCEDURES

Ion exchange resins	Separation conditions and remarks	Ref.
Dowex® 50-X8 (100-200 mesh and 200-400 mesh) and -X12 (50-100 mesh) (H^+ and NH_4^+ forms) and Dowex® 1-X8 (100-200 mesh; chloride form)	A) First column operation Column: 35 × 1.6 cm containing the cation exchanger -X12 in the H^+ form to a height of 30.2 cm a) Water sample adjusted to pH 1.5—1.8 (adsorption of the rare earths and accompanying elements) b) 40% (300 mℓ) and 80% (200 mℓ) acetone (removal of organic matter) c) Water (100 mℓ) (removal of acetone) d) 1.6 *M* HCl (1500 mℓ) (elution of Na, K, Ca, Mg, and Fe[III]) (flow rate: 1 mℓ/min) e) 6 *M* HCl (1100 mℓ) (elution of the rare earths) (coeluted are Sr, Ba, Al, Zr, and Hf) (flow rate: 0.5 mℓ/min) B) Second column operation (anion exchange resin) Column: 7 × 0.727 cm From 12 *M* HCl (2 mℓ) coeluted elements such as residual Fe(III) and also Zr and Hf are adsorbed on a column of this resin, while the rare earths pass into the effluent (after washing the resin bed with an additional 10 mℓ of 12 *M* HCl); for the separation of the individual lanthanides the following procedure is used C) Third column operation Column: 70 × 0.734 cm containing the cation exchanger -X8 (NH_4^+ form) to a height of 59 cm a) 0.1 *M* HCl (1 mℓ) (adsorption of the rare earths, Sr, and Ba, followed by rinses with 0.1 *M* HCl [2 mℓ], and 2 mℓ of eluent [b]) b) 0.08 *M* α-HIBA of pH 4.43 (60 fractions of 5 mℓ each) (elution of Sc, Lu, Yb, Tm, and Er) c) 0.1 *M* α-HIBA of pH 4.43 (40 fractions of 5 mℓ each) (elution of Ho, Y, and Dy) d) 0.15 *M* α-HIBA of pH 4.43 (50 fractions of 5 mℓ each) (elution of Tb, Gd, and Eu) e) 0.2 *M* α-HIBA of pH 5.00 (50 fractions of 5 mℓ each) (elution of Sm and Nd) f) 0.29 *M* α-HIBA of pH 5.00 (45 fractions of 5 mℓ each) (elution of Pr, Ce, and La) (Sr and Ba are further retained by the resin) D) Fourth column operation Column of 0.727 cm ID containing the cation exchange resin -X8 (100-200 mesh; H^+ form) to a height of 7 cm (separation of Dy and Y contained in eluate [c] obtained by the third column operation) a) Eluate (c) acidified to pH 1.8 (adsorption of Dy and Y) b) 0.01 *M* HCl (20 mℓ) (as a wash) c) 0.6 *M* HCl (25 mℓ) (elution of NH_4^+) d) 6 *M* HCl (20 mℓ) (elution of Dy and Y) E) Fifth column operation Column: 130 × 0.727 cm containing the cation exchange resin -X8 (200-400 mesh) to a height of 126 cm a) 0.1 *M* HCl (3 mℓ) (adsorption of Dy and Y) b) Water (20 mℓ) (as a rinse) c) 0.02 *M* ammonium citrate (mutual separation of the two rare earth elements; Dy is eluted before Y) (flow rate: 0.4-0.48 $mℓ/cm^2 \cdot min$); the first 735 mℓ of the effluent are discarded and two 250-mℓ fractions are collected	65

Table 10 (continued)
DETERMINATION OF RARE EARTH ELEMENTS IN MINERAL WATERS AFTER SEPARATION BY MULTICOLUMN ION EXCHANGE PROCEDURES

Ion exchange resins	Separation conditions and remarks	Ref.
	The fraction containing the Dy is acidified to pH 1.8 and Dy is adsorbed on a column of the same resin (H^+ form); after washing the resin bed with 0.01 *M* HCl, the NH_4^+ form of the resin is produced by passing 0.1 *M* aqueous ammonia solution, and finally Dy is eluted with 0.1 *M* ammonium citrate	
	Dy and Y were determined spectrophotometrically	
	F) Sixth column operation	
	Column: 7 × 0.727 cm containing the cation exchange resin -X8 (100-200 mesh; H^+ form) (isolation of Sc from eluate [b] obtained by the operation [C])	
	a) Sample solution of pH 1.8 (adsorption of Sc and other elements)	
	b) 0.01 *M* HCl (20 mℓ) (as a wash)	
	c) 0.6 *M* HCl (25 mℓ) (as a wash)	
	d) 6 *M* HCl (50 mℓ) (elution of Sc)	
	The Sc was determined spectrophotometrically using xylenol orange; before the determination of the other rare earth elements except Dy and Y (see above under [E]) a cation exchange separation procedure (H^+ form resin) was applied to each eluate fraction separately, which was obtained by the third column operation, using the following eluents (removal of α-HIBA)	
	Column of 0.727 cm ID containing Dowex® 50W-X8 (100-200 mesh) to a height of 7 cm (Lu → Tb) or 6 cm (Gd → La)	
	a) Rare earth fraction acidified to pH 1.8 (adsorption of the rare earths contained in this fraction)	
	b) 0.1 *M* HCl (20 mℓ) followed by water (10 mℓ) (removal of α-HIBA)	
	c) 0.1 *M* aqueous ammonia solution (conversion of the resin to the NH_4^+ form; followed by water until neutral reaction is obtained)	
	d) 0.1 *M* ammonium citrate of pH 4.3 (7.5 mℓ) (elution of the rare earth elements)	
	Spectrophotometric procedures were also used for the determination of these rare earth elements utilizing xylenol orange as the color reagent	
	Before carrying out the above set of ion exchange separations the water sample (10 ℓ) is acidified with 6 *M* HCl to pH 1.5—1.8 and boiled	
	A shorter but less efficient procedure using 0.02, 0.04, and 0.06 *M* ammonium citrate solutions of pH 4.19 as eluents for the rare earth elements has also been applied in connection with the spectrophotometric determination of the rare earths in mineral waters[505]	

Table 11
RADIOANALYTICAL DETERMINATION OF Ce AND OTHER ELEMENTS IN NATURAL WATERS AFTER SEPARATION BY COMBINED CATION AND ANION EXCHANGE

Ion exchange resins	Separation conditions and remarks	Ref.
SA-2 cation exchange resin paper and Dowex® 2-X8 and Dowex® 1-X1	Radioactive Ce and other radionuclides are separated (with addition of carriers if necessary) by filtration of the sample through a thickness of at least two SA-2 cation exchange resin papers on a Büchner funnel, and the total β activity of the dried papers is measured; a 3-mm layer of a 1:1 mixture of the two anion exchange resins is then placed on a glass-fiber mat in the funnel and washed with conc HCl; subsequently, the ion exchange papers containing the adsorbed Ce are placed on the layer of the resin mixture and Ce is eluted with conc HCl, whereby coeluted multivalent elements, e.g., Zr and Te, are adsorbed on the anion exchange resins; from the eluate, Ce is isolated by precipitation as an iodate and determined radiometrically	71
1:1 Mixture of Powdex® PAO (anion exchange resin) (OH^- form) and Powdex® PCH (cation exchange resin; H^+ form)	The radionuclides, e.g., ^{141}Ce, ^{144}Ce, ^{131}I, ^{7}Be, ^{103}Ru and ^{106}Ru, ^{140}Ba, ^{137}Cs, ^{95}Zr, ^{95}Nb, and ^{140}La, are concentrated by passing the untreated sample (3 ℓ of rain water) through a bed (4.5 × 1 cm) containing 10 g of the mixed ion exchange resins at 200 mℓ/min; subsequently, the radioactivity adsorbed on the resin and that remaining in the effluent was examined by γ-ray spectrometry	76

Table 12
DETERMINATION OF Ce IN SEAWATER AFTER SEPARATION BY CATION EXCHANGE USING A ONE-COLUMN PROCEDURE

Ion exchange resin	Separation conditions and remarks	Ref.
Dowex® 50	a) ≃0.08 *M* HCl solution containing sufficient 0.1 *M* Na_2EDTA to complex 90% of the Fe present (adsorption of Ce; elution of Fe and other elements) b) 4 *M* HCl (elution of Ce) This separation is repeated to further purify the Ce which is determined spectrophotometrically Prior to the ion exchange separation, Ce is isolated from the sample (200 ℓ) by coprecipitation with ferric hydroxide	55

Table 13
DETERMINATION OF Ce, Y, AND Sc IN BIOLOGICAL MATERIALS AFTER CATION EXCHANGE SEPARATION USING ONE-COLUMN PROCEDURES

Material	Ion exchange resin, separation conditions, and remarks	Ref.
Marine plankton	Dowex® 50 (H^+ form) a) 0.2 *M* HCl (adsorption of rare earths, Fe, and other elements; phosphate passes into the effluent) b) 8 *M* HCl (elution of rare earths and coadsorbed elements) From this eluate, the rare earths are isolated by precipitation as the hydroxides and separated by extraction chromatography on a HDEHP-cellulose column (after removal of Fe by extraction with tri-*n*-octylamine-benzene) Before the ion exchange separation, the plankton sample is ashed, the ash subjected to carbonate fusion (in the presence of carriers, i.e., Y, Ce, Zr, Nb, and Ru) and after removal of silica, Zr, Nb, and Ru the rare earths are isolated by hydroxide precipitation	31
Bone and egg shells	Amberlite® IRA-120 (16-50 mesh; H^+ form) Column with a cross-sectional area of 1.54 cm^2 and a height of 22 cm, giving a total bed volume of ≃35 mℓ a) 0.1 *M* HCl (100 mℓ) (adsorption of Y, Ca, and other elements) (flow rate: 3 mℓ/min) b) 0.8 *M* HCl (1500 mℓ) (elution of mono- and divalent ions) (flow rate: 6—8 mℓ/min) c) 6 *M* HCl (1200 mℓ) (elution of Y at 50°C) (same flow rate as with eluent [b]) The Y was determined spectrographically; before the separation, the sample is dry-ashed at 600°C and the ash (~1 g) is dissolved in HCl; the recovery of Y is ≃94% for concentrations of 2—35 ppm in the ash	15
Plants and animals	Dowex® 50-X8 (H^+ form) (removal of neutron-induced P) Column: 20 × 0.8 cm operated at a flow rate of 10—15 drops/min a) 0.1 *M* HCl (60 mℓ) (adsorption of Sc and other elements [into the effluent pass ^{32}P, ^{124}Sb, and other elements] followed by a rinse [20 mℓ] with the same acid) b) 2 *M* HNO_3 (200 mℓ) (elution of Sc and coadsorbed elements) Sc was determined radiometrically following its isolation from the eluate by coprecipitation on Zr-phytate Before the ion exchange separation, the sample is subjected to neutron irradiation, fused with Na_2O_2, and dissolved in 4 *M* HNO_3 The sensitivity of the method, which is also applicable to the analysis of marine sediments, sands, soils, rocks, and minerals, is ≃0.005 μg of Sc	16

Table 14
DETERMINATION OF STABLE Ce IN BIOLOGICAL MATERIALS AFTER SEPARATION BY A TWO-COLUMN METHOD USING CATION AND ANION EXCHANGE RESINS

Ion exchange resins	Separation conditions and remarks	Ref.
Dowex® 50 (H^+ form) and Dowex® 1 (chloride form)	A) First column operation (cation exchange resin) a) 0.5 *M* HCl (adsorption of rare earths, Fe, and other elements) b) 2 *M* HCl (elution of Fe and other impurities) c) 8 *M* HCl (elution of the rare earths) B) Second column operation (anion-exchange resin) This separation, which is based on the adsorption of trivalent Fe and other elements on this resin from 6 *M* HCl, serves for the further purification of the rare earths which after irradiation with neutrons are separated into individual elements using extraction chromatography on a HDEHP-celite column; before the cation exchange separation, the sample (swordfish bone, shell, and edible part of clam) is ashed, decomposed with HNO_3-$HClO_4$-H_2O_2, and the lanthanides are coprecipitated with ferric hydroxide This preirradiation technique is also applicable to the analysis of rocks[75] and minerals;[73,74] it has later been amended by other investigators,[75,506] in connection with studies concerning the rare earth abundances in Mn nodules[506] and the rare earth fractionation in Hawaiian volcanic rocks[75] They use coprecipitation on Ca oxalate followed by coprecipitation on Fe hydroxide and the same anion exchange separation step	28

Table 15
DETERMINATION OF Ce AND OTHER RARE EARTH ELEMENTS IN STEEL AND CAST IRON AFTER SEPARATION BY CATION EXCHANGE IN HCl, ACETONE, OR PURE AQUEOUS MEDIA

Material	Ion exchange resin, separation conditions, and remarks	Ref.
C steel	Cation exchange resin 743 (made in China) or Zerolit® 225 (H^+ form) (200-400 mesh) a) 4:2 Mixture of acetone and conc HCl (adsorption of the rare earths; elution of Fe) b) 7:3 Mixture of acetone and 4 *M* HCl (elution of bulk of the Fe) c) 6 *M* HCl (elution of the rare earths) The separation is carried out in a high pressure apparatus in which the solutions are subjected to a forced flow using nitrogen at $\simeq$20 kg cm^{-2} The rare earth elements were determined spectrophotometrically with arsenazo III Before the separation, the steel sample is dissolved in 6 *M* HCl + H_2O_2; recoveries of 5 μg amounts of Ce (in 1 g of Fe) were between 100 and 106%	56
Steel and cast iron	Bio-Rad® AG50W-X8 (H^+ form) a) 3:1 Mixture of acetone and 3 *M* HCl (adsorption of Ce[III]; into the effluent pass Fe and other elements) b) 5 *M* HCl (elution of Ce) Amperometric titration with $KMnO_4$ was used for the determination of Ce; results for samples containing from 0.015 to 0.56% of Ce showed satisfactory repeatability	57
Cast iron and steel	Dowex® 50W-X8 a) Acetone-HCl solution of the sample (adsorption of Ce, Zr, and other elements; Fe passes into the effluent) b) 2 *M* HCl (elution of coadsorbed elements) c) 1 *N* H_2SO_4 (elution of Zr) d) 5 *M* HCl (elution of Ce) The Ce was determined spectrophotometrically using the oxine method; the standard deviation is 8 ppm when determining 944 ppm of Ce; the limit of determination is 2—8 ppm; prior to the separation, the sample (0.5 g) is dissolved in HCl-H_2O_2 The method is also applicable to the analysis of heat-conducting (Ni-Cr) alloys	59
Steel	Zeo-Karb® 225 (H^+ form) Following adsorption of the elements on a column of the resin the following eluents are used a) 2.5 *M* HCl (elution of Fe) b) 2 *M* HNO_3 (elution of Ba) c) 2.5 *M* HNO_3 (elution of Ce and other rare earth elements) The Ce was determined by mass spectrometry measuring the ratio of $^{140}Ce/^{142}Ce$ Before this separation, ^{142}Ce is added to the sample (100 mg) which is dissolved in 6 *M* HCl	63

Table 16
DETERMINATION OF Sm, Eu, Gd, Dy, AND OTHER RARE EARTH ELEMENTS IN NONFERROUS ALLOYS AND METALS AFTER CATION EXCHANGE SEPARATION

Material	Ion exchange resins, separation conditions, and remarks	Ref.
Be, U, Zr, and Ti metals, alloys, and oxides	Composite column consisting of Dowex® 50 (100-150 mesh) (or Amberlite® IR-120; 100-150 mesh) (H^+ form) (upper portion of resin bed) and Dowex® 1-X8 (200 mesh, chloride form) (lower layer of the bed) For the analysis of U and Be samples only the cation exchange resin is used (in a separate column [1 in. ID] containing a resin bed of 10 in. height), while for Ti and Zr samples the composite column (containing from bottom to top; glass wool followed by 3 in. of Dowex® 1, glass wool, and 3 in. of Dowex® 50 or Amberlite® IR-120)], together with the following eluents is employed a) 1 *M* HCl (100 mℓ) (adsorption of Y, rare earths, Zr, and Ti on the cation exchanger; into the effluent [through both resins] pass U, Be, Mg, and Ca [flow rate: 3—5 mℓ/min]) b) 1 *M* HCl (600-800 mℓ) (complete elution of Ca) c) 12 *M* HCl (250 mℓ) (elution of Y and rare earths) (Zr and/or Ti which were retained by the cation exchanger are now adsorbed by the anion exchange portion of the column) The rare earths (Eu, Dy, Gd, and Sm) together with Y were precipitated as oxalates, ignited to the oxides, and weighed Before the cation exchange separation the sample (10—50 g) is dissolved in HF and the rare earths are isolated by coprecipitation with the fluorides of Y (added as a carrier), Ca, and Mg; the collection of the rare earths by the fluorides of the latter two elements is complete even in the absence of Y	37
Al and Al alloys	Dowex® 50-X8 or Zerolit® 225 (H^+ form) a) Very dilute HCl (<0.5 *M*) (adsorption of rare earths and accompanying elements) b) 1.2 *N* H_2SO_4 followed by 1.8 *M* HCl (elution of Fe and other extraneous metals) c) 4.8 *M* HCl (elution of the rare earths, mainly Gd, Sm, Eu, and Dy) The traces of rare earth elements contained in the eluate were determined spectrophotometrically Before the cation exchange separation, the sample (2 g) is dissolved in NaOH, and after addition of Tm as internal standard the rare earths were coprecipitated on ferric hydroxide	33
Cu	Zerolit® 225 (200-400 mesh; H^+ form) The sample (0.1—0.5 g) is dissolved in HNO_3, the solution is evaporated, and the residue dissolved in water (20 mℓ); this solution is stirred for 6 min with amounts of thiourea and the resin that are 5 and 13 times, respectively, the weight of Cu present; the resin, now containing Cu as Cu(I) (cationic thiourea complex), rare earth metals, and possibly Ni or Fe(II), is filtered off and washed with 2 *M* H_2SO_4 (35 mℓ); the Cu remains adsorbed, but the rare earth metals are completely eluted with this acid and can be determined spectrophotometrically with arsenazo III Recoveries of rare earth metals (10 and 2.5 μg) added to Cu (0.1 and 0.5 g, respectively) were 97 and 107% (n = 10)	72

Table 17
DETERMINATION OF RARE EARTHS IN ACTINIDE MATRICES AFTER ION EXCHANGE SEPARATIONS[a]

Material	Ion exchange resin, separation conditions, and remarks	Ref.
Irradiated U	A) First column operation (Zeo-Karb® 225; BSS 52-80 mesh; H^+ form) Column: 12 × 0.5 cm ID containing 2 g of the resin a) 0.3 *M* HCl (20 mℓ followed by a 3 × 10 mℓ wash) (adsorption of rare earths and most other fission products; using a flow rate of 0.5 mℓ/cm^2 min; into the effluent pass Rh, some U, and other accompanying elements—the Rh is further purified by using the column operation [D]) (see below) b) 0.5% Oxalic acid (10 mℓ) (elution of Zr and Nb activities) c) 1 *M* NH_4Cl (10 mℓ) followed by 0.3 *M* HCl (10 mℓ) (elution of alkali and alkaline earth elements) d) 11 *M* HCl (200 mℓ) (elution of rare earths, U, and Pu); flow rate: 0.1 mℓ/cm^2 min B) Second column operation (Deacidite® FF; BSS 52-80 mesh; chloride form) Column: 15 × 1.7 cm ID containing 15 g of the resin (separation of the rare earths from U and Pu) This separation is effected by passing the eluate (d) (see above under [A]) through a column of this resin; U and Pu are adsorbed while the rare earths pass into the effluent; complete elution of the rare earths is achieved by the subsequent passage of 13.3 *M* HCl (30 mℓ); after several precipitation stages (in the presence of added Y) the individual lanthanides are separated by means of the following procedure C) Third column operation (Zeo-Karb® 225; NH_4^+ form) Column: 9 × 0.3 cm ID a) 0.5 *M* HNO_3 (100 μℓ) (adsorption of rare earths plus added Eu serving as a column ''marker'' to follow the separation) b) 0.5 *M* α-HIBA solution of pH 3.2 (this eluent is used until Y is eluted) c) Same eluent as (b) but of pH 3.37 (elution until Gd and Eu are eluted) d) Same eluent as (b) of pH 3.75 (elution of Sm and Nd) In a parallel operation with an equal portion of the rare earth solution (0.5 *M* in HNO_3) the same separation is carried out in the presence of added natural Nd, Gd, and Sm in order to be able to determine the required isotopes by isotope-dilutions mass-spectrometry; from all the eluates, the lanthanides are isolated by oxalate precipitation and determined by mass spectrometry D) Isolation of Rh (Deacidite® FF, BSS 52-80 mesh; chloride form) Column: 20 × 1.7 cm ID containing 20 g of the resin a) 0.3 *M* HCl (50 mℓ) (= effluent a) from (A), (adsorption of Rh; U and other accompanying elements pass into the effluent) (flow rate: 0.3 mℓ/cm^2·min) b) 0.3 *M* HCl (3 column volumes) (complete elution of U) c) 4 *M* HCl (125 mℓ) (elution of Rh) (flow rate: 0.5 mℓ/cm^2·min) To make absolutely certain that all of the U is removed, step (a) of the column operation outlined above under (A) is repeated (on a small column: 8 × 0.4 cm ID containing 1 g of the resin) after separation (D), and then Rh is determined spectrophotometrically; before the first column operation, the sample is dissolved in HNO_3 and nitrate is removed by repeated evaporations with 6 *M* HCl	32
Irradiated U	Dowex® 50-X8 (H^+ form) a) 0.1 *M* HCl - 0.2 *M* oxalic acid (adsorption of Ce, Sr; elution of Ru and Rh) b) 1 *M* HNO_3 (elution of Cs) c) 2 *M* HNO_3 (elution of Sr) d) 8 *M* HNO_3 (elution of Ce)	53,54

Table 17 (continued)
DETERMINATION OF RARE EARTHS IN ACTINIDE MATRICES AFTER ION EXCHANGE SEPARATIONS[a]

Material	Ion exchange resin, separation conditions, and remarks	Ref.
	Before this ion exchange step, U is removed by extraction into TBP-kerosine and Zr + Nb are separated by adsorption on silica gel	
	Radiochemical purities of >99% are attainable with this method	
U mill products	Amberlite® IR-120 resin (Na^+ form) (Reeve Angel Resin Circle — Grade SA-3, 3 cm)	93
	Following adsorption of the rare earths on a cation exchange resin filter from 0.1 *M* HNO_3, the filter is dried and the rare earths are determined by X-ray fluorescence; before this separation, the U is removed from the sample by TBP extraction, Y is added as a carrier, and the rare earths are precipitated as oxalates	
Pu	The sample (~100 mg) is dissolved in HCl-$HClO_4$ and from a solution consisting of 90% acetone and 10% 1 *M* HCl the Sc is adsorbed on Reeve-Angel SA-2 cation exchange resin paper; under these conditions Pu passes into the effluent; subsequently, the paper is dried and Sc is determined by X-ray spectrometry; the coefficient of variation was ≈2% when determining 50—400 μg of Sc	60
Th metal	Dowex® 50-X12 (100-200 mesh; H^+ form)	10
	a) Sorption solution which is <4 *M* in HCl (110 ± 10 mℓ) (adsorption of Th and rare earths)	
	b) 4 *M* HCl (first fraction) (0—2250 mℓ) (elution of common impurity elements)	
	c) 4 *M* HCl (second fraction) (2250—6500 mℓ) (elution of the rare earths)	
	d) 4 *M* H_2SO_4 (elution of Th)	
	The rare earth impurities were determined spectrophotometrically; before the separation, the sample (10 g) is dissolved in HCl and Y is added as a carrier	
Radioactive wastes	Dowex® 50-X4 (100-200 mesh; H^+ form)	52
	Column: 5 × 0.5 cm ID operated at a pumping rate of 6.9 mℓ/hr	
	a) Sample plus 0.1 *M* HCl - 0.02 *M* oxalic acid (6 mℓ) (adsorption of rare earths, Cs, Co, and Sr; into the effluent pass Ru, Sb, Zr, and Nb)	
	b) 0.5 *M* HNO_3 (10 mℓ) (elution of Cs)	
	c) 0.45 *M* HCNS (10 mℓ) (elution of Co)	
	d) 1.0 *M* HNO_3 (10 mℓ) (elution of Sr)	
	e) 2 *M* HNO_3 (10 mℓ) (elution of the rare earths [including Y])	
	For these sequential elutions a semiautomatic machine is used which feeds sample and eluents into a battery of ion exchange columns; one operator can separate 12 samples and count the resulting 60 fractions in an 8-hr day	

[a] See also Table 144 in the chapter on Actinides, Volume II and Table 19 in the chapter on Cadmium, Volume IV.

Table 18
CATION EXCHANGE SEPARATION OF Sc FROM OTHER ELEMENTS IN ORGANIC SOLVENT-HCl MEDIA

Elements separated	Ion exchange resin, separation conditions, and remarks	Ref.
Rare earths and many other elements	Dowex® AG 50W-X8 (100-200 mesh; H^+ form) Column of 0.5 cm ID containing 1 g of the resin pretreated with the THF-HCl eluent (5—10 mℓ) and operated at a flow rate of ~0.5 mℓ/min From 95% tetrahydrofuran (THF) - 5% 6 *M* HCl (10 mℓ) which is 0.1 *M* in trioctylphosphine oxide (TOPO), Sc is not retained by this resin, while, however, many other elements are strongly adsorbed; to these belong rare earth metals, alkali and alkaline earth metals, Al, Cr(III), Ni, and the trivalent trans-Pu elements, i.e., elements which do not form anionic chloride complexes; elution of the adsorbed rare earths and other elements can be effected with 6 *M* HCl following passage of 50—150 mℓ of the THF-HCl eluent to completely remove Sc This method has been used to separate milligram amounts of Sc from comparable quantities of the other elements; similarly, ion exchange paper chromatography on Amberlite® SA-2 (H^+ form) paper has been utilized for microscale separations of the RE using 95% THF - 5% 6 *M* HCl which is 0.5 *M* in TOPO;[62] Sc was found to migrate with R_f limits of 0.70—0.48, while the rare earths remain at or near the origin	61
Sc from Mn	Cationite KU-2 (H^+ form) a) 60% Acetone - 40% 2 *M* HCl (adsorption of Sc and elution of Mn) b) 4 *M* HCl (elution of Sc) The method was satisfactory in the tests with ratios of Mn/Sc of 10×10^3:1—1:7×10^3	58

Table 19
CATION EXCHANGE SEPARATION OF RARE EARTH ELEMENTS FROM ACCOMPANYING METAL IONS IN HCl SYSTEMS

Elements separated	Ion exchange resin, separation conditions, and remarks	Ref.
Rare earths from Al, Ga, In, Tl, Fe(III), Ti(IV), Be, U(VI), Mn(II), Co, Cu, Ni, Zn, and Cd	Dowex® 50W-X8 (H^+ form) a) ~0.1 *M* HCl (adsorption of rare earths and of other elements) b) 1.75 *M* HCl (elution of all elements listed in the first column except the rare earths) c) 4 *M* HCl (elution of the rare earths) (coeluted are Zr, Hf, Sr, and Ba) In the eluate, the rare earths including Sc were determined by gravimetry and titrimetry, respectively; the method has been used for the separation of milligram amounts of the elements	2
Rare earths from Al, Fe(III), Ti(IV), and other elements	Bio-Rad® AG 50W-X8 (200-400 mesh; H^+ form) Column of 2 cm ID containing 60 mℓ of the resin a) 0.1 *M* HCl (adsorption of the rare earths and other elements except Ti[IV], Tl[III], and Bi[III] which are adsorbed from 0.5 *M* HCl; in the case of Ti the sorption solution contained H_2O_2) (flow rate of 1.5 ± 0.3 mℓ/min) b) 3 *M* HCl (300 mℓ) containing 50% ethanol (elution of Al, Fe[III], Ti, Ga, In, Be, Mg, U[VI], Bi, Tl, Pb, Zn, Cu, Mn, and Ni) (for the complete elution of Al 500 mℓ of this eluent are required), same flow rate as with (a) c) 4 *M* HCl (500 mℓ) (elution of the rare earths) (Th is further retained while Zr, Hf, Ba, and Sr are coeluted) (flow rate 3.0 ± 0.3 mℓ/min) Zr and Hf can be separated from the rare earths and Sc by eluting the latter group with 0.25 *M* sulfuric acid from a column of Bio-Rad® AG1-X8 anion exchange resin (sulfate form) on which Zr and Hf are retained; Ba and Sr (as well as Ca, K, and Rb which may also be present partially in the rare earth eluate) can be eluted from a column of the cation exchange resin Bio-Rad® AG 50W-X8 with 1.75 or 2.0 *M* HNO_3, the rare earths and Sc being retained; the method has been used for the separation of milligram amounts of the elements; a less satisfactory separation of Al from the rare earths is achieved when 100-200 mesh cation exchange resin is used	8

Table 20
CATION EXCHANGE SEPARATION OF RARE EARTH METALS FROM ACCOMPANYING ELEMENTS IN CHLORIDE, NITRATE, SULFATE, AND OTHER MEDIA

Elements separated	Ion exchange resin, separation conditions, and remarks	Ref.
Y from Rb, Ce(IV), Be, Mg, Bi, Hg, V(IV), Cd, U(VI), In, Co, Ga, Ca, Sr, Ba, Zn, Sc, Fe(III), Al, Ni, and other elements	Dowex® 50W-X8 (50-100 mesh; Na^+ form) Column: 20 × 1.2 cm a) 0.5 *M* HCl (adsorption of Y and other elements; elution of Rb, Ce[IV], Be, Mg, Bi, Hg, and V[IV]) b) 1 *M* HCl (adsorption of Y; elution of Cd, U, and In) c) 2 *M* HCl (adsorption of Y; elution of Co and Ga) d) 1 *M* HNO_3 (adsorption of Y; elution of Ca, Sr, Ba, and Zn) e) 1 *M* H_2SO_4 (adsorption of Y; elution of Sc, Fe[III], and Al) f) 5% NaCl solution (adsorption of Y; elution of Ni) g) 5% Citric acid solution (adsorption of Y; elution of Ce[IV]) h) 1 *M* $(NH_4)_2SO_4$ (elution of Y after adsorption from any of the systems cited above) With other eluents, Y is eluted first so that the following separations can be carried out Y from Cr a) 1 *M* $(NH_4)_2SO_4$ (adsorption of Cr and elution of Y) b) 4 *M* HCl (elution of Cr) Y from Tl (I) and Pb a) 0.5 *M* ammonium acetate (adsorption of Tl and Pb; elution of Y) b) 4 *M* ammonium acetate (elution of Tl and Pb) Y from Th a) 4 *M* HCl (adsorption of Th; elution of Y) b) 8 *M* HCl (elution of Th) Y can also be separated from Ti, Zr, Hf, and Sb by selective sorption from citrate media; thus, these elements form anionic complexes in 5% citric acid solution of pH 2.2, while Y is not complexed and hence adsorbed on the cation exchanger from which it can be eluted with 1 *M* $(NH_4)_2SO_4$	29
Sc from Y and lanthanides, alkali metals, V, U(VI), Bi, alkaline earths, Fe(III), Zn, Cd, Hg, Cu, In, and Al	Dowex® 50-X8 (20-50 mesh; H^+ form) Column: 20 × 1.4 cm operated at a flow rate of 1 mℓ/min A) Separation of Sc from alkali metals, V, U(VI), and Bi a) 0.5 *M* HCl (200 mℓ) (adsorption of Sc and elution of the other elements) b) 1.5 *M* H_2SO_4 (200 mℓ) (elution of Sc) B) Separation of Sc from alkaline earth metals a) 1 *M* HNO_3 (adsorption of Sc and elution of alkaline earths) b) 1.5 *M* H_2SO_4 (200 mℓ) (elution of Sc) C) Separation of Sc from Y and lanthanides a) 1 *M* H_2SO_4 (250 mℓ) (adsorption of lanthanides and Y; elution of Sc) b) 4 *M* HCl (elution of lanthanides and Y) D) Separation of Sc from Fe, Zn, Cd, Hg, Cu, In, and Al a) 1 *M* HCl (200 mℓ) (adsorption of Sc; elution of the other elements) b) 1.5 *M* H_2SO_4 (200 mℓ) (elution of Sc) E) Separation of Sc from Co, Ni, Mn, Pb, and Ag a) 1 *M* NaCl (adsorption of Sc; elution of Co and Ni) b) 2 *M* NaCl (adsorption of Sc; elution of Mn[II] and Pb) c) 0.5 *M* $NaNO_3$ (elution of Ag) d) 1.5 *M* H_2SO_4 (200 mℓ) (elution of Sc) F) Separation of Sc from Th a) 4 *M* HCl (350 mℓ) (adsorption of Th; elution of Sc) b) 8 *M* HCl (300 mℓ) (elution of Th)	9

Table 20 (continued)
CATION EXCHANGE SEPARATION OF RARE EARTH METALS FROM ACCOMPANYING ELEMENTS IN CHLORIDE, NITRATE, SULFATE, AND OTHER MEDIA

Elements separated	Ion exchange resin, separation conditions, and remarks	Ref.
	G) Separation of Sc from Ti, Zr, and Sb(III) a) 5% Citric acid solution of pH 2.2 (adsorption of Sc; the other elements [as anionic citrate complexes] pass into the effluent) b) Water (to remove citric acid) c) 1.5 *M* H_2SO_4 (200 mℓ) (elution of Sc) These methods have been used to separate milligram amounts of the elements	
Rare earths from H_3PO_4	Bio-Rad® AG 50W-X8 (200-400 mesh; H^+ form); the volume of the resin bed is 45 mℓ a) 0.05 *M* HCl up to 6 *M* in H_3PO_4 (100 mℓ) (adsorption of the rare earths) b) 1.85 *M* HCl (300 mℓ) (elution of H_3PO_4) c) 4 *M* HCl (400 mℓ) (elution of the rare earths) The rare earths (Yb and La) were determined by AAS and spectrophotometry and the recoveries were >97 and 100%, respectively	30
La from Ti	Cation exchange resin a) Dilute H_2SO_4 solution ($\not>$1 *N*) containing a small amount of H_2O_2 (30%) (adsorption of La and Ti) b) 2% Oxalic acid solution (elution of Ti) c) HCl (1:2) (elution of La) The method was used to separate milligram amounts of elements	67

Table 21
DETERMINATION OF RARE EARTH ELEMENTS IN GEOLOGICAL MATERIALS AFTER CATION EXCHANGE SEPARATION IN NITRIC ACID MEDIA

Material	Ion exchange resin, separation conditions, and remarks	Ref.
Rocks, minerals, coal, and soil	A) First column operation (Bio-Rad® AG 50W-X8 (100-200 mesh; H^+ form) Column: 20 × 1 cm conditioned with 1 *M* HNO_3 and operated at a flow rate of ≃1 mℓ/min a) 1 *M* HNO_3 (50 mℓ) (adsorption of the rare earths and other elements) b) 2 *M* HNO_3 (130 mℓ) (elution of coadsorbed elements such as Al, Be, Ca, Mn, and U) c) 6 *M* HNO_3 (50 mℓ) followed by 8 *M* HNO_3 (50 mℓ) (elution of the rare earths) (Fe is coeluted) B) Second column operation (Bio-Rad® AGl-X8; 100-200 mesh; chloride form) (removal of Fe) From 8 *M* HCl (10 mℓ), Fe(III) is adsorbed on a column (5 × 1 cm) (flow rate: 3 mℓ/min) of this resin, while the rare earths pass into the effluent (following a wash with 30 mℓ 8 *M* HCl), in which they are determined by inductively coupled argon plasma-atomic emission spectrometry Before the cation exchange separation, the sample is dissolved either by treatment with HF-HNO_3-$HClO_4$ or by fusion with $LiBO_2$ (if the rare earths are suspected to be present in resistant minerals, such as zircons) For most determinations 95% confidence limits of 1—4% were observed In place of eluents (a), (b), and (c) solutions of 1 *M* HCl (50 mℓ), 2 *M* HCl (100 mℓ), and 8 *M* HCl (160 mℓ), respectively, can be employed for the separation; in this case complete separation of the rare earths from Fe(III) is achieved with the 2 *M* HCl eluent, but not from Sr, Ba, and some Ca which are coeluted with the rare earths using the 8 *M* HCl eluent	95,96
Water samples	Dowex® 50 (H^+ form) a) Water sample (1 ℓ) containing Ce-carrier (adsorption of Ce) b) 3 *M* HNO_3 (25 mℓ) (elution of ^{144}Ce) Then, the radioactive Ce is obtained by successive precipitations as the fluoride, the hydroxide, and finally as the oxalate; the Ce-oxalate is weighed to determine carrier loss and is then counted The sensitivity of detection by this procedure is $\simeq 10^{-8}$ μCi/mℓ	99

Table 22
DETERMINATION OF THE RARE EARTH ELEMENTS IN INDUSTRIAL MATERIALS AFTER CATION EXCHANGE SEPARATION IN NITRIC ACID MEDIA[a]

Material	Ion exchange resin, separation conditions, and remarks	Ref.
U mill products	Amberlite® IR-120 (Na^+ form) (in the form of Reeve Angel Resin Circle—Grade SA-2,3 cm) The rare earths are adsorbed by filtering the solution (0.1 *M* in HNO_3) seven times through the resin paper on which these elements are then determined by X-ray fluorescence Before carrying out this ion exchange separation, U is removed by TBP extraction, Y is added as a carrier, and the rare earths are precipitated as oxalates	93
Thin films of Gd-Fe-oxide ($Gd_3Fe_5O_{12}$) and Gd-Fe (Gd_3Fe_5)	Amberlite® IR-120 (Na^+ form) The sample of thin film is dissolved in a 1:1:3 mixture of acetic acid, nitric acid, and water and after dilution to 50 mℓ, the Gd and Fe(III) are adsorbed on disks of paper impregnated with the cation exchange resin; to ensure complete adsorption, the solution is stirred for 3 hr and this collection step is repeated twice; each disk is then sandwiched between 0.02-mm sheets of Mylar® film for the X-ray fluorescence determination of Gd and Fe; this method of analysis allows the preparation of standards by normal solution techniques without the problem of matrix effects	97
Gd used as a moderator poison for a water reactor	Zeo-Karb® 225 (H^+ form) From a dilute nitric acid solution of pH 5, Gd is adsorbed on this resin and then it is determined by the measurement of the absorption of thermal neutrons from a Po-Be source; direct neutron-absorption determination of Gd at concentration down to 0.1 ppm in aqueous media is less suitable	92

[a] See also Table 10 in the chapter on Gallium, Volume VI.

Table 23
SEPARATION OF THE RARE EARTHS FROM OTHER ELEMENTS USING CATION EXCHANGE IN NITRIC ACID MEDIA

Elements separated	Ion exchange resin, separation conditions, and remarks	Ref.
Sc, Y, and lanthanides from Ba, Sr, and other elements	Bio-Rad® AG 50W-X8 (200-400 mesh; H^+ form) Column of 2.4 cm ID containing 90 mℓ of the resin and operated at a flow rate of 2.5 ± 0.3 mℓ/min a) 0.1—0.5 *M* HNO_3 (adsorption of the rare earths, Ba, Sr, and other elements) b) 2 *M* HNO_3 (500 mℓ) (elution of Mg, Ca, Sr, Ba, Pb, Bi, Zn, Mn, and U) c) 4 *M* HNO_3 (400 mℓ) (elution of rare earths) The rare earths were determined gravimetrically or by atomic absorption spectrophotometry; the method was used to separate milligram amounts of the elements	94
Rare earths from divalent metal ions	Dowex® 50W-X8 (100-200 mesh; H^+ form) Column of 1.2 cm ID containing the resin to a height of 16 cm a) 0.05 *M* solution of the nitrate or perchlorate salts of the metal ions (25 mℓ) (adsorption of the rare earths and other elements) b) 1.5 *M* HNO_3 (200 mℓ) (elution of divalent ions such as Ca, Mg, Sr, and Ni) (flow rate: 2 mℓ/min) c) Water (as a rinse) d) 3 *M* HNO_3 (200 mℓ) (elution of the rare earths and other trivalent ions such as Al and Fe[III]) In the eluates, the rare earths and the other elements were determined by EDTA-titration	98

Table 24
DETERMINATION OF RARE EARTH ELEMENTS IN INDUSTRIAL MATERIALS AFTER SEPARATION BY CATION EXCHANGE IN SULFURIC ACID AND AMMONIUM SULFATE MEDIA

Material	Ion exchange resin, separation conditions, and remarks	Ref.
Metallic U	A) First column operation (cationite KU-2; 300-350 mesh; H^+ form) Column: 20 × 0.8 cm operated at a flow rate of ~1.0 mℓ/min a) 1 *N* H_2SO_4 (10 mℓ) (adsorption of the rare earths and elution of U) b) 1 *N* H_2SO_4 (five to six times the free volume of the column) (elution of U) c) Water (removal of H_2SO_4) d) 7 *M* ascorbic acid solution of pH 4—4.5 (~15 mℓ) (elution of residual U) e) Water (removal of ascorbic acid) f) 4—5 *M* HCl (elution of the rare earths) Subsequently, the rare earth impurities are irradiated with neutrons and after dissolution of the sample in HCl the rare earths are isolated by coprecipitation on ferric hydroxide B) Second column operation (Dowex® 1-X8, ~400 mesh; chloride form) The ferric hydroxide (see above) is dissolved in conc HCl (few drops) and Fe is adsorbed on a column (2 × 0.25 cm) of this resin while the rare earths pass into the effluent C) Third column operation (cationite KU-2, ~400 mesh; NH_4^+ form) (separation of individual rare earths) Microcolumn: 12 × 0.2 cm a) Dilute acid solution of pH 3—4 (adsorption of the rare earths) b) 0.17 *M* ammonium α-HIBA of pH 4.65 (elution at room temperatures of Yb + Lu, Ho + Dy, Gd, Eu, Sm, etc.) In the various fractions the rare earths were determined radiometrically	105
Metallic U	Cationite KU-2 (H^+ form) a) 0.5 *M* H_2SO_4 (adsorption of the rare earth impurities and elution of U) b) 4—5 *M* HCl (elution of the rare earths) Further purification was achieved on a column of Dowex® 1-X8 and finally Gd, Eu, and Sm were determined using a luminescence method; in a similar method the isolated Eu impurity was determined spectrophotometrically[107]	106
High-purity Th	Dowex® 50-X8 (100-200 mesh; H^+ form) Column: 20 × 1.2 cm a) 0.5 *M* H_2SO_4 - 0.12 *M* $(NH_4)_2SO_4$ (120 mℓ) (adsorption of the rare earths; the bulk of Th passes into the effluent) b) Water (as a rinse) c) 6.5 *M* HNO_3 (~100 mℓ) (elution of the rare earths) The rare earths were determined spectrophotometrically; before the separation, the sample (1 g) is dissolved in 6 *M* HNO_3 (in the presence of Sc added as internal standard); the detection limits range from 0.01 μg for Yb to 0.5 μg for Pr and Nd	108
Zr-rare earth alloys	Dowex® 50W-X8 (100-200 mesh; H^+ form) Column: 30 × 2.5 cm filled with the resin to a height of 18—20 cm and operated at a flow rate of ~4 mℓ/min a) 10% H_2SO_4 (250 mℓ) (adsorption of the rare earths and elution of Zr) b) 2 *M* $(NH_4)_2$ SO_4 of pH 1.2 (200 mℓ) (elution of the rare earths)	109

Table 24 (continued)
DETERMINATION OF RARE EARTH ELEMENTS IN INDUSTRIAL MATERIALS AFTER SEPARATION BY CATION EXCHANGE IN SULFURIC ACID AND AMMONIUM SULFATE MEDIA

Material	Ion exchange resin, separation conditions, and remarks	Ref.
	The rare earths were determined gravimetrically in the range of 0.10—90%; the rare earth recovery (Eu) averaged 97.6%; before the separation, the sample is dissolved in 10% H_2SO_4 containing a few drops of HF	10
Th, U, and Zr metals and compounds	Dowex® 50-X12 (H^+ form) a) Ammonium sulfate solution of the sample (of pH 3.0 ± 0.5 in the case of Th and U samples and of pH 2.0 ± 0.5 when Zr is present) (adsorption of the rare earths; into the effluent pass Th, U, and Zr as anionic sulfate complexes) b) 0.75 *M* $(NH_4)_2SO_4$ (pH 3 ± 0.5) (elution of residual Th, U, and Zr) c) 2 *M* $(NH_4)_2SO_4$ (elution of rare earth elements) The rare earth impurities were determined spectrographically	112
Al alloys	Cationite KU-2 (Na^+ form) (3—5-g column) a) Ammonium sulfate solution of the sample (pH 6) (15 mℓ)(adsorption of Ce, Mn, and Al) b) Ammonium acetate buffer solution of pH 6.2 containing 10% of H_2O_2 (50 mℓ) (formation of a peroxy complex of Ce with the resin; see also Table 118) c) Water (10—20 mℓ) (as a rinse) d) 1 *M* KCl or KNO_3 (150 mℓ) (elution of Mn) e) 1:10 H_2SO_4 (elution of Ce and Al) Coeluted Al does not interfere when Ce is determined titrimetrically using the Ag-persulfate method; for the titrimetric determination of Mn, EDTA was used utilizing murexide as indicator; before the cation exchange separation, the sample (≃0.4 g) is dissolved in dilute H_2SO_4 and ammonia is used to adjust the pH of the solution to the required value of 6; errors are $<1.7\%$ for Ce and $<\pm 0.73\%$ for Mn	111
U hexafluoride	Strongly acidic cation exchange resin (80-100 mesh; H^+ form) Column: 7 × 0.5 cm a) Sample solution (5 mℓ) which is <0.2 *M* in HCl (adsorption of rare earths and other impurities; the anionic fluoride complex of U passes into the effluent) (flow rate: ≃1 mℓ/min) b) 0.5 *M* H_2SO_4 (10 mℓ) (elution of most of the U) (flow rate: 0.5 mℓ/min) c) Water (3—4 mℓ) (as a rinse) d) 4 *M* HCl (8 mℓ) (elution of the rare earths and impurities) The eluate (d) is evaporated to dryness, the residue is dissolved in HCl (1 drop) and water (1 mℓ), and the solution is transferred to the same column Impurities are washed through with 1.2 *M* HCl (15 mℓ), the rare earth metals are eluted with eluent (d) and then determined spectrographically; before the separation the aqueous solution of the sample (5 g) is treated with internal standard (0.5 μg Tm) and evaporated to crystals with HCl in three stages (20, 5, and 5 mℓ); the residue is dissolved in water (50 mℓ) to prepare solution (a) Contents of Gd, Dy, Sm, and Eu between 0.01 and 0.25 ppm can be determined with a mean coefficient of variation of <26%. The average recovery is 100—112%	

Table 25
CATION EXCHANGE SEPARATION OF Sc FROM THE RARE EARTHS AND OTHER ELEMENTS IN DILUTE SULFURIC ACID MEDIA

Elements separated	Ion exchange resin, separation conditions, and remarks	Ref.
Sc from the rare earths, Al, Ca, Cd, Co and many other elements	Dowex® 50-X8 (100-200 mesh; H^+ form) Column: 0.9 cm ID containing 5 g of the resin and operated at a flow rate of 0.5—1.5 mℓ/min a) Dilute mineral acid (adsorption of Sc and accompanying elements) b) 0.025 *M* H_2SO_4 - 0.3 *M* $(NH_4)_2SO_4$ (~50 mℓ) (elution of Sc) c) 10% H_2SO_4 (10 mℓ) (elution of Cr[III]) d) 0.025 *M* H_2SO_4 - 1 *M* $(NH_4)_2SO_4$ (40 mℓ) (elution of Fe[III] and In) e) 3—6 *M* HCl (elution of Al, Ca, Cd, Co, Cu, Ga, Mg, Mn, Ni, Zn, and the lanthanides) The method has been used for the separation of milligram amounts of the elements which were determined titrimetrically or gravimetrically	110
Sc from Y, La, Ce, Sm, Er, and Yb	Bio-Rad® AG 50W-X8 (100-200 mesh; H^+ form) Column: 50 × 2 cm containing 20 g of the resin and operated at a flow rate of 3.0 ± 0.2 mℓ/min a) Dilute mineral acid solution (<0.3 *N* free acid) (adsorption of Sc, Y, and the lanthanides) b) 2 *N* H_2SO_4 (450 mℓ) (elution of Sc) c) 0.1 *M* HCl (50 mℓ) (removal of H_2SO_4) d) 4 *M* HCl (500 mℓ) (elution of Y and the lanthanides) Sc and the other rare earths were determined by titration with EDTA; the method was used for the separation of milligram amounts of the elements By using volumes of 70 and 150 mℓ of the eluents (b) and (d), respectively, the separation of smaller quantities of the elements can be performed in a small column (35 × 1.15 cm) containing 5 g of the resin and operated at a flow rate of 0.6 ± 0.1 mℓ/min	104
Sc from Y and La	Dowex® 50W-X4 (50-100 mesh; H^+ form) Column: 30 × 0.95 cm containing 11.1 g of the resin a) Very dilute HNO_3 solution (adsorption of Sc, Y, and La) b) 0.2 *M* H_2SO_4 or 1 *M* monochloroactic acid of pH 2.0—2.2 (elution of Sc) c) 1 *M* monochloroactic acid of pH 3 (elution of Y) d) 15% HCl solution (elution of La) e) 1 *M* H_2SO_4 (elution of Y or La) The Sc_2O_3 obtained by treatment of the eluate has a purity of 99.99%	378

Table 26
DETERMINATION OF RARE EARTH ELEMENTS IN INDUSTRIAL MATERIALS AFTER CATION EXCHANGE SEPARATION IN HF MEDIA[a]

Material	Ion exchange resin, separation conditions, and remarks	Ref.
U compounds	Dowex® 50-X8 (50-100 mesh; H^+ form) Column of 0.5 in. ID containing 10 g of the resin conditioned with 2.5 *M* HF (200 mℓ) From 2.5 *M* HF, the rare earths (Eu, Gd, Dy, Yb, Er, Nd, and Sm) are adsorbed on this resin while U (20 g) passes into the effluent as anionic fluoride complex; after this separation the resin is ashed at 800°C, mixed with graphite, and the analysis is completed spectrographically; 0.01—100 ppm of these rare earths in U may be determined (essentially complete removal of U is achieved after the passage of 15—20 column volumes of 2.5 *M* HF) (one column volume = 9.15 mℓ); a similar technique has been utilized for the determination of rare earth impurities in unirradiated U oxide nuclear fuel[507]	115
Zr for nuclear reactors	Dowex® 50 (50-60 mesh; H^+ form) Column of 1 in. ID containing a 36-in. resin bed a) Dilute HF solution of the sample (2000 mℓ) (adsorption of rare earths and Y-carrier; Zr as ZrF_6^{2-} passes into the effluent together with HfF_6^{2-} and TiF_6^{2-}) (flow rate: 10 mℓ/min) b) water (removal of acid) (until the effluent is neutral) c) 6 *M* HCl (300 mℓ) (elution of the rare earths + Y-carrier) After additional separation steps utilizing solvent extraction and precipitation, the rare earths (Gd, Tb, Dy, Ho, and Sm <2 ppm each) were determined spectrographically; quantitative recoveries are obtained within the limits of experimental error of $\pm 10\%$; before the separation the sample (100 g) is dissolved in HF and Y-carrier is added; the procedure is applicable to the determination of rare earths in any element that dissolves in HF to form a stable anionic complex (U, Be, B, Ge, Ti, etc.)	116
Complex fission product mixtures	Cationite KU-2-X6 (H^+ form) a) 1 *M* HF (adsorption of the rare earths, alkali and alkaline earth metals, and other elements; into the effluent pass U, Ta, Hf, and W) b) 5 *M* HF (elution of Cr, Fe, Ca, As, Cd, and In) c) 10 *M* HF (elution of Na, Rb, Cs, Mn, Pb, and Bi) d) 15 *M* HF (elution of Ni, Zn, Tc, and Hg) e) 20 *M* HF (elution of Co, Cu, and Tl) f) Conc HNO_3 + H_3BO_3 (elution of rare earths and alkaline earth elements) The complex fission product mixtures to which this method was applied were obtained by using Te, Hf, W, and U as proton targets	117

[a] See also Table 2 in the chapter on Tungsten, Volume IV.

Table 27
CATION EXCHANGE SEPARATIONS OF RARE EARTH METALS FROM EACH OTHER AND FROM OTHER ELEMENTS IN PHOSPHATE MEDIA

Elements separated	Ion exchange resin, separation conditions, and remarks	Ref.
Sc from Y and other elements	Dowex® 50W-X4 (50-100 mesh; H^+ form) Column: 30 × 0.95 cm containing 11.1 g of the resin and operated at a flow rate of 1 mℓ/min a) Very dilute HNO_3 or H_2SO_4 solution of the elements to be separated (adsorption of Sc, Y, and accompanying elements) b) 0.5 or 1 *M* H_3PO_4 (390 and 200 mℓ, respectively) (selective elution of Sc) c) 0.1—0.5 *M* HNO_3 (removal of H_3PO_4) d) 15% HCl solution (elution of Y) The method has been used to separate milligram and gram amounts of the elements	508
Ce(IV), Ni, Co, and Cu	Dowex® 50W-X12 (50-100 mesh; H^+ form) Column of 1 cm ID containing 10—12 g of the resin and operated at a flow rate of 5—10 mℓ/min a) <0.4 *M* H_2SO_4 containing the polyphosphates of Ce(IV), Ni, Co, and Cu (adsorption of Ni, Co, and Cu; the Ce passes into the effluent) b) 10% HCl (100 mℓ) (elution of Ni, Co, and Cu) This method was used for the separation of milligram amounts of the elements which were determined titrimetrically The polyphosphates are formed in the eluent (a) by the addition of an excess of 1 *M* Na polyphosphate	509

Table 28
CATION EXCHANGE SEPARATION OF Sc FROM RARE EARTHS AND OTHER ELEMENTS IN THIOCYANATE MEDIA

Elements separated	Ion exchange resin, separation conditions, and remarks	Ref.
Sc from rare earths, Th, Fe, Al, Ti, and Zr	Diaion® SK1 (H^+ form) (equivalent to Dowex® 50-X8) (100-200 mesh) Column of 1.1 cm ID containing 10 g of the resin (height of resin bed: 16 cm); a flow rate of 1.0 ± 0.2 mℓ/min is maintained throughout a) Sample solution (adsorption of Sc, rare earths, and the other elements) b) 1 *M* NH_4SCN - 0.5 *M* HCl (~380 mℓ) (elution of Sc) c) 2 *M* NH_4SCN - 0.5 *M* HCl (200 mℓ) (elution of the rare earths) Al, Ti, Fe, and Zr can be eluted before the Sc by using the following eluents a) 1 *M* NH_4SCN - 0.5 *M* HCl (170 mℓ) (elution of Al) b) 0.5 *M* NH_4SCN - 0.5 *M* HCl (80 mℓ) (elution of Ti and Fe) c) 1 *M* NH_4SCN - 0.5 *M* HCl (60 mℓ) or 1 *M* NH_4SCN - 1 *M* HCl (elution of Zr) These procedures have been used for the separation of microgram and milligram amounts of the elements which were determined by spectrophotometry or titrimetry	120,150

Table 29
CATION EXCHANGE SEPARATION OF Sc FROM OTHER ELEMENTS IN ACETONE-HBr MEDIA

Ion exchange resin	Separation conditions and remarks	Ref.
Bio-Rad® AG 50W-X8 (200-400 mesh; H^+ form)	Column: 19 × 2.0 cm containing 60 mℓ of the resin and operated at a flow rate of 2.0 ± 0.5 mℓ/min (for Cd, In, and Al) and 3.0 ± 0.5 mℓ/min for the other elements a) 0.1 *M* HBr (~200 mℓ) (adsorption of Sc, Cd, In, Ga, and Al; elution of Au[III]) b) 50% Acetone - 0.2 *M* HBr (~200 mℓ) (elution of Cd) c) 60% Acetone - 0.5 *M* HBr (~200 mℓ) (elution of In) d) 90% Acetone - 0.5 *M* HBr (~250 mℓ) (elution of Ga) e) 30% Acetone - 3.0 *M* HBr (~300 mℓ) (elution of Al) f) 4 *M* HCl (~300 mℓ) (elution of Sc) In the eluates, the elements were determined spectrophotometrically or by atomic absorption measurements The method was used for the separation of millimole amounts of the elements	126

Table 30
COMPARISON OF SEPARATION FACTORS OF RARE EARTHS AND TRANS-Pu ELEMENTS IN HYDROXYCARBOXYLIC ACID SOLUTIONS (DOWEX® 50)

Elements	α-HIBA (25°C)	Lactic acid (90°C)	Citric acid (87°C)
Lu-Yb	1.54	1.31	1.31
Yb-Tm	1.70	1.43	1.53
Tm-Er	1.71	1.47	1.60
Er-Ho	1.73	1.52	1.50
Ho-Dy	1.80	1.38	1.37
Dy-Tb	2.30	1.56	1.40
Tb-Gd	2.40	1.73	1.37
Gd-Eu	1.65	1.22	1.12
Eu-Sm	1.88	1.28	1.19
Sm-Pm	2.13	1.32	1.38
Pm-Nd	1.54	1.33	1.09
Nd-Pr	1.68	1.36	1.18
Pr-Ce	1.85	1.73	1.69
Ce-La	1.97	1.83	—
Am-Cm	—	1.21	1.20
Cm-Bk	—	1.54	1.63
Bk-Cf	—	1.58	1.42
Cf-Es	—	1.24	1.2
Es-Fm	—	1.77	1.2
Fm-Md	—	1.3	1.18

Table 31
DETERMINATION OF RARE EARTH ELEMENTS IN GEOLOGICAL MATERIALS AFTER CATION EXCHANGE SEPARATION IN α-HIBA MEDIA

Material	Ion exchange resin, separation conditions, and remarks	Ref.
Rocks	Dowex® 50W-X12 (−400 mesh) Column: 2.8 × 0.2 cm a) 0.03 *M* HCl (~50 μℓ) (adsorption of the rare earths) b) 0.15—0.30 *M* α-HIBA at controlled, but increasing pH (3.95—4.92) (concentration and pH-gradient elution of individual lanthanides) (flow rate: 3 drops/min) Good separation between rare earths with atomic numbers that differ by three or more can be accomplished in ~10 min; in the case of adjacent rare earths, less complete but usable separation could be obtained in ~16 min The eluted fractions were analyzed by spectrometry Prior to the cation exchange separation step, the sample (0.1—1 g) is irradiated with neutrons and fused with Na_2O_2 in the presence of La-carrier and ^{152}Eu and ^{154}Eu tracer for determining chemical yield; then the rare earths are purified by a series of fluoride and hydroxide precipitations This method has been used for the determination of rare earth elements in samples such as monazite, granite G-1, and diabase W-1	187, 188
Rocks (basalt, shale, and chondrites)	Dowex® 50W-X4 (200-400 mesh; NH_4^+ form) The rare earths are adsorbed on a slurry (about 2 mℓ) of the resin which is centrifuged out (after equilibrating it with the solution [few milliliters] of the rare earths for 2 min) and transferred to the top of the resin bed in the ion exchange column (60 × 0.8 cm ID) using 0.5 *M* α-HIBA of pH 3.1 (2 × 2 mℓ) as a rinse; subsequently, the rare earths are separated by pH-gradient elution using 0.5 *M* α-HIBA solutions of pH 3.1 and 4.05 to cover this pH range (flow rate: 6—8 drops/min); the elution of La is complete at the end of 16 hr In the fractions obtained, the rare earths are precipitated as oxalates and determined radiometrically Before this ion exchange separation, the sample (1 g) is irradiated with neutrons, fused with Na_2O_2 in the presence of rare earth carriers, and the rare earths are preconcentrated and purified by a sequence of hydroxide and oxalate precipitations; the average precision for duplicate analysis is better than ±4% mean deviation for a shale with a total rare earth content of 200 ppm	144
Marine sediments	Dowex® 50 On a column of this resin individual lanthanides are separated by gradient elution with α-HIBA following neutron irradiation of the sample and radiochemical purification based on hydroxide and oxalate precipitations	241

Table 32
DETERMINATION OF RARE EARTHS IN MONAZITE AND OTHER MATERIALS AFTER CATION EXCHANGE SEPARATION IN α-HIBA MEDIA

Material	Ion exchange resin, separation conditions, and remarks	Ref.
Monazite	Bio-Rad® AG 50W-X8 (NH_4^+ form) Separations of the rare earths are effected by elution chromatography on a column (19 × 1 cm) of this resin using 0.3 *M* α-HIBA at a flow rate of 0.3—0.5 mℓ/min; gradient elution was also carried out by changing the pH from 3—5; all the rare earths except Dy and Y could be separated; before the separation, Ce, which comprises 40—50% of the total rare earth metals, is removed by solvent extraction	145
Monazite sand	LiChromosorb® KAT cation exchange resin (NH_4^+ form; 10 μm) In a stainless-steel column (25 × 0.46 cm ID) containing this resin (~3 g) pretreated with the eluent (35 mℓ) 14 rare earth elements (about 4 μg each) are separated within 40 min by HPLC using concentration gradient elution with 0.04—0.48 *M* α-HIBA; a flow rate of 1 mℓ/min is maintained throughout The rare earths were determined by UV spectrophotometry following reaction with arsenazo III; after dissolution of the sample Th is separated by the hexamine or urea method and finally the rare earth solution is diluted with water to ~500 ppm; the same method is applicable to the determination of rare earth impurities in high-purity oxides of La, Ce, Sm, and Y	174

Table 33
RADIOANALYTICAL DETERMINATION OF RARE EARTHS IN FERROUS MATERIALS AFTER CATION EXCHANGE SEPARATION IN α-HIBA MEDIA

Material	Ion exchange resin, separation conditions, and remarks	Ref.
High-purity Fe	Dowex® 50-X8 (H^+ form) a) Aqueous solution of the rare earths and other elements (adsorption of the rare earths and accompanying elements) b) 0.5 *M* NH_4Cl - 0.25 *M* NaCl (elution of Na, K, Rb, and Ce; the individual elements of this group are separated on a separate column of this resin using 0.1 *M* HCl as the eluent) c) 0.48 *M* α-HIBA of pH 4.15 (elution of the rare earths; if necessary, the individual rare earth elements can be fractionated on a separate column of the resin using ammonium lactate of pH 3.2—5.6 as the eluent) d) 0.8 *M* α-HIBA of pH 4.15 (elution of Ca) e) 1 *M* α-HIBA of pH 5 (elution of Sr) f) 6 *M* HCl (elution of Ba) In the various elutates, the elements were determined radiometrically Prior to the ion exchange separation step, the sample (1 g) is irradiated with neutrons, dissolved in HNO_3, and Fe as well as many other elements are removed by solvent extraction procedures To isolate Ni, Co, and Cd the diethyldithiocarbamates of these elements are first extracted with chloroform and then the elements are separated from each other using the following anion exchange separation step (on a column of Dowex® 1-X8; chloride form) a) 12 *M* HCl (adsorption of Co and Cd; elution of Ni) b) 4 *M* HCl (elution of Co) c) Water (elution of Cd) By using the same solvent extraction procedure at pH 8, Mn and Zn pass into the chloroform and subsequently can be separated on a column of the same anion exchange resin using the following eluents a) 6 *M* HCl (adsorption of Zn and elution of Mn) b) Water (elution of Zn)	242
Steel	Cationite KU-2 (~400 mesh; NH_4^+ form) Column: 10 × 0.25 cm operated at room temperature and at a flow rate of 1.2 mℓ/cm² min a) 0.1 *M* HCl (1—2 mℓ) (adsorption of the rare earths carried out under static conditions with a small portion of the resin; then the resin is washed with 1 *M* NH_4Cl and water and transferred to the top of the resin bed in the column) b) 0.28 *M* ammonium α-HIBA of pH 4.75 (elution of Nd and Pr) c) 0.35 *M* ammonium α-HIBA of pH 4.75 (elution of Ce) d) 0.7 *M* ammonium α-HIBA of pH 4.75 (elution of La) In the collected fractions the rare earths (Ce, La, Pr, and Nd) were determined radiometrically Before this cation exchange separation, the sample (0.2—0.5 g) is irradiated with neutrons and dissolved in aqua regia in the presence of La-carrier; the Fe is then removed by extraction into diethyl ether from 6 *M* HCl and the rare earths are purified by the fluoride-hydroxide precipitation cycle, repeated twice The method is suitable for steels with individual rare earth metal contents from 0.1 ppm to 0.01%; the coefficients of variation are <8—14%	175

Table 34
DETERMINATION OF RARE EARTH FISSION PRODUCTS IN NUCLEAR FUELS AFTER CATION EXCHANGE SEPARATION IN α-HIBA MEDIA

Ion exchange resin	Separation conditions and remarks	Ref.
Dowex® 50-X4 (−400 mesh; NH_4^+ form) and Dowex® 50-X8 (100-200 mesh; H^+ form)	A) First column operation (−X4 resin) A slurry (~0.1 mℓ) of this resin is added to the 1 *M* HCl solution of the rare earths, and after a few minutes the resin is transferred to a column (15 × 0.8 cm) of the same resin; subsequently, individual lanthanides are separated with 0.5 *M* α-HIBA first at pH 3.5 and then at pH 3.9, collecting the fractions (<0.25 mℓ) on an automatic collector that incorporates a γ-ray monitor; the retained fractions of each element are then combined, diluted with water, and after acidification with 6 *M* HCl (1 drop) the following separation is carried out B) Second column operation (−X8 resin) (1.5 × 0.5-cm microcolumn; e.g., in a small eye dropper) a) Weakly acid fraction (adsorption of the rare earth element) b) Water (1 mℓ) followed by 1 *M* HCl (3 mℓ) (removal of α-HIBA) c) 6 *M* HCl (3 mℓ) (elution of the rare earth element) The rare earth elements (lanthanide fission products) thus isolated and purified were determined by mass analysis; filament loading is greatly facilitated in the absence of α-HIBA so that this second cation exchange separation serves the purpose of removing this complexing agent Before the first cation exchange separation, the lanthanides (in the presence of carriers) are coprecipitated with $Ni(OH)_2$, separated from other elements by extraction chromatography on HDEHP columns, and finally reprecipitated with Ni-hydroxide which is dissolved in 1 *M* HCl (minimum amount) The method has been used to determine the concentration and distribution of microgram levels of fission product Ce, Nd, and Sm in ^{233}U, ^{235}U, and ^{239}Pu nuclear fuels; standard deviations for Ce, Nd, and Sm were <0.5 and 0.8% for the 1—5- or 0.1—1-μg ranges, respectively.	146

Table 35
SEPARATION OF RARE EARTHS FROM NEUTRON-IRRADIATED INDUSTRIAL MATERIALS USING CATION EXCHANGE IN α-HIBA MEDIA

Material	Ion exchange resin, separation conditions, and remarks	Ref.
Neutron irradiated Sm_2O_3	A) First column operation (Bio-Rad® AG 50W-X12; 200-400 mesh; NH_4^+ form) From 0.1 *M* HCl the rare earth activities are sorbed on a small batch of the resin which is then transferred to the top of the resin column (7.0 × 1.25 cm ID); subsequently, the rare earths are eluted at 80°C with 0.4 *M* α-HIBA of pH ~4 using a flow rate of ~4 drops/min; approximately 35 fractions are collected, and those containing ^{145}Pm are combined and ^{60}Co contamination is removed by the following separation step B) Second column operation (Bio-Rad® AG1-X8; 100-200 mesh; chloride form) From 9 *M* HCl, Co is adsorbed on a column of this resin, while ^{145}Pm passes into the effluent; a ^{145}Pm fraction virtually free from any ^{145}Sm is obtained by an additional cation exchange separation on a 34 × 0.5-cm column Prior to the cation exchange separation step, the sample of Sm-oxide (isotopically enriched to >94% in ^{144}Sm) is irradiated with neutrons and allowed to decay for 1 year in order to permit the growth of the ^{143}Pm daughter activity; afterwards the target was dissolved in 0.1 *M* HCl This technique was utilized to produce millicurie amounts of isotopically pure ^{145}Pm for Mössbauer applications	199
Neutron-irradiated Nb and Zr metals	Dowex® 50-X8 (200-400 mesh; H^+ form) Column: 14 × 0.6 cm preconditioned with eluent (b) and operated at a flow rate of 0.3 mℓ/min a) Very dilute acid solution of the sample (1 mℓ) (adsorption of Y) b) 0.5 *M* α-HIBA of pH 1.80 (6 mℓ) (elution of Nb or Zr) c) 0.25 *M* α-HIBA of pH 3.8 (20 mℓ) (elution of Y) The Y was determined radiometrically Before this separation, the sample (40 mg) is irradiated with neutrons and dissolved in HNO_3-HF in the presence of Y-carrier	243

Table 36
DETERMINATION OF RARE EARTHS IN MIXTURES OF FISSION PRODUCTS AFTER CATION EXCHANGE SEPARATION IN α-HIBA MEDIA

Material	Ion exchange resin, separation conditions, and remarks	Ref.
U fission products	Dowex® AG50W-X4 (−400 mesh; NH_4^+ form) Column: 27 in. × 8 mm ID containing the resin to a height of ~ 25 in. and operated at room temperature a) Very dilute HNO_3 solution of the sample (adsorption of the rare earths on a slurry [1 mℓ] of the resin which is then transferred to the top of the resin bed in the column) b) 0.5 *M* α-HIBA solution (pH-gradient elution with pH increasing from 3.4— 4.0 at a rate of 0.025 pH unit per hour); with an elution rate of 1 drop every 17—19 sec (~9 mℓ/hr) the elution is complete in ≃30 hr, each element requiring about 2 hr for the elution of 10 mg In the fractions obtained, the individual rare earth metals are located by addition of oxalic acid and determined radiometrically Before the cation exchange separation, the rare earths are isolated, as a group (in the presence of carrier for each element and Pm as a tracer), using a series of radiochemical operations; these include, in this order, fuming with $HClO_4$, a Zr-phosphate scavenging precipitation, two precipitations of rare earth fluorides, a $BaSO_4$ scavenging precipitation, precipitation of rare earth hydroxides, use of anion exchange in conc HCl, and reprecipitation of rare earth hydroxides For the determination of only one or two adjacent rare earths, it is convenient to use an eluent of a single pH; for example, in the determination of ^{147}Nd, 0.5 *M* α-HIBA of pH 3.73 was used	147
Rare earth fission products	Dowex® 50-X8 (100-200 mesh; NH_4^+ form) The very dilute HNO_3 solution (30 mℓ) of the rare earths is equilibrated (5 min) with a small batch (1 mℓ) of the resin which is then transferred to the top of the cation exchange column (30 in. × 7 mm ID); then the rare earths are eluted at a rate of 1 drop/7 sec by means of concentration gradient elution; the eluent is ammonium α-HIBA of pH 5.2, the molarity of which is kept at 0.1 *M* for 5 hr, then increased at 0.0085 *M*/hr until a value of 0.4 *M* is reached; in the eluate fractions the rare earths were determined radiometrically; good separations (less than 0.01% of contamination) of Y, Tb, Gd, Eu, Sm, Pm, Nd, Pr, Ce, and La are achieved, with peak-to-trough ratios of 10^2—10^4	176
	Dowex® 50-X8 (−400 mesh, NH_4^+ form; and 100-200 mesh; H^+ form A) First column operations (NH_4^+-form resin) Column: 23 × 0.3 cm ID a) 0.1 *M* HNO_3 (~1.5 mℓ) containing H_2O_2 (3 drops of 30% solution) to reduce Ce to the trivalent state (adsorption of ^{148}Nd and other rare earth elements) b) Water (~6 mℓ) (as a rinse) c) 0.4 *M* α-HIBA (pH gradient elution from pH 3.72—4.33 using a flow rate of ≃1 drop/50 sec and collecting fractions of ~0.4 mℓ each) From the fraction containing ^{148}Nd (which is obtained after 13 hr), the α-HIBA is removed by use of the following cation exchange separation	148

Table 36 (continued)
DETERMINATION OF RARE EARTHS IN MIXTURES OF FISSION PRODUCTS AFTER CATION EXCHANGE SEPARATION IN α-HIBA MEDIA

Material	Ion exchange resin, separation conditions, and remarks	Ref.
	B) Second column operation (H^+-form resin) Column: 1.6 × 0.6 cm a) Nd fraction adjusted to pH 2 with 6 *M* HNO_3 (adsorption of Nd and elution of α-HIBA) b) 0.1 *M* HNO_3 (5 mℓ) (elution of residual α-HIBA) c) 6 *M* HNO_3 (5 mℓ) (elution of Nd) Following evaporation of the eluate, the Nd is irradiated with neutrons and the γ-activity of the ^{149}Nd is measured	
Fission products from, e.g., ^{252}Cf	Aminex® A-9 cation exchange resin (11.5 ± 0.5 μm) In a HPLC column (15 × 0.32 cm ID) (stainless-steel tube) containing this resin, the rare earths are eluted at a flow rate of 1.0 mℓ/min using a gradient elution technique; a gradient is formed on the column using two pumps which continuously vary the concentration and pH of an α-HIBA solution from 0.65 *M* and pH = 3.6 to 0.95 *M* and pH = 4.8; the column is operated at 95°C which improves the resolution between adjacent rare earth elements by ~50%; the rare earths were determined radiometrically; before the cation exchange separation step, the rare earths are separated from other fission products by extraction chromatography in a HPLC column; both separations are carried out in a microprocessor-controlled system and the whole process takes ≃3 min, and isotopes of half-lives >30 sec can be observed	189
Fission products	Dowex® 50-X12 (fraction with a settling rate of <0.5 cm/min in water) Column: 2.8 × 0.2 cm With 0.23 *M* α-HIBA of pH 5.5 as the eluent, U and most of the fission products appear in the first few drops of effluent and then the rare earths elute in the reverse order of atomic number; flow rate: ~3 drops/min In the various fractions obtained, the rare earths were determined radiometrically Prior to the cation exchange separation, ^{235}U (300 μg in 50 μℓ of 0.01 *M* HCl) is irradiated with neutrons and then the irradiated solution is loaded onto the column	244

Table 37
DETERMINATION OF RARE EARTH IMPURITIES IN RARE EARTH OXIDES AFTER CATION EXCHANGE SEPARATION IN α-HIBA MEDIA

Material	Ion exchange resin, separation conditions, and remarks	Ref.
High-purity rare earth oxides	Diaion® SK-1 (NH_4^+ form; 100-200 mesh) Column: 45 × 1.2 cm ID; the elutions with eluents (c) and (d) are performed at a flow rate of 30—40 mℓ/hr a) 0.1 *M* HCl (20 mℓ) (adsorption of the rare earths) b) Water (10 mℓ) (removal of acid) c) 0.12—0.35 *M* ammonium α-HIBA of pH 3.77 (concentration gradient elution of the rare earths from Lu to Eu) d) 0.5 *M* ammonium citrate (elution of the other rare earths) The rare earths were determined radiometrically; before the separation, the sample (50—100 mg) is irradiated with neutrons and dissolved in HCl or HNO_3 together with a carrier element having an atomic number of one less than that of the target material; the amount of carrier added is almost the same as that of the sample By means of this method, the impurities Gd, Dy, Ho, and Eu in Tb-oxide, Dy, Er, and Yb in Ho-oxide, and Sc, Eu, Tb, and Tm in Yb-oxide can be determined	177
Gd oxide	Dowex® 50W-X8 (NH_4^+ form; 200-400 mesh) The separation of the rare earths and Co is carried out on a column (16 cm × 0.4 cm^2) of this resin using elution with an exponential ligand concentration gradient The ligand is α-HIBA, the initial concentration being 0.119 *M* in the mixing chamber (constant volume 50 mℓ) and 0.156 *M* in the reservoir; the desired ligand concentration is obtained by adjustment of the pH of 0.4 *M* α-HIBA with ammonia solution; the eluate may be collected in 2-mℓ fractions in which the rare earths are determined radiometrically; Co is eluted after Sm has been eluted; before this separation, the sample was subjected to neutron irradiation and then transferred to the cation exchange resin using 0.01 *M* HCl (few milliliters); in the case of the irradiated standards, an anion exchange separation step is employed to separate Co (added as a neutron flux monitor) from the rare earths; for this purpose, a solution (few milliliters) of the standards in 9 *M* HCl is passed through a column (5 cm × 0.5 cm^2) of Dowex® 1-X8 (100-200 mesh) on which Co is adsorbed while the rare earths pass into the effluent Co is eluted with 0.5 *M* HCl; in the 9 *M* HCl effluent the rare earths are determined radiometrically; the method has been used for the determination of traces of rare earth impurities (Lu, Tb, Eu, and Sm)	178
High-purity Lu oxide	Diaion® SK-1 (NH_4^+ form; 100-200 mesh) The sample of Lu_2O_3 (~23 mg) is irradiated with neutrons, mixed with nonirradiated Yb_2O_3 (as a carrier), dissolved in HCl, and the rare earths are isolated on a column (40 cm × 10 mm) of this resin using the following eluents a) 0.1 *M* HCl (10 mℓ) (adsorption of the rare earths) b) Water (20 mℓ) (removal of HCl) c) 0.12 *M* α-HIBA of pH 3.77 (900 mℓ) (successive elution of Lu, Yb, and Tm) d) 0.25 *M* citrate solution (7:3 mixture of citric acid and ammonium citrate) (150 mℓ) (elution of other rare earths lighter than Tm; at room temperature using a flow rate of 21 mℓ/hr) In the various fractions the rare earth impurities (i.e., ppm amounts of Sc, Yb, Tm, Tb, and Eu) were determined radiometrically	245

Table 37 (continued)
DETERMINATION OF RARE EARTH IMPURITIES IN RARE EARTH OXIDES AFTER CATION EXCHANGE SEPARATION IN α-HIBA MEDIA

Material	Ion exchange resin, separation conditions, and remarks	Ref.
Intermediate product of rare earth metal smelting	Sulfonic acid-type resin YSG-SO_3Na (grain size: 15—20 μm) Columns: 14.6 × 0.8 cm or 9.6 × 0.8 cm linked to a coulometric detector Complete separation of Lu, Tm, Er, Ho, Dy, and Yb is effected within 50 or 42 min on the respective columns by isocratic elution with 0.11 *M* α-HIBA (pH 4.3) at a forced flow rate of 1.7 mℓ/min Before the separation, the sample (600 mg) is dissolved in HNO_3, the solution evaporated nearly to dryness, and the residue diluted with water to 25 mℓ A portion of this solution is applied to the slurry-packed column	511

Table 38
DETERMINATION OF RARE EARTH ELEMENTS IN GARNETS, GLASSES, AND Co-BASED ALLOYS AFTER CATION EXCHANGE SEPARATION IN α-HIBA MEDIA

Material	Ion exchange resin, separation conditions, and remarks	Ref.
Gd-Ga garnet	Cation exchange resin (NH_4^+ form) Column: 10 × 1.6 cm a) Dilute HCl sample solution of pH 1.2 (adsorption of Nd, Gd, and Ga) b) 0.2 *M* α-HIBA (elution of Gd and Ga) c) 2 *M* acetic acid (elution of Nd) The Nd was determined titrimetrically Before the separation, the sample (0.2 g) containing ≃3% Nd_2O_3 is fused with Na_2CO_3-$Na_2B_4O_7$	246
Standardized glasses	Dowex® 50 A pH-gradient technique utilizing an α-HIBA solution as eluent is employed for the isolation of individual lanthanides By using eluent fractions of narrow pH range, rapid elution of short-lived activity and separation of adjacent elements is achieved; before the cation exchange separation, the sample is irradiated with neutrons and decomposed with H_2SO_4-HF	144
Co-based alloys	Analysis is carried out on an SY-202 high performance ion exchange chromatograph (Sichuan Analytical Apparatus Works, China) with a column (8 cm × 8 mm) of YSG-SO_3H (20—25 μm) cation exchange resin (a DVB-ethylstyrene copolymer) and 2-hydroxy-2-methylpropionic acid (α-HIBA) solution as eluent (1.6 mℓ/min) The sample is dissolved in HNO_3 (1:1), the solution is diluted with H_2O to 25 mℓ, and a suitable aliquot containing ≤45 μg of any individual rare earth metal is injected into the column through which the mobile phase is already flowing; for solutions containing Co, Er, Gd, and Sm, the mobile phase is 0.18 *M* α-HIBA adjusted to pH 4.5 with aq NH_3 and separate collections of eluate are made at 5.5, 12.5, and 21 min for Er, Gd, and Sm, respectively (Co begins to appear after 60 min); for solutions containing Co, Sm, and Pr the mobile phase is 0.26 M α-HIBA adjusted to pH 5.27, and collections are made from 45 sec to 7.25 min for Sm and then up to 13.75 min for Pr (Co begins to appear after 13.75 min); the rare earth metal in each of the fractions is determined spectrophotometrically with arsenazo I	194

Table 39
MUTUAL SEPARATION OF RARE EARTH ELEMENTS BY CATION EXCHANGE IN α-HIBA MEDIA OF VARYING pH USING THE HPLC TECHNIQUE

Ion exchange resin	Separation conditions and remarks	Ref.
TSK LS-212 (high performance cation exchange resin) (6 μm)	Following introduction of the sample solution (0.22 μℓ) (containing ≃0.05—0.5 Ci/μℓ of each radioactive rare earth metal) into the column (7.5 × 0.5 mm ID), stepwise gradient elution is carried out at a rate of 8 μℓ/min with 0.4 *M* α-HIBA in the pH range 3.1—6 (adjusted with aqueous ammonia solution); in the eluate, the rare earths are determined radiometrically; by the use of this technique, 16 rare earth metals are separated in 38 min for a total injection of 6—10 μg of the metals; if only lanthanides are present, the gradient elution is simpler and 14 lanthanides can be separated in 30 min	149
Hitachi® 2611 (strongly acidic cation exchange resin -X10; 8—11 μm)	On a column (15 × 0.9 cm) of this resin (at 80°C) 16 rare earths are separated by HPLC using 0.4 *M* α-HIBA solution as the mobile phase and a pH gradient from 3.5—7.0 (flow rate: 1.3 mℓ/min); the retention factors of each of the rare earths were determined at three flow rates; the limits of detection (coulometric detector) ranged from 0.78—4.8 n*M*; separation was complete in 75 min, and resolutions were 1.38—3.88, except for the separation of Dy from Y, for which resolution was 0.91	150-152

Table 40
MUTUAL SEPARATION OF RARE EARTH ELEMENTS BY CATION EXCHANGE IN α-HIBA MEDIA OF VARYING CONCENTRATIONS USING THE HPLC TECHNIQUE

Ion exchange resin	Separation conditions and remarks	Ref.
Dowex® 50W-X8 (NH_4^+ form; 25—60 μm)	For a mixture containing 5 mg of each lanthanide (except Pm, which is present in a tracer amount) good separation in about 100 min is achieved by use of a column (33 × 0.9 cm) of this resin operated at 80°C and with a pressure drop of 500 lb/in.^{2}, a flow rate of 10 mℓ/min, and a concentration gradient of 0.115—0.5 *M* of α-HIBA (adjusted to pH 4.4 with aqueous ammonia solution) as eluent; for 20 mg of each stable lanthanide adequate separation can be attained on a column (11.7 × 1.5 cm) of the same resin at room temperature with a flow rate of 5 mℓ/min (pressure drop ≃50 lb/in.^{2})	135
Bio-Rad® AG 50W-X12 [(25 ± 7) – μm particles]	The solution of 14 rare earths (milligram amounts) in 1 *M* HCl (1 mℓ) is applied to a column (24 × 0.63 cm) of this resin and the rare earths are eluted at 23°C with 0.05—0.42 *M* α-HIBA at pH 5.2—5.3 using a forced flow rate of 10.3—13.5 mℓ/cm^2 min; the rare earths were detected by flow-through electrical conductivity measurement or by adding trace radioactivity	179

Table 41
MUTUAL SEPARATION OF RARE EARTH ELEMENTS BY CATION EXCHANGE IN α-HIBA MEDIA OF VARYING CONCENTRATIONS AND pH VALUES USING THE HPLC TECHNIQUE

Ion exchange resin	Separation conditions and remarks	Ref.
Bio-Rad® AG 50W-X12 (NH_4^+ form)	A) Medium flow rate system (−400 mesh resin) In a column (100 cm × 10 mm ID) utilizing a pressure of 6—8 bar and a flow rate of 190 mℓ/hr, the rare earths from Ce to Tm can be separated by gradient elution using 0.2—0.5 *M* α-HIBA at the pH gradient of 3.2—4.0 (to separate the rare earths from Ho to Lu and Y, 0.2 *M* α-HIBA is used isocratically); the separations are performed at room temperature and can be completed in 7—10 hr B) High flow rate system (resin of grain size 25 ± 5 μm) For this rapid separation of the rare earths which can be completed in 4 hr, a pressure of 180 bar damped to 40 bar at the column inlet is applied to a column (100 cm × 7 mm) of the resin; the rare earths are eluted at 90°C using a flow rate of 500 mℓ/hr employing a molarity gradient of 0.2—0.5 *M* α-HIBA and a pH gradient of 3.2—4.5	190
YSG-SO_3Na (grain size: 15—20μm)	Traces of lanthanoids are separated on a column (8.5 × 0.8 cm) of the resin by gradient elution with 0.1—0.3 *M* α-HIBA (pH 3.5—6.7) at 50°C; the rare earth elements are determined polarographically	512

Table 42
MUTUAL SEPARATION OF RARE EARTH ELEMENTS BY CATION EXCHANGE IN α-HIBA AND 2-HYDROXYBUTYRIC ACID MEDIA

Elements separated	Ion exchange resin, separation conditions, and remarks	Ref.
13 Lanthanides	Dowex® 50W-X8(NH_4^+ form; 250—325 mesh) Column: 93.5 cm × 2.8 mm ID Gradient elution with 0.4 *M* α-HIBA of pH 2.9—4.5 is employed to separate 5-μg amounts of the lanthanides using a forced flow rate of 0.4 mℓ/min; the order of elution is the same as that usually observed, i.e., in inverse order of atomic number For the continuous detection of the rare earths in the eluate, the method of in-stream color addition is employed which eliminates the need for fraction collecting, radioisotope handling, or continuous attention; as the color reagent PAR (4-[2-pyridylazo]resorcinol) is used The Lu-Yb and Nd-Pr-La groups are not completely separated	153
Lu, Yb, and Tm	Dowex® 50-X8 (NH_4 form) The separation of 0.001—1-mg amounts of Lu, Yb, and Tm is effected on a column (15—25 cm × 3 mm ID) of this resin by elution with 0.1 *M* ammonium α-HIBA; the separation takes 23 hr at pH 3.83, but if a pH gradient from 3.79—4.05 is applied, the time is reduced to 13 hr	154
Dy, Tb, Gd, Eu, Sm, Pm, Cf, and Cm	Dowex® 50-X12 (+400 mesh) From a column (5 × 0.2 cm) of this resin the rare earths and actinides are eluted with 0.3 *M* α-HIBA at 87°C using a flow rate of 1 mℓ/cm² min	247
Tb from a mixture of light lanthanides	Cation exchange resin (20—45 μm) The sample (0.5 mℓ) of pH 2—3 (total metal concentration of 1.4 mg) is applied to a column (10 cm × 9 mm) operated at 70°C; Tb is eluted with 0.2 *M* 2-hydroxybutyric acid (adjusted to pH 5.2—5.3 with ammonia solution) at a forced flow rate of 1—1.1 mℓ/min and determined coulometrically	139

Table 43
DETERMINATION OF RARE EARTH ELEMENTS IN GEOLOGICAL MATERIALS AFTER CATION EXCHANGE SEPARATION IN LACTATE MEDIA

Material	Ion exchange resin, separation conditions, and remarks	Ref.
Monazite, allanite, gadolinite, xenotime, fergusonite, euxenite, and yttrialite	Diaion® SK 1 (corresponding to Dowex® 50) Column: 20 × 0.5 cm operated at 87°C a) Dilute HCl solution of the sample (adsorption of the rare earths) b) 0.5 *M* ammonium lactate (of pH 3.6 and 4.6) (elution of the rare earths at a flow rate of 6—7 mℓ/hr) In order to improve the separation of the elements of the Y family from each other, the same eluent is used at pH 3.5 and 4.3; in the fractions obtained, the rare earth elements were determined spectrophotometrically using arsenazo I or 5,7-dichloro-8-hydroxyquinoline	155, 156
Surface waters and atmospheric water precipitations	Dowex® 50W-X8 (NH_4^+ form; 50-100 mesh) Column filled with 110 mℓ of settled resin a) Water sample of pH 2—3 containing carriers of the elements to be determined (adsorption of radioelements, Ca, Mg, and other elements) b) 1.0 *M* ammonium lactate of pH 7.5 (elution of ^{144}Ce and ^{90}Y which are followed by Ca and ^{137}Cs) (flow rate: 5 ± 1 mℓ/min); the first 250 mℓ of eluate (= Ce fraction), the next 150 mℓ eluate are rejected, and the following 350 mℓ contain the Cs c) 0.15 *M* ammonium citrate of pH 7.5 (elution of ^{90}Sr-^{89}Sr followed by ^{140}Ba) (flow rate: same as with [b]) The initial 300 mℓ contain the Sr, the next 250 mℓ are discarded, and the following 600 mℓ contain the Ba In the various eluates Ce, Cs, Sr, and Ba were determined radiometrically; the limit of detection amounts to 0.005 pCi/ℓ for ^{144}Ce, 0.06 pCi/ℓ for ^{137}Cs, 0.004 pCi/ℓ for ^{90}Sr, and 0.006 pCi/ℓ for ^{140}Ba with a 30-ℓ sample; these senstivities can be improved by using a battery of columns; the method can be applied to water samples with total Ca-Mg contents up to 110 meq; in place of eluent (b) Ce and Cs can also be eluted with 0.6 *M* ammonium glycollate - 0.2 *M* HCl (pH 5)[386]	272, 386
Rainwater	Dowex® 50-X8 (100-200 mesh; NH_4^+ form) Column: 10 × 2.2 cm containing 21 mℓ of the settled resin a) Water sample (containing carriers for the elements to be determined) (adsorption of Ce, Cs, Sr, Ba, and other elements) (flow rate: 15 mℓ/min) b) 0.7 *M* HCl (240 mℓ) (elution of ^{137}Cs and other alkali metals) (the first 120 mℓ containing Na and K are discarded) c) 1.5 *M* ammonium lactate of pH 7.0 (120 mℓ) (elution of ^{144}Ce before the ^{90}Sr) d) 4 *M* HCl (100 mℓ) (elution of ^{140}Ba) From its eluate fraction (50—60 mℓ interval), Ce is isolated by a hydroxide precipitation to separate it from coeluted Ca and finally it is precipitated as oxalate and determined radiometrically as are the radionuclides of Cs, Sr, and Ba; before its determination, the Sr (via measurement of ^{90}Y) is first isolated from the eluate	273

Table 43 (continued)
DETERMINATION OF RARE EARTH ELEMENTS IN GEOLOGICAL MATERIALS AFTER CATION EXCHANGE SEPARATION IN LACTATE MEDIA

Material	Ion exchange resin, separation conditions, and remarks	Ref.
	fraction (80—120-mℓ interval) by adsorbing it on a small column (containing the resin to a height of 6 cm) from a solution of pH 2; then ^{90}Y is eluted with an ammonium citrate solution (5% with respect to citrate) (30 mℓ) of pH 3.8, precipitated as Y-oxalate (following addition of Y-carrier), and determined by radioactive measurements	
	Before the determination of Ba (via radiometric assay of its ^{140}La daughter) the eluate (d) is evaporated to dryness, Ba is adsorbed (from 100 mℓ aqueous solution) on a small column containing the resin to a height of 6 cm, and the ^{140}La activity is eluted with ammonium citrate solution (30 mℓ) of pH 3.8	
	The elutions of Sr and Ba are effected with 1.5 *M* ammonium lactate solutions of pH 7.0	

Table 44
DETERMINATION OF RARE EARTH ELEMENTS IN INDUSTRIAL MATERIALS AFTER CATION EXCHANGE SEPARATION IN LACTATE MEDIA

Material	Ion exchange resin, separation conditions, and remarks	Ref.
Neutron irradiated U_3O_8	Dowex® 50 (250-500 mesh; NH_4^+ form) (at 87°C) Column containing 3.4 mℓ of settled resin a) Dilute $HClO_4$ solution of the rare earths (adsorption of the rare earths) b) 0.85, 0.90, 0.95, and 1 *M* lactic acid solutions of pH 3 (elution of Y, Tb, Gd, Eu, Sm, Pm, and Nd in this order) The eluted fission product rare earths were determined radiometrically; prior to this separation, the sample is irradiated with neutrons, dissolved in 6 *M* HNO_3 in the presence of carriers, and the rare earths are isolated by standard radiochemical precipitation steps	180
Proton-irradiated Yb-acetate or $Hf(OH)_4$	Cationite (NH_4^+ form) For the separation of the radioisotopes of the rare earths, a column of this resin at 90°C was used with a solution of 0.4 *M* ammonium lactate of pH 3.4 as the eluent Before the separation, the proton-irradiated sample (70—80 mg) is dissolved in 5% acetic acid (for Yb) or a mixture of 0.5 *M* HCl and 5% oxalic acid (for Hf) Subsequently, separation of Yb is carried out by electrolysis with a Hg cathode, and that of Hf by passing the solution through the column of the cation exchange resin (H^+ form) which retains the rare earth elements, but not Hf	274
Al	Dowex® 50-X12 (200-400 mesh; NH_4^+ form) Column: 60 × 1 cm operated at 90°C a) Solution of the rare earths (adsorption of the rare earths) (flow rate: 4—5 drops/min) b) Water (15—20 mℓ) (as a rinse) c) 1 *M* ammonium lactate solutions of pH 3.2—4.5 (pH-gradient elution of the rare earths using a flow rate of 4—5 drops/min) For the complete elution a time of 30 hr is required In the various fractions obtained, the rare earths were determined radiometrically following their precipitation as oxalates in the presence of La-carrier; before the cation exchange separation, the sample is irradiated with neutrons, dissolved in 6 *M* HCl, and the rare earths are purified by a number of radiochemical precipitations in the presence of La as a carrier	157, 158

Table 45
MUTUAL SEPARATION OF RARE EARTH ELEMENTS BY CATION EXCHANGE CHROMATOGRAPHY IN LACTATE MEDIA OF VARYING pH VALUES

Ion exchange resin	Separation conditions and remarks	Ref.
Dowex® 50-X12 (−400 mesh; NH_4^+ form)	To a mixture of purified rare earth activities a small amount of the resin (which settled between 1.0 and 1.5 cm/min) is added, and after equilibration (for several min) the exchanger is transferred to the jacketed column (60 × 0.7 cm ID) Subsequently, pH-gradient elution is carried out at 90°C by passing 1 *M* ammonium lactate buffer of pH 3.19 to which is added continuously, with stirring, the lactate buffer, but of the much higher pH of 7; flow rate: 0.4 mℓ/cm² min; in the various eluate fractions, the rare earth elements were determined radiometrically; the rare earth activities are obtained by bombarding a target consisting of pure U and Ta metal foil with protons The target is dissolved in HF-HNO_3 in the presence of La-carrier and carriers for other elements; among the radiochemical separation steps which were then used for the purification of the rare earth activities is an anion exchange procedure (Dowex® 1) involving the use of conc HCl to remove those elements which form anionic complexes under these conditions	159
Dowex® 50 (grain size: 0.04—0.05 mm; NH_4^+ form)	Jacketed column: 200 × 1 cm ID operated at a flow rate of 3.6—5.9 mℓ/min The rare earth are separated at 76°C by stepwise elution using the following eluents which are also 0.1 *M* with respect to phenol a) 1 *M* lactic acid of pH 3.00 (elution of Sc, Lu, and Yb) b) Same eluent of pH 3.05 (elution of Tm, Er, and Ho) c) Same eluent of pH 3.10 (elution of Y, Dy, and Tb) d) Same eluent of pH 3.15 (elution of Gd and Eu) e) Same eluent of pH 3.25 (elution of Sm [Pm]) f) Same eluent of pH 3.30 (elution of Nd and Pr) g) Same eluent of pH 3.44 (elution of Ce) h) Same eluent of pH 3.50 (elution of La)	160
Amberlite® CG-120 type II (200-400 mesh; NH_4^+ form)	Column: 100 × 0.5 cm ID operated at 80°C a) Neutral or slightly acid solution (HCl) of the rare earths (20 mℓ) (adsorption of the rare earths) b) Water (10 mℓ) (to remove HCl) The eluents (a) and (b) are employed on a small sampling column of the resin in the H^+ form which is then transferred to the top of the large chromatographic column on which fractionation of the rare earths is achieved using eluent (c) c) 1 *M* ammonium lactate solution of pH 3.2—4.0 (pH-gradient elution of the rare earths at a flow rate of 4—5 mℓ/hr); the lactic acid solution used to prepare eluent (c) is purified (removal of traces of metal ions) by passage through a column (20 × 1.6 cm) of Dowex® 50-X8 (50-100 mesh; H^+ form); in the collected fractions the rare earths were determined spectrophotometrically; only Dy and Y are not separated The maximum error for any one rare earth element is +14% and the error for total rare earth metals is $<+4\%$	162,164

Table 45 (continued)
MUTUAL SEPARATION OF RARE EARTH ELEMENTS BY CATION EXCHANGE CHROMATOGRAPHY IN LACTATE MEDIA OF VARYING pH VALUES

Ion exchange resin	Separation conditions and remarks	Ref.
Sulfonated polystyrene resin (8—12X) (particle size: 8—11 μm)	Separations of the individual rare earths are effected at 50 or 70°C on a column (10.7 × 0.9 cm) of this resin using pH-gradient elution (from pH 3.4—5) with 0.5 *M* lactic acid - NaOH as eluent Also, by use of a − 10X cross-linked resin in a column (5 × 0.6 cm) operated at 50°C with the pH increasing sharply from 3.4—7.2, 16 rare earth metals, including Sc and Y, were separated in ≈80 min	165
Diaion® SK mixed with Diaion® SA in the ratio of 1:2 (grain size of the resins: ~23 μm)	On a column (height: 20.5 cm) containing 4.02 mℓ of this mixture of strongly acidic and strongly basic anion exchange resins, the rare earths are eluted with 0.5 *M* lactic acid - 0.06 *M* NaCl at various pH values, e.g., 2.63, 2.8, 2.9, 3.0, and 3.1 using flow rates of 30 mℓ/hr (with the eluent of pH 2.63) or 60 mℓ/hr (with all other eluents); the separation of 14 rare earths is achieved in 4 hr; the rare earths were determined coulometrically	161, 163, 166
Dowex® 50-X8 (400 mesh; NH_4^+ form)	On a column (7 × 2 mm) (operated at 20°C) of this resin, Y, Pm, Nd, Pr, Ce, and La are eluted in this order with 0.4 *M* lactic acid - 0.002 *M* EDTA as eluent using pH-gradient elution, the pH being increased from 3.0—3.5	167

Table 46
MUTUAL SEPARATION OF RARE EARTH ELEMENTS BY CATION EXCHANGE CHROMATOGRAPHY IN LACTATE MEDIA USING FORCED FLOW CONDITIONS

Ion exchange resin	Separation conditions and remarks	Ref.
Zeo-Karb® 225 (grain size: 16—27 μm)	Column: 20—30 cm × 3—4 mm ID operated at 87°C a) 0.01—0.1 *M* HCl (minimum volume) (adsorption of 1-mg amounts of the rare earths) b) 1 *M* lactate buffer of pH 3.25 under a pressure of 5—15 lb/in.^{2}, at a flow rate of 10—12 mℓ/hr (elution of Y, Eu, Sm, Nd, Pr, and Ce in this order) The complete separation takes 8—9 hr; the limits of individual fractions are recognized by testing with oxalic acid and also by radioactive tracers, i.e., the rare earth fission products	268, 513
Hitachi® 2611 (cation exchange resin)	On a column 15 × 1 cm of the resin, the rare earths (La, Ce, Pr, Nd, and Sm) are separated by HPLC using 0.5 *M* lactic acid of pH 4.2 as the eluent; the rare earths were detected with arsenazo III solution; detection limits ranged from 13 ng (for Sm) to 90 ng (for La)	195
Cation exchange resin IEX-210 SC (10 μm; Toyo Soda Co.)	On a column (25 × 0.4 cm; operated at 50°C), separation of 14 lanthanoids and Y is achieved by HPLC using lactate buffer of pH 4.2 (flow rate: 1.4 mℓ/min) The rare earths were determined directly in the eluate fractions using ICP-atomic emission spectrometry	514

Table 47
CATION EXCHANGE SEPARATIONS OF RARE EARTHS FROM EACH OTHER AND ACCOMPANYING ELEMENTS IN LACTATE MEDIA

Elements separated	Ion exchange resin, separation conditions, and remarks	Ref.
Y from light and medium lanthanides	Wofatit® KPS-X8 (strongly acidic cation exchange resin) Trace amounts of Y are separated from mixtures with Sm, Nd, and Nd-Sm-Eu-Gd-Tb by chromatography on a column of this resin with 0.25—0.75 *M* ammonium lactate at pH 2.79—3.2 as eluent; similar results are obtained with the chelating exchange resin Duolite® ES-63; in this case Y is adsorbed by the resin and the impurities are collected in the eluate	191
Y-Sr and Sc-Ca	Dowex® 50W-X8 (200-400 mesh; NH_4^+ form) By the use of 0.5 *M* ammonium lactate solutions with the pH values of 3.45, 3.10, 3.2, 3.35, and 3.05, ^{90}Y is eluted ("milked") from a column of this resin while ^{90}Sr is further retained; similarly, ^{46}Sc can be separated from ^{45}Ca using lactate solutions of pH 2.6, 2.7, 2.8, and 2.9; on the same resin (H^+ form) Sr and Y can also be separated employing the following eluents a) 1.4 *M* HNO_3 (elution of ^{90}Sr) b) 3 *M* HNO_3 (elution of ^{90}Y) To separate Ca from Sc the following eluents may be used a) 1 *M* HNO_3 (elution of ^{45}Ca) b) 3 *M* HNO_3 (elution of ^{46}Sc)	276

Table 48
RADIOMETRIC DETERMINATION OF Ce AND OTHER RADIONUCLIDES IN ENVIRONMENTAL MATRICES AFTER CATION EXCHANGE SEPARATION IN CITRATE AND OTHER MEDIA[a]

Ion exchange resins	Separation conditions and remarks	Ref.
Dowex® 50 (H^+ form) and Dowex® 1 (chloride form)	A) First column operation (cation exchanger) a) 0.2 *M* HCL (adsorption of rare earths, Th, U, Zr, Fe, and Ra; into the effluent passes Ru) b) 5% Oxalic acid solution (elution of Th, U, Zr, and Fe) c) 0.5 *M* NH_4Cl (H^+ form of the resin is converted into the NH_4^+ form) d) 1 *M* formic acid - 1 *M* ammonium formate (elution of the rare earths) e) 0.1 *M* EDTA - 2 *M* ammonium acetate - 6 *M* NH_3 solution (elution of Ra) B) Second column operation (anion exchanger) Eluate (d) (see above) is evaporated, ammonium salts are decomposed, carriers for Nd, Sm, and Ce are added, and the solution (12 *M* in HCl) is purified by passing it through a column of this resin; the rare earth hydroxides are precipitated from the effluent, Ce is separated as ceric iodate, followed by a hydroxide and oxalate precipitation; the oxalate is ignited to the oxide and Ce is determined radiometrically; the other rare earth nuclides are separated on Dowex® 50 (using lactate) from the supernate after the ceric iodate precipitation Before the first column operation, the sample of sea water (100 ℓ) is first subjected to five different precipitation steps and then the rare earths, Zr, Ru, U, Th, and Ra are collected on an iron hydroxide precipitate Fe is then removed by extraction with isopropyl ether Marine sediments are either extracted with HCl + $HClO_4$ under reflux (residue discarded), or fused, after removal of organic material by ignition, with NaOH and Na_2CO_3 The fish sample is ashed, the ash treated and leached with HCl, and the different nuclides are separated by using the following ion exchange separation steps I) Cation exchange separation (cation exchange resin in the H^+ form) a) 0.2 *M* HCl (adsorption of rare earths and other nuclides; Ru passes into the effluent) b) 0.5% Oxalic acid (elution of Zr and Fe) c) 5% Oxalic acid followed by water (as a rinse) d) 0.5 *M* NH_4Cl - 0.1 *M* HCl (elution of Cs) e) Water (as a rinse) f) 5% Ammonium-citrate solution of pH 3.5 (elution of rare earths, Mn, Co, and Zn) g) 5% Ammonium-citrate solution of pH 5.1 (elution of Cd) h) 5% Ammonium-citrate solution of pH 5.1 (elution of Sr) II) Anion exchange separation (anion exchange resin) The eluate (f) (see above) is evaporated to dryness, Ce, Nd, Sm, Zn, Co, and Mn carriers are added, and the rare earths are separated as a group by using the anion exchange procedure outlined above under (B); from the 12 *M* HCl, Zn, Co, and Mn are retained by the resin, while the rare earths pass into the effluent from which they are isolated as described above for seawater	277

Table 48 (continued)
RADIOMETRIC DETERMINATION OF Ce AND OTHER RADIONUCLIDES IN ENVIRONMENTAL MATRICES AFTER CATION EXCHANGE SEPARATION IN CITRATE AND OTHER MEDIA[a]

Ion exchange resins	Separation conditions and remarks	Ref.
	In connection with the determination of radio-Ce only in fish samples, 18 different separation steps are used utilizing, among others; hydroxide, oxalate, iodate, and fluoride precipitations; the ash is treated with HCl and HNO_3 and any undissolved residue is subjected to alkali fusion; Ce is added as a carrier and the solution is then purified from U and Pu by using the anion exchange procedure described above under (B) (U and Pu are retained by the resin), from Ra by a $BaSO_4$ scavenging and from Th and Pa by adsorption of these two actinides on Dowex® 1 from 7.5 *M* HNO_3; in the effluent Ce is separated from the other rare earths as Ce-iodate	
Anion exchange resin SB-2 filter and cation exchange resin SA-2 filter	Two anion exchange filters (upper layer) and four cation exchange filters (lower layer) in a Büchner funnel are utilized to adsorb all cations and anions from the tapwater sample containing hold-back carriers; subsequently, the retained radionuclides are eluted by using the following eluents a) 3% Citric acid solution adjusted to pH 3.5 with ammonia solution (elution of rare earths and other trivalent ions as well as Co and Cs, which were all adsorbed on the cation exchange resin filters) (divalent cations such as Mn, Zn, Sr, and Ba are further retained by this resin) b) 0.1 *M* Na chromate solution (containing acetic acid, ammonium acetate, and Sr nitrate) (elution of Mn and Zn); Ba is precipitated on the resin filter as $BaCrO_4$ and determined radiometrically; from the effluent Mn and Zn are isolated by hydroxide and sulfide precipitations, respectively, and then their radioactivities are measured; the radioactive rare earths and Y contained in eluate (a) are precipitated with oxalic acid in the presence of Y- and Ce-carriers and determined radiometrically; in the filtrate, Co is also precipitated as oxalate following the addition of carrier and then determined radiometrically; to determine Ce, the oxalates are dissolved in HNO_3 and this element is separated from the other rare earths by precipitation as ceric iodate c) 4.5 *M* HCl - 0.06 *M* HF (elution of Zr and Nb from SB-2 resin filters) In the eluate, these elements were determined radiometrically	279, 280

[a] See also Table 13 in the chapter on Zinc, Volume IV.

Table 49
DETERMINATION OF RARE EARTHS AND OTHER ELEMENTS IN INDUSTRIAL MATERIALS AFTER ION EXCHANGE SEPARATION IN CITRATE AND OTHER MEDIA

Material	Ion exchange resin, separation conditions, and remarks	Ref.
Neutron-irradiated U	A) First column operation (Diaion® SK-1; H^+ form) Column: 32 × 1.0 cm containing 25 mℓ of the resin a) 0.3 *M* HCl (~20 mℓ) (adsorption of rare earths, U, Np, Zr, Nb, Cd, Te[IV], Sn, Cs, Rb, Zn, Be, Ru, Sr, and Ba; into the effluent pass Pd, Mo, Ru, Rh, Tc, and Te[VI]) b) 0.3 and 0.5 *M* HCl (~5 mℓ each) (removal of residual non-adsorbed elements) c) 0.75 *M* HCl (50 mℓ) followed by water (20 mℓ) (elution of Cd, Te[IV], and Sn) d) 0.2 *M* oxalic acid (25 mℓ) and water (20 mℓ) (elution of Zr-Nb, Np, and part of the U) e) 1 *M* oxalic acid (25 mℓ) and water (20 mℓ) (elution of U[VI]) f) 1 *M* HCl (200 mℓ) (elution of Cs, Rb, and Zn as well as Be and Ru) g) 1 *M* NH_4Cl (65 mℓ) and water (20 mℓ) (the H^+ form of the resin is converted into the NH_4^+ form) h) 6:4 Mixture (pH 4.2) (25 mℓ) of 0.25 *M* ammonium citrate and 0.25 *M* citric acid (elution of the rare earths) i) Water (20 mℓ) (as a rinse) j) 0.25 *M* ammonium citrate - 0.225 *M* ammonia (pH 6.9) (25 mℓ) followed by water (20 mℓ) (elution of Sr) k) 0.5 *M* ammonium citrate - 0.45 *M* ammonia (pH 6.9) (25 mℓ) followed by water (20 mℓ) (elution of Ba) B) Second column operation (isolation of rare earths from eluate [h] using Diaion® SK-1 in the NH_4^+ form) Column: 10.5 × 0.4 cm containing 1.3 mℓ of the resin a) Eluate (h) acidified with an amount of HCl equivalent to the ammonium citrate (or ammonium-α-HIBA) contained in the solution and diluted with water to 0.2 *M* in ammonium ion (adsorption of the rare earths; citric acid [or α-HIBA] passes into the effluent) b) 1 *M* HCl (10 mℓ) (elution of ammonium ions) c) 4—6 *M* HCl (20 mℓ) (elution of rare earths) C) Third column operation (mutual separation of the rare earths on Diaion® SK-1; NH_4^+ form) Column: 32 × 1 cm containing 25 mℓ of the resin a) 0.1—0.3 *M* HCl (adsorption of rare earths) b) 3:7 Mixture (290 mℓ) (pH 3.7) of 0.3 *M* ammonium α-HIBA and 0.3 *M* α-HIBA acid (elution of Y and Tb) (because of its small fission yield compared to that of Y, Tb has to be purifed further by repeating the separation process on another cation exchange column of the same dimensions) c) 3:7 Mixture (525 mℓ) (pH 3.7) of 0.5 *M* ammonium-α-HIBA and 0.5 *M* α-HIBA acid (elution of Eu, Sm, and Pm successively, in this order) Subsequently, a fresh column (32 × 1 cm) containing 25 mℓ of the same resin is connected in series to the end of the old one before Nd and Pr are eluted with a 3:7 mixture (680 mℓ) (pH 3.3) of 0.25 *M* ammonium citrate and 0.25 *M* citric acid Ce is eluted with a 4:6 mixture (130 mℓ) (pH 3.6) of 0.25 *M* ammonium citrate and 0.25 *M* citric acid, and finally La with a 6:4 mixture (75 mℓ) (pH 4.2) of 0.25 *M* ammonium citrate and 0.25 *M* citric acid; the flow rate of the eluent is kept at	248

Table 49 (continued)
DETERMINATION OF RARE EARTHS AND OTHER ELEMENTS IN INDUSTRIAL MATERIALS AFTER ION EXCHANGE SEPARATION IN CITRATE AND OTHER MEDIA

Material	Ion exchange resin, separation conditions, and remarks	Ref.
	25 mℓ/hr or less; from every eluate obtained by means of this third column operation, the rare earth fraction is concentrated and simultaneously freed from the organic complexing agent by means of a cation exchange separation identical to that described above under (B); in the final eluates (4—6 *M* in HCl) the various rare earths were determined radiometrically	
	Isolation of Mo (Diaion® SA-100; chloride form)	
	Column: 13 × 0.7 cm containing 5 mℓ of the resin	
	Following removal of Pd from effluent (a) (first column operation) by precipitation, the Mo is separated by using the following eluents	
	a) 6 *M* HCl (adsorption of Mo; elution of accompanying elements)	
	b) 0.1 *M* HCl - 0.06 *M* HF (50 mℓ) and 0.1 *M* HCl (20 mℓ) (removal of residual amounts of other elements)	
	c) 12 *M* HNO_3 (30 mℓ) (elution of Mo)	
	Isolation of Cd and Te (from eluent [c] obtained by the first column operation) (Diaion® SA 100; chloride form)	
	Column: 13 × 0.7 cm containing 5 mℓ of the resin	
	After coprecipitation of Te with ferric hydroxide, Cd is collected on ZnS; from the latter, the Cd is isolated by adsorbing it on a column of this resin from 3 *M* HCl; subsequently, the resin is washed with 3 *M* HCl (20 mℓ) followed by 0.1 *M* HCl (50 mℓ) and Cd is eluted with water (35 mℓ) and determined radiometrically	
	The ferric hydroxide on which Te was coprecipitated is dissolved in HCl and Te is deposited on Cu metal and then separated from the Cu (after dissolution in HNO_3) by using the following cation exchange separation step (Diaion® SK-1; H^+ form)	
	Column: 20 × 0.8 cm containing 10 mℓ of the resin	
	a) 0.3—0.5 *M* HCl (adsorption of Te and Cu)	
	b) 1 *M* HCl (elution of Te prior to that of Cu)	
	To separate the components contained in fraction (d) (see first column operation [A]) the resin Diaion® SA-100 (chloride form) (column 13 × 0.7 cm containing 5 mℓ of the resin) is utilized, employing the following eluents	
	a) 0.2 *M* oxalic acid (= eluate [d]) (adsorption of Np and Zr; elution of Nb)	
	b) 12 *M* HCl (20 mℓ) (removal of oxalic acid)	
	c) 0.1 *M* HNO_3 (100 mℓ) (elution of Np)	
	The same resin (chloride form in a column 20 × 0.8 cm containing 10 mℓ of the resin) is used to isolate U from fraction(e) (see first column operation [A]); for this purpose the U is adsorbed directly from fraction (e) (1 *M* oxalic acid), then the resin is washed with 6 *M* HCl (50 mℓ) and finally U is eluted with the 0.1 *M* HCl (30 mℓ)	
	Similarly, Zn is separated from fraction (f) (see first column operation [A]) by adsorbing it on this resin (Column: 13 × 0.7 cm containing 5 mℓ of the resin) from 3 *M* HCl; elution of Zn is effected with 0.1 *M* HCl (50 mℓ); from the 3 *M* HCl effluent, Cs and Rb are isolated by adsorbing them on a column (13 × 0.7 cm containing 5 mℓ of the resin) of Diaion® SK-1 resin (NH_4^+ form) from an EDTA solution containing an ammonium ion concentration of ≈0.2 *M*	

Table 49 (continued)
DETERMINATION OF RARE EARTHS AND OTHER ELEMENTS IN INDUSTRIAL MATERIALS AFTER ION EXCHANGE SEPARATION IN CITRATE AND OTHER MEDIA

Material	Ion exchange resin, separation conditions, and remarks	Ref.
	Subsequently, the following eluents were used	
	a) 1:1 Mixture of acetic acid and ammonium acetate (20 mℓ) (elution of Al and Be)	
	b) 2 *M* ammonium acetate (20 mℓ) (elution of Cs and Rb)	
	To obtain Cs and Rb individually, the following eluents were used to separate them on a column of the same exchanger	
	Column: 40 × 0.8 cm containing 20 mℓ of the resin in the H^+ form	
	a) 0.1 *M* HCl (adsorption of Rb and Cs)	
	b) 0.6 *M* HCl (chromatographic elution of Rb and Cs in this order)	
	Prior to the first column operation (A), purified U_3O_8 (about 300 mg) is irradiated with neutrons and dissolved in HNO_3–HCl in the presence of carriers for the elements to be determined	
	By means of the analytical scheme outlined above La, Ce, Pr, Nd, Pm, Sm, Eu, Tb, Y, Zn, Cd, Rb, Cs, Sr, Ba, Mo, Pd, and Te can be separated simultaneously from the irradiated sample; of these elements Zn, Mo, Pd, Cd, and Fe are not separated quantitatively; a polarographic method is used to determine their recoveries	
U	Amberlite® IR-120 (grain size: 0.1—0.3 and 0.5—1 mm; H^+ form)	281, 282
	Column of 1 cm ID containing the resin to a height of 25 cm and operated at a flow rate of 1—1.5 mℓ/min	
	a) Dilute acid solution of the sample (adsorption of U, rare earths, Fe, Cu, Cd, Ni, Co, and Mn)	
	b) 0.5 *N* oxalic acid (~240 mℓ) (elution of U, Fe, and Cu at 40—50°C)	
	c) 1 *M* HCl (~240 mℓ) (elution of Cd, Ni, Co, and Mn)	
	d) 5% Ammonium citrate solution of pH 4 (160 mℓ) (elution of the rare earths)	
	Before the determination of the rare earths, citrate is removed from eluate (d) using a small column (5 × 0.4 cm) of the same resin on which the rare earths are adsorbed from this eluate, acidified with HCl to a pH of <1 (flow rate: 3 mℓ/min)	
	Subsequently, the rare earths are eluted with 6 *M* HCl (70 mℓ)	
	Similarly, traces of the rare earth elements are separated from 10-g samples of U nitrate by adsorption on a column (20 × 2 cm) of this resin and elution of the U with 1 *N* oxalic acid; finally, the rare earths are eluted with 5 *M* HCl; both methods can be used in connection with the determination of rare earth impurities (and of other metal impurities) in U samples	
"Pure" Y	Cationite KU-2 (grain size: 0.09—0.12 mm)	259
	Column: 44 × 5 cm	
	A) First column operation (on NH_4^+ form of the resin)	
	a) Solution of pH 2—3 (adsorption of the rare earths)	
	b) 0.25 *M* citric acid adjusted to pH 2.86 with ammonia solution (elution of ≃95% of the Y with the first 1.9 ℓ of the eluent and then the impurity rare earths, i.e., Ce, Pr, and Nd, are eluted with the next 2.8 ℓ of this eluent)	
	To remove citric acid and to concentrate the impurity rare earths contained in the eluate, the following separation is performed	

Table 49 (continued)
DETERMINATION OF RARE EARTHS AND OTHER ELEMENTS IN INDUSTRIAL MATERIALS AFTER ION EXCHANGE SEPARATION IN CITRATE AND OTHER MEDIA

Material	Ion exchange resin, separation conditions, and remarks	Ref.
	B) Second column operation (on H^+ form of the resin) a) Eluate (b) containing the impurity rare earths is passed through a column of this resin to adsorb the rare earths (elution of citrate) b) 4 *M* HCl (elution of the rare earths) The rare earths were determined by conventional spectrography; the method has been used for determining La (18 ppm), Ce (8 ppm), Pr (≃10 ppm), and Nd (26 ppm), and for the analysis ≃14 hr is required Before the cation exchange separation, the sample (1 g) is decomposed with 20% HCl	

Table 50
SEPARATION OF RARE EARTHS FROM EACH OTHER AND ACCOMPANYING ELEMENTS BY CATION EXCHANGE SEPARATION IN CITRATE MEDIA

Ion exchange resin	Separation conditions and remarks	Ref.
Dowex® 50W-X8 (treated with Cl_2) (100-200 mesh; NH_4^+ form)	Column: 17 × 1.2 cm ID containing the resin to a height of ≈10.5 cm and operated at a flow rate of 2.3—2.4 mℓ/min (eluents [f] through [n]) a) Sample solution (20 mℓ) ≈2% in citric acid with the pH adjusted to 3.0 with NH_3 and treated with Cl_2-gas at 95—97°C (adsorption of rare earths and many other elements; into the effluent pass Sc, Fe[III], Cr, Hf, W, Zr, Nb, Mo, Tc, Ta, Sb, Bi, Ru, Os, Ir, Re, Pd, Ga, Au, and Hg) b) 2% Citric acid solution (30 mℓ) with the pH adjusted to 3.0 with NH_3 (elution of residual nonadsorbed elements) c) Water (2 × 5 mℓ) (as a rinse) d) 1% HEDTA solution (55 mℓ) (0.036 *M*) with the pH adjusted to 3.0 with NH_3 (elution of Pd, Ga, Cu, Zn, Co, Cd, In, and Hg) e) Water (3 × 5 mℓ) (as a rinse) f) Buffer wash with 0.08 *M* formic acid (adjusted to pH 4.0 with ammonia) (20 mℓ) followed by water (5 mℓ) g) 1% Citric acid solution (15 mℓ) with the pH adjusted to 5.0 with NH_3 (flow is stopped for 30 min and then resumed with more of the solution) (35 mℓ) (elution of La, Ce, Pr, Nd, Pm, Sm, Eu, and Y) h) Water (2 × 5 mℓ) (as a rinse) i) 1% EDTA solution (55 mℓ) (0.031 *M*) adjusted to pH 6.5 with NH_3 (elution of Pd, Hg, Ag, Sr, and Ca) j) Water (5 mℓ) (as a wash) k) 2.5% EDTA solution (30 mℓ) adjusted to pH 9.0 with NH_3 (elution of Na, Ba, and Tl) l) Water (3 × 5 mℓ) (as a rinse) m) 3 *M* HCl (30 mℓ) (elution of K, Rb, and Cs as well as residual amounts of Pd and Au) n) 5 *M* HCl (100 mℓ) (elution of the residual amounts of Pb, Au, and Ce) Dilute HF is needed to elute residual amounts of Ta (~15%) remaining on the resin Excellent separations (less than 0.1% of element contaminating any other fraction) are obtained for Sc, Fe, Cr, Hf, W, Cu, Zn, Co, Cd, La, Pr, Nd, Pm, Sm, Eu, Sr, Ca, Na, Ba, K, Rb, and Cs; for Nb, Mo, Tc, Ta, Sb, Bi, Ru, Os, Ir, Re, In, Ce, and Y the separations are good (corresponding to 0.1—1% contamination of any other fraction); poor separations which correspond to 1—10% contamination are obtained for Ga and Tl, while to over 10% contamination is observed in the case of separations involving the elements Pd, Au, Hg, and Ag This analytical scheme is recommended for use on a tracer level Resin chlorination, which is used to hold as many elements as possible in single, well-defined oxidation states throughout the entire resin bed, causes a slight decrease in strong acid capacity which, however, has little practical effect on the separations; the chlorine concentration in the sorption solution (a) must be kept small to minimize undesirable side reactions with ammonium ion and with citrate; with the former, chlorine reacts to liberate nitrogen gas, proceeding through a hy-	283

Table 50 (continued)
SEPARATION OF RARE EARTHS FROM EACH OTHER AND ACCOMPANYING ELEMENTS BY CATION EXCHANGE SEPARATION IN CITRATE MEDIA

Ion exchange resin	Separation conditions and remarks	Ref.
	drazine intermediate; therefore, the resin must be chlorinated while in the H^+ form; in solution the chlorination of NH_4^+ ions is slow, so that no gas evolution takes place in the sorption solution; on passing chlorine through the sorption solution at 95°C the pH drops from 3.0 to 2.0—2.2 and must be readjusted by adding ammonia; this pH drop is an indication for the reaction of chlorine with the complexing agent of which 30% are destroyed at 85°C accompanied by a pH drop from 3 to 1.1—1.3	
Dowex® 50-X8 (100-200 mesh; H^+ form)	Column of 2.2 cm ID containing the resin to a height of ≃22.5 cm and operated at a flow rate of 0.7 cm/min a) Solution of mixed chlorides of the rare earths (adsorption of La, Ce, Pr, Nd, and Sm) b) Water (500 mℓ) (as a rinse) c) 50:1 Mixture (3000 mℓ) of 1.25% diammonium H citrate solution with a solution prepared by diluting 40 mℓ of 28% aq NH_3 to 160 mℓ plus 100 mℓ of anhydrous acetic acid diluted to 300 mℓ (elution of Sm, Nd, and Pr as separate peaks) d) 20:1 Mixture (~1000 mℓ) of the two solutions (see above under [c]) (elution of Ce) e) 5:1 Mixture (500 mℓ) of the two solutions (see above under [c]) (elution of La) Following isolation from the eluates by precipitation as oxalates the rare earths were determined gravimetrically	284

Table 51
DETERMINATION OF RARE EARTH ELEMENTS IN GEOLOGICAL MATERIALS AFTER CATION EXCHANGE SEPARATION IN EDTA MEDIA

Material	Ion exchange resin, separation conditions, and remarks	Ref.
Rocks	Dowex® 50-X8 (100-200 mesh) Column: 5 × 0.8 cm operated at a flow rate of 1—2 mℓ/min a) EDTA solution (20 mℓ) adjusted to pH 2.1 with ammonia solution (adsorption of light rare earths, i.e., La → Tb; elution of heavy lanthanides (Ho → Lu) which form more stable EDTA complexes under these conditions) (this EDTA solution contains 4.5 mℓ of 0.01 *M* EDTA) b) 0.05 *M* NH_4Cl solution of pH 2.1 (20 mℓ) (removal of EDTA) c) 7 *M* HNO_3 (15 mℓ) (elution of La → Tb) From the two fractions (a) and (c) thus obtained, the rare earths were isolated by precipitation as fluorides, which was followed by a hydroxide precipitation, and finally the rare earths were determined radiometrically Before the cation exchange separation, the sample (500 mg) is irradiated with neutrons, fused with Na_2O_2 in the presence of rare earth carriers, and the rare earths are purified by repeated use of the hydroxide-fluoride precipitation cycle	357
Monazite	Dowex® 50W-X8 (40-50 mesh; NH_4^+ form) a) Sample solution of pH 3.5 (adsorption of lanthanides) b) 0.025 *M* EDTA (diammonium salt) (elution of Sm, Nd, and Pr) c) 1 *M* NH_4Cl (elution of La) At pH 3.5, Th and Ce were found to be essentially insoluble and can be filtered off (their hydrated oxides were dried at 120°C) In the various eluates, the rare earths were determined gravimetrically The method has been used for the recovery of the rare earths from Egyptian monazite following decomposition with H_2SO_4 and precipitation of double sulfates	358

Table 52
ISOLATION OF RARE EARTH IMPURITIES FROM INDUSTRIAL MATERIALS USING CATION EXCHANGE IN EDTA MEDIA

Material	Ion exchange resin, separation conditions, and remarks	Ref.
U	Amberlite® IRC-50 (Na^+ form) From the sample solution containing EDTA (disodium salt), the U is adsorbed on a column of this resin while the rare earths pass into the effluent This method can be used for the analytical separation of traces of rare earth elements (in the order of 10^{-5} g and less)	359
Th compounds	Dowex® 50 (20-50 mesh; NH_4^+ form) Column of 2.2 cm ID containing the resin to a height of 30 cm; the resin is pretreated with eluent (b) a) Sample solution (70—80 mℓ) of pH 2—2.3 containing the equivalent of 1.5 g of Th-oxide as nitrate or perchlorate and slightly less than an equivalent amount of EDTA (adsorption of the rare earths, e.g., ^{90}Y and ^{140}La; Th passes into the effluent) (flow rate: 2—3 mℓ/min) b) 10% Ammonium formate buffer of pH 2.1 (~50 mℓ) (removal of EDTA) c) Water (350—400 mℓ) (removal of formate buffer) Eluent for the rare earths not stated in original paper! The preferential dissociation of the EDTA complexes with pH makes possible the retention on the resin of the traces of rare earths along with a small portion of the Th which is uncomplexed	360
High-purity La-oxide	Cationite KU-2 (NH_4^+ form) a) Solution of La-chloride (adsorption of the rare earths) b) 0.5% Solution of EDTA of pH 4.2 (elution of the rare earth impurities Ce, Pr, Nd, and Sm) From the eluate, the rare earths are isolated by precipitation as oxalates; the precipitate is ignited, weighed, and analyzed spectrographically	361

Table 53
MUTUAL SEPARATIONS OF THE RARE EARTH ELEMENTS BY DISPLACEMENT CHROMATOGRAPHY ON CATION EXCHANGE RESINS USING POLYAMINO POLYACETATES AS ELUENTS

Ion exchange resin	Separation conditions and remarks	Ref.
Dowex® 50W-X2 (200-400 mesh; Cu^{2+}/H^+ form)	This mixed form of the resin is prepared by treatment of the H^+ form with 0.4 *M* $CuSO_4$ - 1.3 *N* H_2SO_4 or by mixing it with resin in the Cu^{2+} form; in case that too much of the metal form of the resin is present the sparingly soluble compound Cu(Cu-EDTA)$5H_2O$ is formed; on the other hand, precipitation of H_4 EDTA takes place if the H^+ form is present in great excess over the Cu(II) form due to the much higher acidity in the resin bed; the Cu^{2+}/H^+ form of the resin is used in a capillary column (70 cm × 2 mm) (tube of uniform bore and packed uniformly with the resin); before use of this column, the rare earths (40—50 mg) are adsorbed on a small column of this resin in the NH_4^+ form which is then transferred to the top of the resin bed in the capillary column; subsequently, 0.04 *M* EDTA (triammonium salt) buffered to pH 4.6 with ammonium acetate or formate is passed at a flow rate of 4 mℓ/day; the Cu(II)/H^+ resin acts as an "ion barrier", i.e., the EDTA complexes of the rare earths which are formed by the reaction of EDTA with the rare earths adsorbed on the NH_4^+-form resin, are decomposed by Cu^{2+} (= retaining ion), and readsorbed on the cation exchanger, while the complexing agent passes into the effluent as H_2Cu-EDTA; this readsorption of the rare earths on the resin (which should be only slightly colored) occurs in the order of the stability of their complexes, as sharply defined zones visible in UV light; the length of each zone is a measure of the amount of rare earth element present in it; the length of each zone is measured and expressed as a percentage of the total zone length; the rare earth elements that can be separated are (from top of the resin bed downwards) La, Ce, Pr, Nd, Sm + Eu, Gd + Tb + Y, Dy, Ho, Er, Tm, and Yb + Lu; Sc does not appear, since its very stable EDTA complex is not decomposed by the Cu on the resin; interferences are caused by elements such as Zn, Cd, Pb, Mn, and Co; these elements can be removed by elution from the first portion of the resin with 0.08 *M* Na glycinate of pH 8.8;[212] this technique of displacement chromatography can be improved if the resin contains a small amount of morin adsorbed on it[212,213]	210, 211
	Under these conditions any rare earth element of the La series (except Eu) as well as Y can be determined on such a resin; a similar method was described[214] for the separation of the pair Eu-Gd using a capillary column packed with Dowex® 50W-X4 or X5 (Cu^{2+}/H^+ form) containing adsorbed morin; elution was carried out at 80°C and a pressure of 100 torr with 0.04 *M* EDTA containing sufficient sodium acetate to give a pH of 3 (elution of Eu and Gd at this temperature gives a separation factor of 1.4, while at 25°C the factor is only 1.1)	
	After 10 hr the lengths of the fluorescent zones were measured; the ratio of the length of a single band (zone) to the total length of the rare earth metal zone gives the	

Table 53 (continued)
MUTUAL SEPARATIONS OF THE RARE EARTH ELEMENTS BY DISPLACEMENT CHROMATOGRAPHY ON CATION EXCHANGE RESINS USING POLYAMINO POLYACETATES AS ELUENTS

Ion exchange resin	Separation conditions and remarks	Ref.
	ionic fraction for that element; results on a mixture of Sm, Eu, and Gd showed relative errors of −1.6 to +0.9%; the method has been applied to the analysis of solutions containing up to 20 g of the rare earth-oxides/liter	
Dowex® 50W-X4 or -X5 (400 mesh; H^+ form)	This resin is treated with 0.5 *M* ammonium citrate and 3 *M* HCl and impregnated with morin and then packed into a capillary column; the sample solution (0.1—0.13 mmol of rare earth metals) is passed through a small column of the same resin in the Na^+ form, which is then transferred to the capillary column; subsequently, 0.04 *M* Na_2H(HEDTA) (disodium salt of *N*-(2-hydroxyethyl) ethylene-diaminetriacetic acid) - 0.04 *M* Na acetate is passed through the column at a pressure of 280 torr; after a steady state is reached the lengths of the separated rare earth metal zones (located under UV radiation) are measured; the volume occupied by each zone is proportional to the amount of each rare earth metal; Th and Al interfere; since the acid HEDTA is readily soluble in water (which is not the case with H_4-EDTA), the H^+ form of the resin can be utilized as the retaining ion)	215
Zerolit® 225-X8 (Fe^{3+} form)	On a column packed with the resin, the eluent is a 1.5% solution of NTA (ammonium nitrilotriacetate) at pH 2.7—4.0; when the cation exchanger is in the NH_4^+ form the Fe can be used as the displacing ion utilizing 1.5% Fe(III)-nitrilotriacetate solution at pH 2.8—4.2 as the eluent	216
	Good separation is attained for Sm, Nd, Pr, and La; in the chromatography of Y and the heavy rare earths, Yb, Ho, Dy, Tb, and Gd and part of Y are eluted together; a pure Y fraction is also obtained	
Dowex® 50W-X2 (200-400 mesh; VO^{2+} form)	Column of 2.5 cm ID containing the resin to a height of 93 cm and operated at flow rates of 1.3 and 0.9 mℓ/min a) Solution of lanthanides (0.2 *M*) containing a stoichiometric amount of HIDA (*N*-[2-hydroxyethyl]-iminodiacetic acid and adjusted to pH 5.2 with ammonia solution) (adsorption of the rare earths) b) 0.2 *M* HIDA adjusted to pH 8.7 with ammonia solution (elution of rare earths) In the eluate, the rare earths were determined by complexometric titration after removal of VO(II)	217
Dowex® 50W-X2 (200-400 mesh; Ni^{2+} or VO^{2+} forms)	The sample solution adjusted to pH 5—6 is first passed through a column (11 × 1.1 cm) of the resin in the NH_4^+ form to adsorb trivalent cations and others, then the washed resin is transferred to the top of a calibrated capillary column (65 × 0.2 cm) packed with the same resin (Ni[II] or VO[II] form); this column is then washed with an ammoniacal solution of HIDA at pH 8.8 to remove Cu(II), Sc(III), In(III), Fe(III), Ga(III), and other interfering species, leaving the rare earth metals separated as uniform adjacent zones, according to the stability of their HIDA complexes, and in the descend-	218

Table 53 (continued)
MUTUAL SEPARATIONS OF THE RARE EARTH ELEMENTS BY DISPLACEMENT CHROMATOGRAPHY ON CATION EXCHANGE RESINS USING POLYAMINO POLYACETATES AS ELUENTS

Ion exchange resin	Separation conditions and remarks	Ref.
	ing order La, Ce, Pr, Nd, Y, Sm, Eu, and Gd; the zones are revealed in different colors under UV radiation and the concentrations are obtained from the measured volumes of the zones by reference to calibration graphs prepared from standard solutions of each element (up to 0.25 meq); the efficiency of separation depends mainly on selection of the retaining ion (Ni[II] or VO[II]) and the eluent (0.08 *M* HIDA - 0.139 *M* aqueous ammonia or 0.2 *M* HIDA - 0.365 *M* aqueous ammonia), also on the optional addition of Mg^{2+} to the sample solution as auxiliary cation or the addition of a rare earth metal to improve the sharpness of the boundary between two faintly colored zones	
	Each complete analysis takes 5—6 days and the maximum error is ±3% for 0.15—0.26 meq of each element in prepared mixtures; duplicate results for a commercial sample of mixed rare earth oxides were also in good agreement, but other results showed that the presence of ions, e.g., those of Sn, Ag, Cd, Pb, Co, Cr, and the alkali metals, may cause appreciable errors	

Table 54
SEPARATION OF RARE EARTH ELEMENTS BY ELUTION CHROMATOGRAPHY ON CATION EXCHANGE RESINS USING EDTA MEDIA

Elements separated	Separation conditions and remarks	Ref.
La, Sm, Y, Sc, Th, and Fe	Dowex® 50-X4 (50-100 mesh; Na^+ form) Column of 2 cm ID containing 5 g of the resin and operated at flow rates in the range of 8—15 mℓ/min (eluents [a] and [b]) A) Separation of La from Th a) 10:30:15 Mixture (~55 mℓ) (adjusted to pH 2.2) of 0.05 *M* metal ion solution, buffer solution (0.2 *M* glycine containing ≃50 mℓ of 0.05 *M* EDTA per liter and adjusted to pH 2.2), and 0.05 *M* EDTA (adsorption of La and elution of Th) b) Buffer solution of pH 2.2 (90—120 mℓ) (see above) (elution of residual Th) c) 4 *M* HCl (90 mℓ) (elution of La) (flow rate: 1.5—2 mℓ/min) B) Separation of Sm from Fe a) Same mixture (~55 mℓ) as above of pH 1.8 (adsorption of Sm and elution of Fe) b) Glycine buffer solution of pH 1.8 (90—120 mℓ) (elution of residual Fe) c) 4 *M* HCl (90 mℓ) (elution of Sm) (flow rate: 1.5—2 mℓ/min) C) Separation of Y from Sc a) Same mixture (~55 mℓ) as above under (A) adjusted to pH 1.35 (adsorption of Y and elution of Sc) b) Glycine buffer solution of pH 1.35 (90—120 mℓ) (elution of residual Sc) c) 4 *M* HCl (90 mℓ) (elution of Y) (flow rate: 1.5—2 mℓ/min) In the various eluates, the elements were determined by EDTA titration following wet ashing of organic materials with HNO_3-$HClO_4$	356
Er from Fe, Al, and Ti	Cationite KU-2 X20 (Na^+ form) a) EDTA solution of pH 2 (adsorption of Er; EDTA-complexed Fe, Al, and Ti pass into the effluent) b) HCl, HNO_3, or H_2SO_4 solutions (elution of Er) The method was used to separate milligram amounts of the elements which were determined titrimetrically	363
^{90}Y from ^{90}Sr	Dowex® 50-X2 (NH_4^+ form) Quantitative separation of ^{90}Y from ^{90}Sr is effected on a column of this resin by using 0.01 *M* EDTA at pH 5.11 as eluent	365
Rare earth fission products	Dowex® 50 (grain size 74—36 μm) Column: 33 cm × 1 cm^2; operated at 80°C a) 0.025 *M* EDTA solution of pH 3.05 (elution of Y, Eu, and Pm) b) Same solution of pH 3.65 (elution of Ce)	362
Ho or Tm	Dowex® 50-X8 (Na^+ form) Column: 10 × 0.4 cm ID A substoichiometric amount of EDTA is added to the sample solution (containing Ho or Tm down to 0.4 μg/5 mℓ and labeled with ^{166}Ho or ^{175}Yb) and the anionic complexes are formed by heating the mixture (at pH 5.5) at 50°C for ≃30 min; the solution is then passed at a rate of 0.3 mℓ/min through a column of the resin and the activity of the effluent is measured Elements interfering with this isotope dilution method such as Bi, Cd, Co, Fe, Ga, Hg, In, Ni, Pb, and Zr must first be removed by extractions with cupferron-diethylammonium diethyldithiocarbamate	364

Table 55
CATION EXCHANGE SEPARATION OF THE RARE EARTHS IN SULFOSALICYLIC ACID MEDIA

Elements separated	Ion exchange resin, separation conditions, and remarks	Ref.
Rare earths from Be, Al, Fe(III), Th, Mo, U, Ti, Zr, V, and W	Dowex® 50W-X8 (100 mesh; NH_4^+ form) Column of 1 cm ID containing the resin to a height of 8 cm a) 0.02 *M* sulfosalicylic acid adjusted to pH 6.5 (1500 mℓ) (adsorption of rare earths except Sc; elution of U, Th, Mo, Fe, Al, and Be) b) 0.02 *M* sulfosalicylic acid of pH 6.5 (50 mℓ) followed by water (as a rinse) c) Saturated NH_4Cl solution (25 mℓ) (elution of rare earths) Following coprecipitation with Be-cupferronate, the rare earths were determined spectrographically; La, Yb, Y, and Gd in 16-μg amounts were thus determined in Th (1 g) and U (1 g) with recoveries of 14—18 μg Separation from Ti, Zr, V, and W is carried out similarly, but at pH <4.0	367
Ce(IV) from U(VI), Ti, and Zr	Strongly acidic cation exchange resin (NH_4^+ form) a) 5% Sulfosalicylic acid solution of pH 4 (adsorption of Ce; elution of U, Ti, and Zr) b) 2 *N* H_2SO_4 (elution of Ce) Ce was determined spectrophotometrically using catechol violet	368
Ce(IV) from Fe(III) and Al	Merk I cation exchange resin (NH_4^+ form) a) 2 *M* sulfosalicylic acid (pH 2—4.5) (adsorption of Ce [IV]; elution of Fe [III] and Al) b) 4 *M* HCl (elution of Ce) In the eluate, Ce was determined spectrophotometrically	369
Ce(III,IV) and La	Merk I cation exchange resin (NH_4^+ form) a) 5% Sulfosalicylic acid of pH 4 (adsorption of Ce and elution of La) b) 10% Sulfosalicylic acid solution of pH 4 (elution of Ce[III]) c) 30% Sulfosalicylic acid of pH 4 (elution of Ce[IV]) Ce was determined spectrophotometrically	370

Table 56
CATION EXCHANGE SEPARATION OF THE RARE EARTHS IN ACETATE MEDIA

Elements separated	Ion exchange resin, separation conditions, and remarks	Ref.
Sc and lanthanides	Dowex® 50-X12 (50-100 mesh; H^+ form) Column of 0.7 cm ID containing 3.75 mℓ of the resin a) 0.5 *M* ammonium acetate buffer of pH 4.6 (adsorption of lanthanides and elution of Sc) b) 1 *M* ammonium acetate buffer of pH 5.0 (elution of lanthanides) The method was used to separate milligram amounts of the elements	371
Dy and Y	Dowex® 50W-X8 or -X12 (50-100 mesh; NH_4^+ form) Column: 20 × 3 cm operated at a flow rate of 10^{-3} mℓ/sec a) 0.5 *M* HCl (1—2 mℓ) (adsorption of Dy and Y) b) 24:1 Mixture (1400 mℓ) of ammonium acetate and acetic acid (0.49 *M* in acetate) (Dy is eluted before the Y) In the fractions obtained, the rare earths were determined by EDTA titrations	372
Y from Ce	Cationite KU-2 (H^+ form) Column: 22 cm × 15 mm a) Sample solution of pH 2 (adsorption of Y and Ce) b) 1 *M* acetic acid adjusted to pH 4.4 (Y eluted before the Ce)	373

Table 57
CATION EXCHANGE SEPARATION OF THE RARE EARTHS IN OXALIC ACID MEDIA

Elements separated	Ion exchange resin, separation conditions, and remarks	Ref.
Sc from rare earth elements	Dowex® AG 50W-X8 (100-200 mesh; H^+ form) Column of 0.8 cm ID containing 2 g of the resin pretreated with 0.1 *M* oxalic acid (10—15 mℓ) a) 20:180 Mixture (~200 mℓ) of saturated oxalic acid solution and water (adsorption of the rare earths; elution of Sc) (flow rate: 10 mℓ/min) b) 0.1 *M* oxalic acid (80 mℓ) (elution of residual Sc) c) 0.2 *M* HNO_3 (10—15 mℓ) (removal of remaining oxalic acid) d) 5 *M* HNO_3 (or 3 *M* HCl) (elution of the rare earths) Sc was quantitatively recovered as the oxide by evaporation of the solution and ignition of the residue; before the separation, the sample containing Sc (100 mg) is dissolved, the solution is evaporated to near dryness, and the acid-moist residue is taken up in eluent (a); a modification of the above procedure also allows the separation of small amounts of the Sc from large quantitites of the rare earths; to prevent precipitation of the latter as insoluble oxalates the eluents (a) and (b) are made 1 *M* in HNO_3 in addition to 0.1 *M* oxalic acid; a separation factor of >800 between Sc and the rare earth metals was measured	281
Sc from La and Th	Diaion® SK1 (H^+ form) (100-200 mesh) Columns: 5 × 1 cm and 24 × 1 cm With 2% oxalic acid solution (15 mℓ) Sc is eluted from the small column of this resin while La is further retained By using 3% ammonium-oxalate - 1% ammonium acetate as the eluent the Sc is eluted before the Th (large column) These procedures have been used for the separation of milligram amounts of the elements	382
^{90}Y from ^{90}Sr	Dowex® 50-X8 (H^+ form) Column: at least 20 cm long and 0.8 cm ID a) Sorption of alkaline earth fraction of fission products b) 1% Ammonium oxalate solution (elution "milking" of ^{90}Y) The method has been used for the preparation of carrier-free ^{90}Y from ^{90}Sr; it can also be utilized for the separation of the rare earths, as a group, from other fission products; the advantages of the use of the ammonium oxalate solution instead of citrate or lactate solution are as follows In the preparation of the eluent, no pH adjustment is neccessary There is no risk of growth of bacteria in the solution The volume required to elute the rare earths is less than that needed in the case of citrate or lactate; hence, one can obtain an eluate having a higher specific activity The oxalate in the effluent is easily decomposed or converted into another anion	383

Table 58
CATION EXCHANGE SEPARATION OF THE RARE EARTHS IN GLYCOLLIC ACID MEDIA

Elements separated	Ion exchange resin, separation conditions, and remarks	Ref.
La, Ce, Pr, Nd, Sm, and Y	Amberlite® IR-120 (Na^+ form) (Amberlite® SA-2 cation exchange paper) On a resin paper rotated at 1500 rpm (centrifugally accelerated) with 0.25—0.4 *M* glycollic acid (pH 3.0—4.5) as eluent, the elements named in column 1 can be separated within 3—9 min, with the exception of Pr and Nd (which are only partially separated), and Sm and Y; the latter pair can be separated with 0.4 *M* lactic acid	385
Tm, Tb, Lu, Er, and Yb	Dowex® 50-X12 (−400 mesh) The rare earths can be separated within 30 min by selective elution with 0.25 *M* glycollic acid (buffered to pH 3.48 or 3.58) from a column (<1 cm × 2 mm) of this resin Separation of Y from Yb and of Am from Cm is poor on this short column The Ca in commercial glycollic acid must be removed initially by adsorption on Dowex® 50 (H^+ form)	77

Table 59
CATION EXCHANGE SEPARATION OF THE RARE EARTHS IN FORMIC ACID MEDIA

Elements separated	Ion exchange resin, separation conditions, and remarks	Ref.
Fission products	Dowex® 50-X8 (H^+ form) A) First column operation Column: 8 × 1 cm containing 100-150 mesh resin; flow rate: 0.3—0.4 mℓ/min a) 0.2 *M* HCl (20 mℓ) (adsorption of the rare earths and other fission products; elution of Ru) b) Water (as a rinse) c) 0.5% Oxalic acid solution (50-70 mℓ) (elution of U, Pu, Zr, and Nb) d) Water (as a rinse) e) 1 *M* formic acid containing 1 *M* ammonium formate (in the ratio of 7:3) (= formate buffer solution) of pH 3.2 (110 mℓ) (elution of Cs) f) Formate buffer solution (170 mℓ) of pH 3.8 (ratio 4:6) (elution of rare earths) g) Formate buffer solution (100 mℓ) of pH 4.2 (ratio 2:8) (elution of Sr) h) 2 *M* ammonium formate (100 mℓ) (elution of Ba) B) Second column operation (separation of light from heavy rare earths) Column: 20 × 1 cm containing 200-250 mesh resin; flow rate 0.3—0.4 mℓ/min a) Formate buffer solution of pH 3.8 (the Y group is eluted before the Ce group) b) Formate buffer solution of pH 3.6 (ratio 1:1) (partial separation of the elements of the Ce group); for better separation, a buffer solution of pH 3.3 ~ 3.4 or a larger column must be used In the various eluates, the elements were determined radiometrically The method may be useful for the determination of fission products in natural waters; furthermore, it can be utilized for the preparation of pure and carrier-free radioactive elements from mixed fission products	379

Table 60
CATION EXCHANGE SEPARATION OF THE RARE EARTHS IN ASCORBIC ACID AND MALATE MEDIA

Elements separated	Ion exchange resin, separation conditions, and remarks	Ref.
^{141}Ce, ^{153}Sm, ^{160}Tb, and ^{170}Tm	Dowex ® 50W-X8 (20-50 mesh; Na^+ form) Column of 0.5 cm ID containing the resin to a height of 9 cm a) 2 *M* HCl (adsorption of rare earth tracers) b) 0.5% Na ascorbate solution of pH 6.5—7.0 (the rare earths are eluted in the general sequence of elution, i.e., Tm first and then followed in this order by Tb, Sm, and Ce)	387
Yb, Eu, Sc, Y, Dy, Ho, and Tb	Dowex® 50W-X8 (NH_4^+ form; 200-400 mesh) Column: 22 × 1.2 cm ID for the separation of Yb from Eu; and column: 11 × 1.2 cm ID for all other separations A) Separation of Yb from Eu With 0.3 *M* malic acid of pH 3.0 the Yb is eluted before the Eu B) Separation of Sc from Y Sc is eluted before Y using 0.2 *M* malic acid of pH 3.25 C) Separation of Sc, Yb, and Dy a) 0.1 *M* malic acid of pH 3.5 (Sc is eluted before Yb) b) 0.1 *M* malic acid of pH 3.75 (elution of Dy) D) Separation of Tb, Yb, and Ho a) 0.1 *M* malic acid of pH 3.5 (Tb is eluted before Yb) b) 0.2 *M* malic acid of pH 3.5 (elution of Ho) The methods were used for the separation of 0.05-mmol quantities of the elements	388

Table 61
CATION EXCHANGE SEPARATION OF THE RARE EARTHS IN AQUEOUS MEDIA CONTAINING ORGANOPHOSPHORUS COMPOUNDS

Elements separated	Ion exchange resin, separation conditions, and remarks	Ref.
Ce from Y	Cation exchange resin a) 0.1% Ethylenedinitrilotetrakis (methylphosphonic acid) solution of pH 4.35 (adsorption of Ce and elution of Y) b) 2 *M* HCl (elution of Ce) With 1-hydroxyethylidenebis (phosphonic acid) and ethylenedinitrilotetrakis (iso-propylphosphonic acid) maximum separation of Ce from Y is attained by elution with a 0.1% solution of the organophosphorus compound at pH 4 or 6.5, respectively	389

Table 62
DISTRIBUTION COEFFICIENTS OF RARE EARTH ELEMENTS IN SYSTEMS CONSISTING OF 90% ORGANIC SOLVENT AND 10% NITRIC ACID OF VARYING MOLARITY (DOWEX® 1-X8)

Metal ion	Molarity of HNO_3	Organic solvent						
		Methanol[407]	Ethanol[408]	Isopropanol[408]	Methylglycol[408]	Tetrahydrofuran[409]	Acetone[409]	Acetic acid[410]
Ce(III)	0.15	$>10^3$	$>10^3$	$>10^3$	$>10^3$	45	$>10^3$	$>10^3$
Yb(III)	0.15	12	23	104	<1	<1	9	54
Y(III)	0.15	8	16	109	3	4	13	33
Sc(III)	0.15	<1	10	22	<1	<1	5	13
Ce(III)	0.3	$>10^3$	$>10^3$	$>10^3$	$>10^3$	92	$>10^3$	$>10^3$
Yb(III)	0.3	13	34	181	<1	<1	11	54
Y(III)	0.3	9	29	190	4	4	22	42
Sc(III)	0.3	<1	11	44	<1	<1	10	15
Ce(III)	0.6	$>10^3$	$>10^3$	$>10^3$	$>10^3$	270	$>10^3$	$>10^3$
Yb(III)	0.6	13	75	370	4	<1	13	54
Y(III)	0.6	11	68	400	4	5	27	73
Sc(III)	0.6	3	12	48	<1	<1	11	17
Ce(III)	0.9	$>10^3$	$>10^3$	$>10^3$	$>10^3$	$>10^3$	$>10^3$	$>10^3$
Yb(III)	0.9	18	103	740	5	5	14	54
Y(III)	0.9	12	105	$>10^3$	4	9	30	73
Sc(III)	0.9	4	14	95	1	<1	12	17
Ce(III)	1.2	$>10^3$	$>10^3$	$>10^3$	$>10^3$	$>10^3$	$>10^3$	$>10^3$
Yb(III)	1.2	20	116	$>10^3$	6	11	14	54
Y(III)	1.2	15	207	$>10^3$	6	16	35	73
Sc(III)	1.2	3	17	110	3	<1	12	17

Table 63
DISTRIBUTION COEFFICIENTS OF RARE EARTH ELEMENTS IN 0.6 *M* NITRIC ACID MEDIA CONTAINING VARYING CONCENTRATIONS OF ORGANIC SOLVENTS (DOWEX® 1-X8)

	Percentages					
Metal ion	**0**	**20**	**40**	**60**	**80**	**90**
			Methanol[407]			
Ce(III)	<1	<1	3	12	$>10^3$	$>10^3$
Yb(III)	<1	<1	<1	<1	4	13
Y(III)	<1	<1	<1	<1	7	11
Sc(III)	<1	<1	<1	<1	<1	3
			Ethanol[408]			
Ce(III)	<1	<1	13	72	240	$>10^3$
Yb(III)	<1	<1	<1	3	10	75
Y(III)	<1	<1	<1	<1	17	68
Sc(III)	<1	<1	<1	<1	4	12
			Isopropanol[408]			
Ce(III)	<1	<1	1	13	$>10^3$	$>10^3$
Yb(III)	<1	<1	3	12	50	370
Y(III)	<1	<1	<1	4	18	400
Sc(III)	<1	<1	<1	<1	10	48
			Methyl Glycol[408]			
Ce(III)	<1	<1	4	12	200	$>10^3$
Yb(III)	<1	<1	<1	<1	3	4
Y(III)	<1	<1	<1	<1	2	4
Sc(III)	<1	<1	<1	<1	<1	<1
			Tetrahydrofuran[409]			
Ce(III)	<1	7	13	14	40	270
Yb(III)	<1	<1	<1	<1	<1	<1
Y(III)	<1	<1	<1	<1	2	5
Sc(III)	<1	<1	<1	<1	<1	<1
			Acetone[409]			
Ce(III)	<1	<1	<1	9	$>10^3$	$>10^3$
Yb(III)	<1	<1	<1	<1	4	13
Y(III)	<1	<1	<1	<1	7	27
Sc(III)	<1	<1	<1	<1	4	11
			Acetic Acid[410]			
Ce(III)	<1	<1	1	4	$>10^3$	$>10^3$
Yb(III)	<1	<1	<1	<1	16	54
Y(III)	<1	<1	<1	<1	12	43
Sc(III)	<1	<1	<1	<1	<1	17

Table 64
DETERMINATION OF RARE EARTH ELEMENTS IN GEOLOGICAL MATERIALS AFTER SEPARATION BY ANION EXCHANGE IN PURE AQUEOUS NITRIC ACID MEDIA

Material	Ion exchange resin, separation conditions, and remarks	Ref.
Marine sediments	Dowex® 1 (100-200 mesh; nitrate form) Column: 10 × 0.9 cm containing 5 g of the resin; a flow rate of 1 mℓ/min is maintained throughout a) 9 *M* HNO_3 - 0.05 *M* $NaBrO_3$ (5 mℓ) (adsorption of Ce[IV]; elution of accompanying elements such as all other rare earths and Ca) b) 9 *M* HNO_3 - 0.05 *M* $NaBrO_3$ (20 mℓ) followed by 9 *M* HNO_3 (5 mℓ) (elution of residual impurities) c) 98% 4.5 *M* HNO_3 - 2% $NH_2OH{\cdot}HCl$ (10% aqueous solution) (25 mℓ) (elution of Ce) The Ce was determined spectrophotometrically using the arsenazo I method; before the ion exchange separation, the sample (2 g) is decomposed with $HClO_4$-HNO_3-HF and Ce (+ other rare earths) is coprecipitated with Ca-oxalate	401
Seawater	A) First column operation (strongly basic anion exchange resin in the nitrate form) a) 8 *M* HNO_3 (adsorption of Pu; elution of Fe, Sb, and the greater part of the rare earths) b) 12 *M* HCl (elution of residual rare earths) After further purification by a cycle of hydroxide, fluoride, and hydroxide precipitations, and removal of Ra-isotopes, ^{210}Pb, and some Th by a double $BaSO_4$ scavenging, the individual rare earths are separated using the following procedure B) Second column operation (Dowex® 50) a) Ammonium lactate solution of pH 3.25 (elution of Y, Eu, Sm, and Nd in this order; Pm is found between Sm and Nd) b) Ammonium lactate solution of pH 3.50 (elution of residual rare earths) The rare earths were determined radiometrically Before the first ion exchange separation step, the radioactive isotopes of Ce, Pm, and other elements are isolated from the seawater (100-ℓ sample) by coprecipitation on ferric hydroxide in the presence of ^{236}Pu tracer and carriers for Sb, Ce, Nd, Sm, Eu, Y, Cs, and Sr; this precipitation is repeated twice	168, 169

Table 65
DETERMINATION OF RARE EARTH ELEMENTS IN GEOLOGICAL MATERIALS AFTER SEPARATION BY ANION EXCHANGE IN METHANOL-NITRIC ACID MEDIA

Material	Ion exchange resin, separation conditions, and remarks	Ref.
Silicate ores	Bio-Rad® AG1-X8 (200-400 mesh; nitrate form) Column: 32 × 1 cm operated at a flow rate of 0.7 mℓ/min a) 95% Methanol - 5% 7 *M* HNO_3 (50 mℓ) (adsorption of the rare earths and Th; elution of Sc and matrix elements) b) Eluent (a) (150 mℓ) (removal of Sc and base metals) c) 55% Methanol - 45% 7 *M* HNO_3 (250 mℓ) (elution of the Y subgroup) d) 8 *M* HNO_3 (180 mℓ) (elution of the Ce subgroup) (Th is further retained) The rare earths were determined by a DC-arc technique or by use of an inductively coupled plasma torch (emission spectrography); the total rare earths in the subgroup can be determined spectrophotometrically with xylenol orange; before the ion exchange separation, the sample is decomposed by fusion with KHF_2 and digestion with HF-HNO_3-$HClO_4$ The method is applicable to a wide range of silicate ores containing rare earth elements from trace amounts to major concentrations; the recoveries range from 92—99%	419
Apatite minerals	Dowex® 1-X8 (100-200 mesh; nitrate form) Column of 1 cm ID containing the resin to a height of 14 cm and operated at a flow rate of ~0.8 mℓ/min a) 95% Methanol - 5% 7 *M* HNO_3 (50 mℓ) (adsorption of the rare earths; Ca and other matrix elements such as Na, Mg, Al, K, Mn, and Fe as well as Sc pass into the effluent) b) Eluent (a) (50 mℓ) (complete elution of Ca) c) 55% Methanol - 45% 7 *M* HNO_3 (200 mℓ) (elution of the heavier rare earths) d) Water (100 mℓ) (elution of the lighter rare earths) The method should be suitable for preparing light and heavy concentrates from mixtures of rare earths obtained from apatite minerals	515
Apatite minerals	Dowex® 1-X8 (100-200 mesh; nitrate form) The rare earth elements (La, Ce, Pr, and Nd) are adsorbed on a batch (0.5 g) of the resin from 95% methanol - 5% 7 *M* HNO_3 (50 mℓ); the resin is recovered by filtration, dried at 120°C, and after spreading it over a disk of self-adhesive foil backed with a cellulose support, the rare earths are determined by X-ray fluorescence spectroscopy Before this separation, the sample (0.5 g) is dissolved in 7 *M* HNO_3 (10 mℓ)	516
Rocks	Dowex® 1-X8 (100-200 mesh; nitrate form) Column of 1 cm ID containing the resin to a height of 8 cm; the resin is preconditioned with eluent (a) a) 90% Methanol - 10% conc HNO_3 (50 mℓ) (adsorption of Lu, Yb, and Tb; into the effluent pass ^{46}Sc and some Lu and Yb) b) 90% Methanol - 10% conc HNO_3 (50 mℓ) (removal of residual activity of ^{46}Sc) c) 90% Methanol - 10% conc HNO_3 (50 mℓ) (elution of main fraction of Lu and Yb) d) 80% Methanol - 20% 7 *M* HNO_3 (3 × 20 mℓ) (elution of Tb) The rare earth elements were determined by γ-spectrometry; before the separation, the sample (100 mg) is subjected to neutron activation, fused with NaOH in the presence of carriers for Lu, Tb, and Yb, and the rare earths are purified by precipitation of the hydroxides; the precision is given as ±5—15%	517
Rocks	Dowex® 1-X8 (50-100 mesh; nitrate form) Column: 30 × 0.8 cm operated at flow rates of 0.5 mℓ/min (sorption step) and 0.5—0.8 mℓ/min (all other eluents) a) 2.5% 7 *M* HNO_3 - methanol (10 mℓ) (adsorption of all rare earths including Y and Sc; elution of accompanying elements)	518

Table 65 (continued)
DETERMINATION OF RARE EARTH ELEMENTS IN GEOLOGICAL MATERIALS AFTER SEPARATION BY ANION EXCHANGE IN METHANOL-NITRIC ACID MEDIA

Material	Ion exchange resin, separation conditions, and remarks	Ref.
	b) Larger volume (30 mℓ) of eluent (a) (elution of Sc) c) 10% 1 *M* HNO_3 - methanol (50 mℓ) (elution of Lu to Tb with Y) (elements lighter than Gd are further retained by the resin) d) Larger volume (40 mℓ) of eluent (c) (elution of all adsorbed elements except Sm and La) e) 10% 0.5 *M* HNO_3 - methanol (50 mℓ) (elution of Sm) f) 0.01 *M* HNO_3 (50 mℓ) (elution of La) The method was applied to determine Sc, Y, Sm, and La in standard rocks following neutron activation of the sample (50 or 200 mg) which were decomposed by treatment with HF and fused with $KHSO_4$ The ion exchange separation step was performed after isolation of the rare earth elements by repeated hydroxide and fluoride precipitations From the eluates, Y, Sm, and La are precipitated as oxalates, and Sc as the 8-hydroxyquinolinate, and then determined radiometrically	

Table 66
DETERMINATION OF RARE EARTH ELEMENTS IN ROCKS AFTER SEPARATION BY ANION EXCHANGE IN ETHANOL-NITRIC ACID MEDIA

Ion exchange resin	Separation conditions and remarks	Ref.
Dowex® 1-X8 (100-200 mesh; nitrate form)	Column of 0.9 cm ID containing a resin bed of 8 cm height, preconditioned with eluent (a) a) 30:10:5 Mixture (~45 mℓ) of ethanol, water, and conc HNO_3 (adsorption of the rare earths; elution of accompanying elements) b) Eluent (a) (2 × 20 mℓ) (as a wash) c) 5:4:1 Mixture (20 mℓ) of ethanol, water, and conc HNO_3 (elution of Eu) d) Water (20 mℓ) (elution of Ce) The rare earths were determined radiometrically Before the separation, the sample (~500 mg) is subjected to neutron irradiation, fused with NaOH in the presence of Ce- and Eu-carriers, and the rare earths are precipitated as the hydroxides	519

Table 67
DETERMINATION OF RARE EARTH ELEMENTS IN ROCKS AFTER SEPARATION BY ANION EXCHANGE IN ACETIC ACID- AND METHANOL-NITRIC ACID MEDIA[a]

Ion exchange resin	Separation conditions and remarks	Ref.
Bio-Rad® AG1-X8 (200-400 mesh; nitrate form)	A) First column opeartion (bulk rare earth separation) Column: 10 × 1 cm containing 2.5 g of the resin a) 80% Glacial acetic acid - 20% 5 *M* HNO_5 (5 mℓ) (adsorption of rare earths; elution of accompanying elements such as Fe, Al, Mg, Ca, etc.) b) 90% Glacial acetic acid - 10% 5 *M* HNO_3 (35 mℓ) (elution of residual elements (mainly Ba) which accompany the rare earths) (flow rate: 0.1—0.3 mℓ/min) c) 0.05 *M* HNO_3 (8 mℓ) (elution of the rare earths) B) Second column operation (separation of the rare earths into three groups Column: 8 × 0.5 cm containing 0.35 g of the resin a) 90% Methanol - 10% 5.25 *M* HNO_3 (4.5 mℓ) (adsorption of light rare earths, i.e., Gd to La; elution of heavy rare earths Lu to Tb) (flow rate: 0.1 mℓ/min) b) 90% Methanol - 10% 0.01 *M* HNO_3 (16 mℓ) (elution of light rare earths [Gd to Ce]) c) 1 *M* HNO_3 (1 mℓ) (elution of La) The rare earths in the various eluates were determined by mass spectrometry The first column operation described above is a modification of a method which has been used to determine minor elements such as U, Th, and the rare earths in marine sediments[524] Before the ion exchange separation steps outlined above, the sample (200 mg) with added tracer isotopes is decomposed by HF-HNO_3, and after evaporation with 6 *M* HCl, the rare earths are leached from the residue with 80% acetic acid - 20% 5 M HNO_3 (5 mℓ) The error of this mass spectometric isotope dilution method is generally <2% and concentrations down to 0.01 ppm are easily determined; the method has been applied to the determination of nine rare earth elements (La, Ce, Nd, Sm, Eu, Gd, Dy, Er, and Yb) in USGS standard rock samples; furthermore, this technique has been utilized in connection with the determination of the rare earth element contents of deep-sea (>4000 m) principally todorokite-bearing ferromanganese nodules and associated sediments from the Pacific Ocean[525]	520—523

[a] See also Tables 52 and 88 in the chapter on Actinides, Volume II.

Table 68
DETERMINATION OF RARE EARTH ELEMENTS IN ACTINIDE MATRICES AFTER ANION EXCHANGE SEPARATION IN PURE AQUEOUS NITRIC ACID MEDIA

Material	Ion exchange resin, separation conditions, and remarks	Ref.
Th	Dowex® 1-X10 (200-400 mesh; nitrate form) Column of 0.6 or 1.0 cm ID containing 3 or 10 g of the resin, respectively From 8 *M* HNO_3 (10 mℓ), Th is adsorbed on the appropriate column of the resin while the rare earths pass into the effluent (for complete elution, the resin bed is washed with 8 *M* HNO_3 [15—35 mℓ]) in which they are determined spectrographically; before this separation, the sample (Th-metal or -oxide) (100 mg or 1 g) is dissolved in 1:1 HNO_3 containing a few drops of 0.04 *M* HF	526
Th compounds obtained from monazite	Strongly basic anion exchange resin (nitrate form) Same separation method as described above; before the separation, the sample of oxides is dissolved with 8 *M* HNO_3 containing a few drops of 5% HF and Y-nitrate is added as a carrier	527
Th-tetrafluoride	Deacidite® FF 7-9X (100-200 mesh; nitrate form) Column: 22 × 1.9 cm Similar separation method as described above From the effluent the rare earths (Eu, Gd, Dy, Sm, and Er) are isolated by TBP-extraction and then determined spectrographically; before the anion exchange removal of Th, the sample (1 g) is dissolved in 7 *M* $Al(NO_3)_3$ (10 mℓ) and Y is added as carrier and internal standard; to prepare the sorption solution an equal volume of conc HNO_3 is added and the mixture is passed through the column at a rate of $\not>$0.25 mℓ/min	528
Th-U and Pu-Th-U alloys	Composite column of 1.5 cm ID consisting of two layers of Dowex® 1-X8 (100-200 mesh; nitrate form) (lower [2-cm-high bed] and upper [10-cm-high bed] layers) and one middle layer of Kel-F or Plaskon® CTFE 2300 coated with TBP (2 g of TBP mixed with 2 g Kel-F) a) 8 *M* HNO_3 - 0.005 *M* HF (10 mℓ) (adsorption of Th and Pu on Dowex® 1 and of U on TBP-coated polymer; into the effluent pass accompanying impurities such as the rare earths, iron, Al, Ni, etc.) b) 8 *M* HNO_3 (50—80 mℓ) (elution of residual impurities) (flow rate: ~1 mℓ/min) The rare earths were determined spectrographically; before the separation the sample (0.2—0.5 g) is dissolved in 10 mℓ of eluent (a)	529
Pu-metal	Dowex® 1-X4 (50—100 mesh; nitrate form) Column: 15 × 1 cm conditioned with 7.2 *M* HNO_3 From 7.2 *M* HNO_3 Pu(IV) is adsorbed on this resin while the rare earth impurities pass into the effluent in which they are determined by means of a spectrographic method; Pu was converted to Pu(IV) by the addition of H_2O_2; coefficients of variation ranged from 6—12% for the various rare earth metals	530

Table 69
DETERMINATION OF RARE EARTH ELEMENTS IN INDUSTRIAL MATERIALS AFTER ANION EXCHANGE SEPARATION IN PURE AQUEOUS NITRIC ACID MEDIA

Material	Ion exchange resin, separation conditions, and remarks	Ref.
Cast steels	Dowex® 1-X10 (200-400 mesh; nitrate form) From 8 *M* HNO_3 (5 mℓ), Th is adsorbed on a column of 1 cm ID containing this resin to a height of 3.5—4 cm, while Y and the rare earths pass into the effluent (after washing the resin bed with an additional 20 mℓ of the acid) in which they are determined spectrophotometrically using the arsenazo I method Before this ion exchange separation, the sample (1 g) is dissolved in HCl-$HClO_4$ (10 mℓ) and after addition of Th-nitrate, the rare earths are first coprecipitated with Th-fluoride to separate them from iron and then with Th-oxalate to achieve further purification	531
High-temperature alloys	Anionite AV-20E (nitrate form) a) 10 *M* HNO_3 containing $KBrO_3$ (adsorption of Ce[IV] and elution of accompanying elements) b) 3 *M* HNO_3 containing hydrazine sulfate (elution of Ce[III]) Arsenazo III was used for the spectrophotometric determination of Ce Before the separation, the sample was dissolved in aqua regia	402
^{90}Sr	Dowex® 1-X4 (100-200 mesh; nitrate form) + PbO_2 (mixed with the resin) Column of 0.3 or 0.4 cm ID with a settled resin bed volume of ~1 mℓ and containing 50 mg of PbO_2 a) 5—9 *M* HNO_3 (sorption solution) followed by 8 *M* HNO_3 (10 mℓ) (adsorption of Ce[IV]; elution of Sr, Y, and trivalent rare earths) b) 0.5 *M* HNO_3 (6 mℓ) or 4 *M* HCl (elution of Ce) For the quantitative Ce adsorption, 50 mg of PbO_2 per milliliter of resin was found to be adequate The method has been applied to the determination of trace amounts of ^{144}Ce activity in ^{90}Sr and to prepare pure carrier-free Ce tracer	400

Table 70
DETERMINATION OF Ce IN Pu ALLOYS AFTER ANION EXCHANGE SEPARATION IN AQUEOUS AND METHANOLIC MEDIA CONTAINING NITRIC ACID[a]

Material	Ion exchange resin, separation conditions, and remarks	Ref.
Ternary and quaternary alloys containing Ce, Pu, Mn, and Co	Dowex® 1-X8 (100-200 mesh; nitrate form) Column: 23 cm × 22 mm ID filled with the resin to a depth of 10 cm A) First column operation (removal of Pu) From 7.2 *M* HNO_3 the Pu is adsorbed on this resin, while Ce, Mn, and Co pass into the effluent from which they are isolated by using the following separation step B) Second column operation (separation of Ce from Mn and Co) a) 95% Methanol - 5% 5 *M* HNO_3 (20 mℓ) (adsorption of Ce, Mn, and Co) b) 95% Methanol - 5% 5 *M* HNO_3 (275 mℓ) (elution of Mn and Co at a rate of 7—15 drops/min) c) 6 *M* HCl (150 mℓ) (elution of Ce at a fast rate) In the eluates, Ce and Mn were determined by potentiometric titrations The second column operation is only required if Mn is present; before the first ion exchange separation, the sample (1—2 g) is dissolved in 12 *M* HCl; the average recovery of Ce from 52 solutions corresponding to 11—36% of Ce in quaternary alloys containing, also, 50% of Pu was 99.9% with a coefficient of variation of 0.4%; very similar results were obtained for Mn	532

[a] See also Table 34 in the chapter on Actinides, Volume II.

Table 71
DETERMINATION OF BURN-UP OF NUCLEAR FUELS AFTER ISOLATION OF Nd BY ANION EXCHANGE IN METHANOL-NITRIC ACID MEDIA

Material	Ion exchange resin, separation conditions, and remarks	Ref.
Nuclear fuel solutions	Bio-Rad® AG1-X4 (200-400 mesh; nitrate form) Column: 6 × 0.4 cm ID operated at a flow rate of <0.1 mℓ/min a) 90% Methanol - 10% 0.4 *M* HNO_3 (0.5 mℓ) (adsorption of fission product Nd + carrier; into the effluent pass heavier rare earths and other accompanying elements) b) 90% Methanol - 10% 0.032 *M* HNO_3 (6 mℓ) (elution of Nd which is contained in the last 3 mℓ) Mass spectrometry was used for the determination of Nd; even small amounts of phosphoric or sulfuric acid prevented the separation	533
Irradiated U	Dowex® 1-X8 (200-400 mesh; nitrate form) Column bed: 6—7 cm operated at a flow rate of 0.5 mℓ/cm² min a) 90% Methanol - 10% 5 *M* HNO_3 (5 mℓ) (adsorption of Nd and other light rare earths; elution of U and fission products) b) 90% Methanol - 10% 0.2 *M* HNO_3 (fractional elution of the rare earths) The Nd fraction obtained is subjected to extraction chromatography on HDEHP-Chromosorb® W to further purify the Nd which is determined by mass spectrometric measurements	534

Table 72
DETERMINATION OF THE BURN-UP OF NUCLEAR FUELS AFTER ISOLATION OF Nd BY ANION EXCHANGE IN METHANOL-NITRIC ACID MEDIA AND CATION EXCHANGE CHROMATOGRAPHY USING α-HIBA

Material	Ion exchange resin, separation conditions, and remarks	Ref.
Nuclear fuel rods	A) First column operation (Wofatit® SBW-X8; 40—80 μm; nitrate form) Column: 6 cm × 0.126 cm^2 operated at a flow rate of 2 mℓ/cm^2 min a) 95% Methanol - 1 *M* HNO_3 (10 mℓ) (adsorption of fission rare earths, trans-Pu elements, U and Pu, followed by a 15-mℓ wash with the same eluent; elution of U and nonadsorbed fission products) b) 75% Methanol - 3 × 10^{-3} *M* HNO_3 (15 mℓ) (elution of the rare earth and trans-Pu elements) B) Second column operation (Wofatit® KPS-X8; 40—80 μm; NH_4^+ form) Column: 10 cm × 0.126 cm^2; operated at a flow rate of 0.4 mℓ/cm^2 min a) 0.1 *M* HNO_3 (adsorption of the rare earth and trans-Pu elements) b) 0.22 *M* α-HIBA of pH 4.6 (chromatographic elution of the rare earth and trans-Pu elements); in the corresponding fractions obtained, the ^{148}Nd was determined by mass spectrometry after isotope dilution with ^{150}Nd and separation from α-HIBA employing cation exchange on the same resin	249, 250
Solutions of U-fission products	A) First column operation (Dowex® 1-X2; 200-400 mesh; nitrate form); resin bed of 3—5 cm in a column of 0.4 cm ID a) 90% Methanol - 10% 5 *M* HNO_3 (5 mℓ) to which ^{142}Nd was added (adsorption of Nd and other light rare earths; elution of fission products) b) 90% Methanol - 10% 0.032 *M* HNO_3 (20 mℓ) (elution of Nd and other rare earths) B) Second column operation (Dowex® 50W-X4; 200-400 mesh; NH_4^+ form) Column: 10 × 0.4 cm ID; a flow rate of 0.4—0.6 mℓ/cm^2 min is maintained throughout The rare earth solution is equilibrated with a small amount (8 mg) of the resin which is then transferred with water to the top of the column filled with the exchanger; subsequently, the rare earths are separated using as an eluent a 125:516:609 mixture of 0.8 *M* α-HIBA (adjusted to pH 4.62 with aqueous ammonia solution), methanol, and water; the Nd fraction (4—5 mℓ) thus obtained can be further purified by application of the following procedure C) Third column operation (Dowex® 50W-X4; H^+ form) (removal of α-HIBA); resin bed of 3 cm in a column of 0.4 cm ID a) 0.05 *M* HNO_3 (adsorption of Nd; α-HIBA passes into the effluent) b) 6 *M* HNO_3 (2—3 mℓ) (elution of Nd) Mass spectrometric analysis was used for the quantitative determination of Nd	200,201

Table 73
DETERMINATION OF Ce AND OTHER RARE EARTHS IN FERROUS MATERIALS AFTER SEPARATION BY ANION EXCHANGE IN METHANOL-(ACETIC ACID)-NITRIC ACID MEDIA

Material	Ion exchange resin, separation conditions, and remarks	Ref.
Ferrous alloys	Dowex® 1-X8 (100-200 mesh; nitrate form) Column of 0.6 cm ID containing the resin to a height of 10 cm and operated at a flow rate of 0.25—0.3 mℓ/min a) 95% Methanol - 5% 5 *M* HNO_3 (20 mℓ) (adsorption of Ce, Pb, and Bi; into the effluent pass Fe and other accompanying elements) b) 95% Methanol - 5% 5 *M* HNO_3 (~70 mℓ or until no more Fe can be detected in the effluent) c) 90% Methanol - 10% 6 *M* HCl (50 mℓ) (elution of Ce) Titration with EDTA was used for the quantitative determination of Ce; the method has been used for the determination of milligram amounts of Ce in steel samples (1 g) which were dissolved in HNO_3; furthermore, it has been suggested to utilize this separation technique for the determination of Ce in monazite sands using a coulometric method for assay[536]	535, 536
Cast iron, steel, and ferro-Si-Mg alloys	Dowex® 1-X10 (50-100 mesh; nitrate form) Column: 50 mℓ buret filled with ~25 mℓ of the resin and operated at a flow rate of 5—10 mℓ/min; the resin is pretreated with eluent (a) a) 200:85:9:≃6 Mixture (~900 mℓ) of methanol, glacial acetic acid, water, and conc HNO_3 followed by 6 bed volumes of the same mixture as a rinse (adsorption of Ce, Pb, and Bi; into the effluent pass Fe and other elements) b) Water (1 bed volume) followed by HCl (1:3) (4 bed volumes) (elution of Ce) In the eluate, Ce was determined spectrophotometrically with 8-hydroxyquinoline Before the separation, the sample (2.5 g or 0.5 g of the ferro alloy) is dissolved in HNO_3; this method can be applied for the determination of Ce in the range of 0—0.008%	537
Plain C and low-alloy steel	Dowex® 1-X10 (nitrate form) This is a modification of the method described above and is suitable for the determination of Ce in the range of 0—0.08% a) 120:50:~10:6 Mixture of methanol, anhydrous acetic acid, water, and conc HNO_3 (adsorption of Ce; elution of Fe and other elements) b) 1200:500:67:33 Mixture of methanol, anhydrous acetic acid, water, and conc HNO_3 (elution of accompanying elements) c) Water followed by HCl (1:3) (elution of Ce) The oxine method was used for the spectrophotometric determination of Ce Before the separation the sample (0.5 g) was dissolved in HNO_3	538
C steels	Bio-Rad® AG1-X10 (50-100 mesh; nitrate form) Column of 1.2 cm OD containing 1.5 g of the resin and operated at a flow rate of ≯15 mℓ/min	539

Table 73 (continued)
DETERMINATION OF Ce AND OTHER RARE EARTHS IN FERROUS MATERIALS AFTER SEPARATION BY ANION EXCHANGE IN METHANOL-(ACETIC ACID)-NITRIC ACID MEDIA

Material	Ion exchange resin, separation conditions, and remarks	Ref.
	From a ~1:5:12 mixture (~885 mℓ) of nitric acid (1:3), glacial acetic acid and methanol, La, Ce, and Pr are adsorbed on a column of this resin while Fe and other elements pass into the effluent; the resin is then drained, dried, and pelletized and the rare earths are determined by X-ray fluorescence; before the separation, the sample (2 g) is dissolved in HNO_3; based on this weight the detection limits are 0.002—0.004%	

Table 74
DETERMINATION OF RARE EARTH IMPURITIES IN RARE EARTH AND OTHER MATRICES AFTER SEPARATION BY ANION EXCHANGE IN METHANOL-NITRIC ACID MEDIA

Material	Ion exchange resin, separation conditions, and remarks	Ref.
La compounds	Dowex® 1-X8 (200-400 mesh; nitrate form) Column: 75 × 0.9 cm (volume 48 mℓ) operated at 50°C and at a flow rate of 0.33 mℓ/cm² min By using 65% methanol - 35% water (made 0.5 *M* in $LiNO_3$ and 0.01 *M* in HNO_3) as the eluent (up to 600 mℓ), the Pr is eluted ahead of the La (in this system the distribution coefficients of Pr and La are 10.5 and 16.0, respectively) Pr was determined by γ-ray spectrometry Before the separation, the sample (La-sulfide or -oxide) is subjected to neutron irradiation and then dissolved in 10 *M* HCl	418
Nd- and Gd-oxides	Dowex® 1-X8 (200-400 mesh; nitrate form) Column operated at 50°C; resin bed: 25 × 0.6 cm; flow rate of 0.5 mℓ/cm² min at sorption, and 1.5 mℓ/cm² min at elution a) 80% Methanol - 0.01 *M* HNO_3 - 0.3 *M* NH_4NO_3 (1 mℓ) followed by washing with the eluent (few milliliters) (for Nd matrices) (adsorption of Nd; elution of rare earth impurities such as Lu, Gd, Eu, Sm, and Pm) b) 85% Methanol - 0.01 *M* HNO_3 - 0.4 *M* NH_4NO_3 or 85% methanol - 0.4 *M* NH_4NO_3 (1 mℓ) (for Gd matrices)(adsorption of Gd; into the effluent pass the heavy rare earths Lu, Tm, Ho, Dy, and Tb when using the first eluent; with the second eluent Gd is eluted ahead of Eu and Sm) c) 0.02 *M* HNO_3 (elution of Nd or Gd) The method has been used to separate microgram amounts of rare earth impurities from milligram quantities of rare earth oxide matrices	416, 417
Am samples	Dowex® 1-X10 (nitrate form) Before the polarographic determination of Eu, the sample (containing ^{243}Am—^{241}Am and isotopes of Cm) is freed from uni- and bivalent cations, Al and various anions by adsorption of the actinides, and Eu on a column of this resin from 1 *M* HNO_3 in 90% methanol; elution of Eu and of the other adsorbed elements is effected with 0.5 *M* HNO_3	540
Glass, Al-foils, and Pr-oxide	Bio-Rad® AG1-X10 (200-400 mesh; nitrate form) A) Isolation of lanthanides from the glass and Al matrices Column of 0.4 cm ID containing the resin to a height of 10 cm and operated at a flow rate of 0.35 mℓ/min a) 90% Methanol - 10% 7 *M* HNO_3 (30 or 70 mℓ) (adsorption of lanthanides; elution of Al, alkali, and alkaline earth metals) b) Same eluent as (a) (50 mℓ) (elution of residual nonadsorbed elements) c) 50% Methanol - 50% 1.5 *M* HNO_3 (30 mℓ) (elution of La, Ce, Pr, and Nd) The eluted lanthanides were determined mass spectrometrically	541

Table 74 (continued)
DETERMINATION OF RARE EARTH IMPURITIES IN RARE EARTH AND OTHER MATRICES AFTER SEPARATION BY ANION EXCHANGE IN METHANOL-NITRIC ACID MEDIA

Material	Ion exchange resin, separation conditions, and remarks	Ref.
	B) Separation of La, Ce, and Nd from the Pr matrix Column of 1 cm ID containing the resin to a height of 17.5 cm and operated at a flow rate of 1 mℓ/min; for this separation, the following eluents are used to achieve fractional elution of the rare earths: 80% methanol - 20% 1 *M* HNO_3 (0—115 mℓ), 75% methanol - 25% 1.4 *M* HNO_3 (115—260 mℓ), 65% methanol - 35% 1.7 *M* HNO_3 (260—305 mℓ), and 50% methanol - 50% 1.5 *M* HNO_3 (>305 mℓ); mass spectrometry was used to determine La, Ce, and Nd in the various eluate fractions Before the anion exchange separation the samples plus ^{142}Ce and ^{145}Nd or ^{150}Nd are irradiated with neutrons and dissolved in HNO_3-HF (glass) or 7 *M* HNO_3 (Pr-oxide and Al-foil) Concentrations in the lower ppm range can be determined with coefficients of variation of 1—14%	

Table 75
DETERMINATION OF RARE EARTH IMPURITIES IN Y OXIDE AFTER SEPARATION BY ANION EXCHANGE IN METHANOL-NITRATE MEDIUM AND CATION EXCHANGE CHROMATOGRAPHY USING α-HIBA

Ion exchange resin	Separation conditions and remarks	Ref.
Dowex® 1-X8 (200-400 mesh; nitrate form) and Dowex® 50-X8 (−400 mesh; NH_4^+ form)	A) First column operation (at 50°C) (anion exchange resin) Column: 25 × 0.6 cm pretreated with 3—4 bed volumes of eluent (b) a) 4 *M* $LiNO_3$ in methanol (0.5 mℓ) (adsorption of Y and rare earth impurities) (flow rate: 0.5 mℓ/cm² min) b) 85% Methanol - 0.01 *M* HNO_3 - 1.5 *M* NH_4NO_3 (28 free bed volumes) (elution of Y) (flow rate: 2 mℓ/cm² min) c) 0.02 *M* HNO_3 (1.5—2 free bed volumes) (elution of the rare earths) Subsequently, the eluted rare earths are subjected to neutron irradiation and fractionated using the following cation exchange procedure B) Second column operation (cation exchange resin) Column: 10 × 0.2 cm a) 0.1 *M* HCl (few drops) (adsorption of the rare earths) b) 0.1 *M* NH_4Cl (3 bed volumes) followed by water (2 bed volumes) (as a wash) c) Ammonium α-HIBA solution (chromatographic elution and separation of the rare earths) In the eluate fractions obtained the rare earths were determined radiometrically Before the anion exchange separation the sample is dissolved in 5 *M* HNO_3	251

Table 76
ANION EXCHANGE SEPARATION OF RARE EARTH ELEMENTS IN PURE AQUEOUS NITRIC ACID MEDIA[a]

Elements separated	Ion exchange resin, separation conditions, and remarks	Ref.
^{144}Pr from ^{144}Ce	Dowex® 1 (100-200 mesh; nitrate form) (before use, the resin is boiled for several minutes in 6 *M* HNO_3 - 0.5 *M* $NaBrO_3$) Column: 7 × 0.5 cm a) 9 *M* HNO_3 containing $NaBrO_3$ (adsorption of ^{144}Ce [IV]) b) 9 *M* HNO_3 (10 mℓ) (under pressure) (elution of carrier-free ^{144}Pr) c) 3 *M* HNO_3 (15 mℓ) containing hydrazine sulfate (elution of Ce[III]) (flow rate: 0.5 mℓ/min) This method has been used for the carrier-free preparation of ^{144}Pr and incorporated in a procedure for the radiochemical determination of ^{144}Ce	403
Sc from Th	Dowex® 1-X8 (50-100 mesh; nitrate form) Column of 1.1 cm ID containing the resin to a height of ~10 cm and pretreated with 8 *M* HNO_3 a) 8 *M* HNO_3 (10 mℓ) (adsorption of Th and elution of Sc) b) 8 *M* HNO_3 (4 × 10 mℓ) (elution of residual Sc) c) 0.5 *M* HNO_3 (elution of Th) Arsenazo I was used for the spectrophotometric determination of Sc	542

[a] See also Table 94 in the chapter on Actinides, Volume II.

Table 77
ANION EXCHANGE SEPARATIONS OF RARE EARTH ELEMENTS IN Mg NITRATE MEDIA

Elements separated	Ion exchange resin, separation conditions, and remarks	Ref.
Eu, Ce, Nd, Sc, U, and Th	Dowex® 1-X8 (100-200 mesh; nitrate form) Column of 1 cm ID containing the resin to a height of ~9.5 cm and operated at a flow rate of 0.3—0.5 mℓ/min A) Mutual separations of the lanthanides a) 3 *M* $Mg(NO_3)_2$ - 0.1 *M* HNO_3 (150 mℓ) (adsorption of Nd and elution of Eu) b) 3 *M* $Mg(NO_3)_2$ - 0.2 *M* HNO_3 (250 mℓ) (adsorption of Ce and elution of Eu) c) 0.2 *M* HNO_3 (50 mℓ) (elution of ^{144}Ce) d) 0.5 *M* HNO_3 (50 mℓ) (elution of Nd) B) Separation of Sc from Th or U a) 1.5, 2, 2.5, and 3 *M* $Mg(NO_3)_2$ (350 mℓ) (adsorption of Th or U; elution of Sc) b) 2.4 *M* HCl (60 mℓ) (elution of Th) c) 1 *M* $HClO_4$ (60 mℓ) (elution of U) C) Separation of Eu from Th a) 2 *M* $Mg(NO_3)_2$ (150 mℓ) (adsorption of Th; elution of Eu) b) 2.4 *M* HCl (50 mℓ) (elution of Th) The methods have been used to separate milligram and tracer amounts of the elements	406
Eu, Nd, Sm, Pr, and La	Dowex® 1-X8 (200-400 mesh; nitrate form) Column of 0.54 cm ID containing 6.4 g of the resin to a height of ~50 cm; operated at a flow rate of 1 mℓ/6 min A) Separation of Eu from Nd By using 1.4 *M* $Mg(NO_3)_2$ (40 mℓ), Eu is eluted before the Nd B) Separation of Sm from Nd a) 2.0 *M* $Mg(NO_3)_2$ (45 mℓ) (adsorption of Nd and elution of Sm) b) 0.5 *M* HNO_3 (~20 mℓ) (elution of Nd) C) Separation of Pr from La a) 1.6 *M* $Mg(NO_3)_2$ (70 mℓ) (adsorption of La and elution of Pr) b) 0.5 *M* HNO_3 (~10 mℓ) (elution of La) The methods were used to separate microgram amounts of the elements	17

Table 78
MUTUAL SEPARATIONS OF RARE EARTH ELEMENTS BY ANION EXCHANGE IN METHANOL-NITRIC ACID SYSTEMS

Elements separated	Ion exchange resin, separation conditions, and remarks	Ref.
Light and heavy rare earths as well as Y and Sc	Dowex® 1-X4 (200-400 mesh; nitrate form) Column: 1/4 in. 10 mℓ buret containing 1 g of the resin; average height of resin bed: 3 in. a) 90% Methanol - 10% conc HNO_3 (1—2 mℓ) (adsorption of the rare earths) (flow rate: 1—2 mℓ/hr) b) Solution of 7 *M* HNO_3 in methanol (gradient elution of the rare earths) (flow rate: 30 mℓ/hr) The rare earth elements are eluted in order of decreasing atomic number, but without sharp separation of those heavier than Dy; emission spectrography was used for the determination of the rare earths; the method is suitable for macroseparations in the preparation of pure rare earth elements, and rapid fractionation of crude mixtures into concentrates for subsequent cation exchange separation	543
Sm from Nd	Bio-Rad® AG1-X4 (200-400 mesh; nitrate form) Column of 2.1 cm ID containing 58 mℓ (20 g dry weight) of the resin a) 90% Methanol - 0.5 *M* HNO_3 (~20 mℓ) (adsorption of Sm and Nd) (flow rate: 1.0 ± 0.1 mℓ/min) b) 85% Methanol - 0.06 *M* HNO_3 (500 mℓ) (elution of Sm) (same flow rate as with [a]) c) 0.1 *M* HNO_3 (200 mℓ) (elution of Nd) (flow rate: 3.0 ± 0.3 mℓ/min) The rare earths elements were determined spectrophotometrically using the chlorophosphonazo III method; for large amounts a gravimetric procedure based on oxalate precipitation was employed	544
Er, Gd, Nd, and La	Bio-Rad® AG1-X4 or -X8 (nitrate form) On a column of this resin, a mixture of equal amounts of Er, Gd, Nd, and La in 80% methanol - 2 *M* HNO_3 is separated by methanol-concentration gradient elution with 40% methanol - 2 *M* HNO_3 The separation of elements heavier than Gd is less favorable than that of those lighter than Gd	545
Tb, Gd, and Eu	Strongly basic anion exchange resin SBW (grain size: <0.063 mm; nitrate form) a) 75% Methanol - 25% 14.8 *M* HNO_3 (Gd is eluted before Eu) Column: 10 × 0.6 cm operated at a flow rate of 0.4 mℓ/cm² min b) 80% Methanol - 20% 14.8 *M* HNO_3 (Tb is eluted first, followed by Gd, and finally Eu is eluted) Column: 20 × 0.6 cm operated at a flow rate of 0.4 mℓ/cm² min) The method has been used to separate milligram amounts of the elements	546
Sm, Pm, and Nd	Strongly basic anion exchange resin SBW (grain size: ~0.063 mm; nitrate form) Column of 0.6 cm ID containing the resin to a height of 20 cm and operated at a flow rate of 0.4 mℓ/cm² min a) 65% Methanol - 35% 14.8 *M* HNO_3 (elution of Sm, Pm, and Nd in this order) b) 50% Methanol - 50% 14.8 *M* HNO_3 (Pm is eluted before Nd)	547

Table 78 (continued)
MUTUAL SEPARATIONS OF RARE EARTH ELEMENTS BY ANION EXCHANGE IN METHANOL-NITRIC ACID SYSTEMS

Elements separated	Ion exchange resin, separation conditions, and remarks	Ref.
Er, Dy, Gd, Eu, Sm, and Y from Ce, Nd, Pr, and La	Deacidite® FF (7-9X) (100-200 mesh; nitrate form) Column 9 × 1.4 cm conditioned with eluent (b) (4—5 column volumes) a) 65% Methanol - 35% 3.5 M HNO_3 (3—4 mℓ) (adsorption of Ce, Nd, Pr, and La) (flow rate: ≯0.1 mℓ/min) b) 65% Methanol - 35% 3.5 M HNO_3 (15 column volumes) (elution of Er, Dy, Gd, Eu, Sm, and Y) (flow rate: 0.2 mℓ/min) c) Water (5 column volumes) (elution of Ce, Nd, Pr, and La) (flow rate: 1 mℓ/min) Spectrography was used for the determination of the rare earths elements The method was used to separate milligram amounts of the elements	548

Table 79
MUTUAL SEPARATIONS OF RARE EARTH ELEMENTS BY ANION EXCHANGE IN DILUTE NITRIC ACID-METHANOL SYSTEMS IN THE PRESENCE OF AMMONIUM AND Li NITRATES

Elements separated	Ion exchange resin, separation conditions, and remarks	Ref.
Carrier-free isotopes of the light rare earths La → Gd	Amberlite® IRA-400 (nitrate form) a) 65% Methanol - 0.01 M HNO_3 - 0.5 M NH_4NO_3 at 20°C (elution of Eu, Nd, Pr, and Ce) b) 65% Methanol - 0.01 M HNO_3 - 0.5 M $LiNO_3$ at 20°C (elution of Gd, Pm, and Nd) c) 65% Methanol - 0.01 M HNO_3 - 2.5 M NH_4NO_3 at 50°C (elution of Gd, Eu, Sm, Pm, Nd, Pr, Ce, and La) d) 65% Methanol - 0.01 M HNO_3 - 1 M $LiNO_3$ at 50°C (gradient elution of Ca, Sr, Y, Ba, Eu, Sm, Pm, Nd, Pr, Ce, and La) The radioisotopes of the rare earths were isolated from a Ta target irradiated with protons, then separated from each other and purified on a column of Dowex® 50-X8 by elution with α-HIBA solution Methanol-ammonium nitrate systems have been applied to the purification of macro amounts of a particular rare earth metal, e.g., Gd or the determination of trace impurities in it; thus, it is feasible to separate tracer amounts of rare earth metals from 100-mg amounts of adjacent members This technique has been utilized to separate light rare earth tracers from macro amounts of Gd, Sm, and Pr	252, 253
Gd, Eu, Sm, Pm, Nd, Pr, Ce, and La	Dowex® 1-X8 or Amberlite® IRA-400 (nitrate form) The elements are separated on a column of the resin at 20°C and at 0.5—0.7 atm pressure by gradient elution with 65% methanol - 2.5 M NH_4NO_3	88

Table 80
MUTUAL SEPARATIONS OF RARE EARTH ELEMENTS BY ANION EXCHANGE IN ETHANOL-(FORMIC ACID)-NITRIC ACID MEDIA

Elements separated	Ion exchange resin, separation conditions, and remarks	Ref.
Gd, Tb, Dy, Ho, Er, Tm, Yb, Lu, and Y — as well as La, Ce, Pr, Nd, Sm, Eu, and Gd	Amberlyst® A-29 (nitrate form) a) 93:3:4 Mixture of ethanol, 26 *M* formic acid, and 13 *M* HNO_3 (elution of Yb and Lu) b) 18:1:1 Mixture of ethanol, 26 *M* formic acid, and 13 *M* HNO_3 (elution of Tm and Yb) c) 18:1:1 Mixture of ethanol, 20 *M* formic acid, and 13 *M* HNO_3 (elution of Er and Tm); with this eluent Y is eluted between Er and Tm d) 18:1:1 Mixture of ethanol, 15 *M* formic acid, and 7 *M* HNO_3 (elution of Ho and Er) e) 85:7:8 Mixture of ethanol, 23 *M* formic acid, and 13 *M* HNO_3 (elution of Tb, Dy, and Ho) f) 8:1:1 Mixture of ethanol, 20 *M* formic acid, and 13 *M* HNO_3 (elution of Gd and Tb) g) 8:1:1 Mixture of ethanol, 15 *M* formic acid, and 10 *M* HNO_3 (elution of Eu and Gd) h) 8:1:1 Mixture of ethanol, 10 *M* formic acid, and 7 *M* HNO_3 (elution of Sm and Eu) i) 8:1:1 Mixture of ethanol, 15 *M* formic acid, and 7 *M* HNO_3 (elution of Sm, Eu, and Gd) j) 15:3:2 Mixture of ethanol, 10 *M* formic acid, and 7 *M* HNO_3 (elution of Nd and Sm) k) 15:4:1 Mixture of ethanol, 5 *M* formic acid, and 7 *M* HNO_3 (elution of Pr, Nd, and Sm) l) 14:5:1 Mixture of ethanol, 5 *M* formic acid, and 7 *M* HNO_3 (elution of Ce[III] and Pr, and of La and Ce[IV]) In general, the heavier elements are eluted first	549, 550
Y, La, and Nd	Dowex® 1-X8 (nitrate form) a) 80% Ethanol - 0.8 *M* HNO_3 (or -1.6 *M* HNO_3) (Y is eluted before La and Nd) b) 80% Ethanol - 0.16 *M* HNO_3 (Nd and Y are eluted before La) The order of elution was always Y, Nd, La	412

Table 81
ANION EXCHANGE SEPARATIONS OF RARE EARTH ELEMENTS IN NITRIC ACID SYSTEMS CONTAINING PROPANOL OR ACETIC ACID

Elements separated	Ion exchange resin, separation conditions, and remarks	Ref.
Lanthanides, Sc, Y, U, Th, Al, Ga, In, Fe, V, and Mo	Dowex® 1-X8 (nitrate form) a) 90 or 95% *n*-propanol or isopropanol - 10 or 5% 5 *M* HNO_3 (adsorption of La, Ce, Pr, Nd, Sm, Eu, Gd, Tb, Dy, Ho, Er, Tm, Yb, Lu, Y, U, and Th; into the effluent pass Sc, Al, Ga, In, Fe, V, and Mo) b) 90 or 95% methanol - 10 or 5% 5 *M* HNO_3 (elution of Y, Gd, Tb, Dy, Ho, Er, Tm, Yb, Lu, and U) c) 0.5 or 1.0 *M* HNO_3 (elution of La, Ce, Pr, Nd, Sm, Eu, and Th) The rare earth elements were determined spectrophotometrically using Solochrome® Fast Red as the color reagent	411
Yb from Al, Ca, Ga, In, Fe(III), Mg, Mn, Cu, Zn, Ni, and V(IV); Sm and Nd from Th, Bi, and Pb	Amberlyst® XN-1002 (60-100 mesh; nitrate form) Column: 16 × 1.2 cm operated at a flow rate of 0.5 mℓ/min From 1.5 *M* HNO_3 - 85% isopropanol, Yb is adsorbed on this resin, while the elements listed in the first column pass into the effluent; when using 1.5 *M* HNO_3 - 45—55% isopropanol as the eluent, Sm and Nd are eluted ahead of Th, Bi, and Pb, which are more strongly adsorbed The methods were used for the separation of milligram amounts of the elements Titrimetric procedures have been employed for the quantitative determinations	550
Nd from Tm (present as nitrates)	Dowex® 1-X8 (nitrate form) a) 17.03 *M* acetic acid (adsorption of Nd and elution of Tm) b) 15 *M* acetic acid (elution of Nd)	414

Table 82
DETERMINATION OF RARE EARTH ELEMENTS IN GEOLOGICAL MATERIALS AFTER ANION EXCHANGE SEPARATION IN HCl SYSTEMS[a]

Geological material	Ion exchange resin, separation conditions, and remarks	Ref.
Rocks	Dowex® 1-X8 (chloride form)	551, 552
	The neutron-irradiated sample is decomposed by carbonate fusion, the rare earth elements (in the presence of carriers for La, Ce, Nd, Gd, Dy, Er, and Yb) are isolated using a conventional hydroxide-fluoride cycle; the rare earth fraction obtained is purified by passing its solution in 6 *M* HCl through a column of this resin; in the effluent the rare earths were determined radiometrically	
	A similar procedure, but using 12 *M* HCl as the sorption solution, has been employed in connection with studies concerning the distribution of La, Eu, and Dy in igneous rocks and minerals	
Feldspars, ilmenite, magnetite, pyroxenes, and silicate rocks	Dowex® 1-X8 (chloride form)	553
	The sample in 9 *M* HCl is passed through a column of this resin to adsorb Fe and other interfering elements; the rare earths pass into the effluent and are determined radiometrically	
	This method was used after neutron activation of the samples	
Uraniferous rocks	Dowex® 1-X8 (100-200 mesh; chloride form) contained in a column of suitable dimensions	554
	a) 10 *M* HCl (50 mℓ) (adsorption of U, Fe, Co, and Zn; the rare earths pass into the effluent together with the alkali and alkaline earth metals)	
	b) 10 *M* HCl (50 mℓ) (elution of residual rare earths and other nonadsorbed elements)	
	Subsequently, the rare earths in the combined eluates (a) and (b) are coprecipitated with Al-hydroxide, the precipitate is irradiated with neutrons, and the rare earths are determined radiometrically free from interferences by fission-produced rare earth metals and by ^{239}Np derived from ^{238}U	
	Before the ion exchange separation, the sample (100 mg) is decomposed by fusion with Na-tetraborate and silicon is removed by treatment with HF-HNO_3	

[a] See also Table 12 in the chapter on Actinides, Volume II and Reference 552a.

Table 83
DETERMINATION OF Sc IN GEOLOGICAL MATERIALS AFTER ANION EXCHANGE SEPARATION IN HCl SYSTEMS

Geological material	Ion exchange resin, separation conditions, and remarks	Ref.
Terrestrial rocks and stony meteorites (chondrites)	Dowex® AG1-X8 (100-200 mesh; chloride form) Column: 15 × 1 cm a) 9 *M* HCl (10 mℓ) (adsorption of Sc, Co, and Zn) (followed by a wash [10 mℓ] with the same acid) b) 3 *M* HCl (60 mℓ) (elution of Sc and Co) c) Water (100 mℓ) (elution of Zn) In the eluates Sc, Co, and Zn were determined radiometrically; the anion exchange separation was carried out after neutron activation of the sample and a number of separation steps involving solvent extraction, distillation, and precipitations in the presence of carriers	423
Meteorites, tektites, and standard rocks	Dowex® 1-X8 (chloride form) From 9 *M* HCl, accompanying elements are adsorbed on a column of the resin while Sc passes into the effluent; Sc was determined by γ-ray measurements Before the ion exchange separation, the sample is irradiated with neutrons, Sc-carrier is added, the sample is dissolved in HNO_3-HCl, and the Sc activity is isolated by coprecipitation with ferric hydroxide and TTA-xylene extraction	555

Table 84
DETERMINATION OF RARE EARTH ELEMENTS IN ROCKS AFTER ION EXCHANGE SEPARATION USING TWO-COLUMN PROCEDURES

Ion exchange resin	Separation conditions and remarks	Ref.
Dowex® 1-X8 (100-200 mesh; chloride and nitrate forms)	A) First column operation (on chloride form of the resin) Column: 8.5 × 1 cm From 9 *M* HCl (10 mℓ), iron and other matrix elements are adsorbed on the resin, while the rare earths, Ba, Sr, Cs, and other elements pass into the effluent; then the column is washed with 9 *M* HCl (4 × 5 mℓ) and the rare earths and Ba are isolated from the combined effluents by the following anion exchange separation step (after removal of interfering nuclides by TBP extraction) B) Second column operation (on nitrate form of the resin) Column: 14 × 1 cm a) 90% Methanol - 10% conc HNO_3 (50 mℓ) (adsorption of the rare earths, Sr, and Ba) (flow rate: 0.8 mℓ/min) b) 85% Methanol - 15% 7.8 *M* HNO_3 (200 mℓ) (elution of Lu, Yb, and Sr) (flow rate: 0.6 mℓ/min) c) 55% Methanol - 45% 7.8 *M* HNO_3 (200 mℓ) (elution of Sr, Tb, Ba, Eu, Sm, and Gd) (flow rate: 0.5 mℓ/min) d) Water (100 mℓ) (elution of Nd, Ce, and La) (flow rate: 0.7 mℓ/min) The rare earths (i.e., Lu, Yb, Tb, Eu, Sm, Nd, Ce, and La) and Ba were determined radiometrically; before the first ion exchange separation step, the sample is irradiated with neutrons, and following the addition of carriers it is decomposed by fusion with NaOH-Na_2O_2; on leaching the melt with water a hydroxide precipitate is obtained which is dissolved in 9 *M* HCl	556, 557
Dowex® 1-X8 (100-200 mesh; chloride form) and Dowex® 50W-X8 (H^+ form; 100-200 mesh)	A) First column operation (on the anion exchange resin) From 6 *M* HCl (50 mℓ), Fe is retained by this resin, while Y and the lanthanides pass into the effluent from which they are isolated using the following cation exchange separation step B) Second column operation (on the cation exchange resin) Column: 105 × 1.63 cm operated at a flow rate of 1.3 mℓ/min a) 0.8 *M* HCl (50 mℓ) (adsorption of the rare earths and other elements, into the effluent pass Li, Na, K, Rb, Mg and Ca) b) 0.8 *M* HCl (5000 mℓ) (elution of Li, Na, K, Rb, Mg, and Ca) c) 4 *M* HCl (270 mℓ) (elution of Al) d) 6 *M* HCl (730 mℓ) (elution of the rare earths) (this eluate also contains Sc and Ba); the rare earth metals were determined by inductively coupled plasma emission spectrography; recoveries of individual RE metals are 90—100%	558

Table 85
DETERMINATION OF RARE EARTH ELEMENTS IN METEORITES AFTER ION EXCHANGE SEPARATION UTILIZING ANION EXCHANGE IN HCl MEDIA FOLLOWED BY CATION EXCHANGE IN THE PRESENCE OF ORGANIC COMPLEXING AGENTS

Ion exchange resin	Separation conditions and remarks	Ref.
Bio-Rad®, Dowex®-2 (100-200 mesh; chloride form) and Dowex® 50W-X12 (−400 mesh; NH_4^+ form)	A) First column operation (anion exchange resin) Column: 10 × 0.6 cm preconditioned with conc HCl (10 mℓ) From conc HCl (2 mℓ), Fe(III) and other elements are retained by the resin, while the rare earths pass into the effluent (after washing the resin bed with conc HCl [2.5 mℓ]) in which they are twice precipitated as hydroxides The final precipitate is dissolved in a minimum amount of dilute HCl and the individual rare earth elements are separated by using the following procedure B) Second column operation (cation exchange resin) Column: 60 × 0.6 cm pretreated with water followed by 1 *M* ammonium lactate (pH 3.2) and operated at 80°C a) Dilute HCl (~10 mℓ) containing the rare earths (adsorption of the rare earths on a small batch [~0.5 mℓ] of the resin which after equilibration is transferred to the top of the column); before transferring this batch of resin, 1 *M* ammonium lactate. (pH 3.2) is run through the column until the effluent has a constant pH of 3.2 (this pretreatment may take as long as 12 hr) b) 1 *M* ammonium lactate solutions of initial pH 3.2 (pH-gradient elution of the rare earths; the pH is increased at a rate of about 0.1 pH unit/hr using a flow rate of about 3 drops/min) In the various fractions the rare earths were determined radiometrically using a scintillation spectrometer; before the anion exchange separation step, the sample (5 g) is irradiated with neutrons, and after addition of carriers to the sample, followed by Na_2O_2 fusion (stony meteorites), or with Fe meteorites by HCl attack, the rare earths are finally separated as hydroxides; the sensitivity of the method is in the fractional ppm region	170
Dowex® 1-X10 (100-200 mesh; chloride form) and Dowex® 50W-X12 (−400 mesh; NH_4^+ form)	A) First column operation (anion exchange resin) Column: 2.5 × 0.5 in. conditioned with conc HCl (10 mℓ) From 12 *M* HCl (6 mℓ), interfering nuclides and other impurities are adsorbed on this resin, while the rare earths pass into the effluent from which they are isolated by a hydroxide precipitation (after washing the resin bed with 15 mℓ of conc HCl to elute residual rare earths); afterwards the individual rare earths are separated by using the following procedure B) Second column operation (cation exchange resin) Column: 60 × 1.5 cm The resin bed is conditioned with 1 *M* ammonium lactate (pH 3.0) and operated at 80°C using a flow rate for gradient elution corresponding to a pH change of ~0.06 pH units/hr a) 6 *M* HCl (5—6 drops) (adsorption of rare earths) b) 1 *M* ammonium lactate of pH 3.0 (elution of Lu fraction) (flow rate: ~10 drops/min) followed by pH-gradient elution of the rare earths by adding 1 *M* ammonium lactate of pH 8.0 to the former; in the various fractions thus obtained, the rare earths were determined radiometrically; before the anion exchange separation step, the sample (~5 g) is irradiated with neutrons, fused with Na_2O_2, and the rare earths are precon-	171

Table 85 (continued)
DETERMINATION OF RARE EARTH ELEMENTS IN METEORITES AFTER ION EXCHANGE SEPARATION UTILIZING ANION EXCHANGE IN HCl MEDIA FOLLOWED BY CATION EXCHANGE IN THE PRESENCE OF ORGANIC COMPLEXING AGENTS

Ion exchange resin	Separation conditions and remarks	Ref.
	centrated using a number of separation steps involving precipitations of hydroxides and fluorides as well as the removal of Fe(III) by solvent extraction	
Dowex® 1-X8 (chloride form) and Dowex® 50-X12 (Na form)	I) Fe meteorites A) First column operation (anion exchange resin) From 6 *M* HCl, residual iron is adsorbed on a column of this resin, and in the effluent Sc, together with other elements, is precipitated as a hydroxide which is dissolved in a 1—2:1 mixture of 0.01 *M* HCl and 0.05 *M* EDTA; the resulting solution is subjected to the following substoichiometric separation step B) Second column operation (cation exchange resin) On this column the negatively charged EDTA complex of Sc is separated from the uncomplexed portion of Sc^{3+}; elution is effected with 0.01 *M* HCl and ^{46}Sc is determined radiometrically; before the first ion exchange separation step the neutron-irradiated sample (+5 mg of Sc-carrier) is dissolved in 6 *M* HCl and the bulk of the Fe is extracted into ethyl ether II) Stony meteorites The irradiated sample is decomposed by fusion with Na_2O_2, Sc-carrier is added, and silica is removed by evaporating the solution with 6 *M* HCl; then the hydroxides are precipitated with ammonia in the presence of tartaric acid to prevent the coprecipitation of Zr; the precipitate is dissolved in HCl, the solution is made 9 *M* in HCl and passed through a column of the anion exchange resin to separate, e.g., Fe and Co, and then Sc is separated in substoichiometric amounts as mentioned above under (I.B) The method is sensitive to 10^{-8}% of Sc, and the results are reproducible to within ± 15%	366

Table 86
DETERMINATION OF RARE EARTHS AND OTHER ELEMENTS IN SILICATES AFTER ION EXCHANGE SEPARATION USING A THREE-COLUMN PROCEDURE

Ion exchange resin	Separation conditions and remarks	Ref.
Amberlite® IRA-400-X8 (chloride form) and Dowex® 50-X8 (H^+ form)	A) First column operation (anion exchange resin) a) 2 *M* HCl (adsorption of Cd, Zn, Bi, Tl, Sn, and Ag; into the effluent pass the rare earths together with alkali metals, alkaline earths, Al, Fe, Ti, Mn, Ni, Co, and V) b) 0.25 *M* HNO_3 (elution of Cd, Zn, Bi, Tl, Sn, and Ag) B) Second column operation (anion exchange resin) (further fractionation of the elements contained in the effluent [a] obtained by the first anion exchange separation) a) 11.3 *M* HCl (adsorption of Ti, Zr, Co, Fe, Ga; into the effluent pass the rare earths, Al, Mn, Ni, V, Pb, and Ag) b) 8 *M* HCl (elution of Ti and Zr) c) 4 *M* HCl (elution of Co) d) 1 *M* HCl (elution of Fe and Ga) C) Third column operation (cation exchange resin) (separation of the rare earths from the accompanying elements contained in eluate [a] of the second column operation) a) 2 *M* HCl (adsorption of Sc, Y, La, Ce, Nd, Sr, and Ba; elution of Ti, Sn, Pb, Zn, Zr, Al, Li, Na, V, Fe, Mg, Mn, Ni, Co, Ga, K, Rb, Cs, and Ca) b) 6 *M* HCl (elution of Sc, Y, lanthanides, and other adsorbed elements) In the various eluates the elements were determined spectrographically This analytical scheme can be utilized for silicate analysis following dissolution of the samples by HF-aqua regia-H_2SO_4	559

Table 87
DETERMINATION OF RARE EARTHS AND OTHER ELEMENTS IN SILICATES AFTER ANION EXCHANGE SEPARATION IN A METHANOL-HCl MEDIUM AND BY A TWO-COLUMN PROCEDURE USING CATION EXCHANGE

Ion exchange resin	Separation conditions and remarks	Ref.
Dowex® 1-X8 (200-400 mesh; chloride form), Dowex® 50-X8 (200-400 mesh; H^+ form), and Dowex® 50-X4 (200-400 mesh; NH_4^+ form)	A) First column operation (anion exchange resin) Column: 10 × 1.2 cm ID pretreated with eluent (a) (20 mℓ) and operated at a flow rate of 2.5 ± 0.3 mℓ/min a) 7 *M* HCl in methanol (80 mℓ) (adsorption of Mn, Cu, Co, Zr, Fe, Nb, and Zn; into the effluent pass the rare earths and all other elements which do not form anionic chloride complexes) b) 12 *M* HCl (150 mℓ) (elution of Mn and Cu) c) 10% Methanol - 6 *M* HCl (150 mℓ) (elution of Co and Zr) d) 1 *M* HCl (100 mℓ) (elution of Fe and Nb) e) 0.05 *M* HCl (250 mℓ) (elution of Zn) From the effluent (a) the rare earths and the other elements are isolated by using the following procedure B) Second column operation (Dowex® 50-X8; H^+ form) Column: 18 × 1.2 cm ID operated at a flow rate of 2.5 ± 0.3 mℓ/min a) 0.05 *M* HCl (50 mℓ) (adsorption of rare earths and all other elements which passed into effluent [a] obtained by the first column operation, except phosphoric acid which was not retained) b) 0.3 *M* HCl - 0.15% H_2O_2 (200 mℓ) (elution of Li and V) c) 0.3 *M* HCl (550 mℓ) (elution of Na + K) d) 0.3 *M* HCl - 0.5% H_2O_2 (700 mℓ) (elution of Ti) e) 0.8 *M* HCl (350 mℓ) (elution of Mg and Ni) f) 1 *M* HCl (1100 mℓ) (elution of Ca, Sr, and Al) g) 4 *M* HCl (100 mℓ) (elution of rare earths and Ba) C) Third column operation (Dowex® 50-X4; NH_4^+ form) (fractionation of the rare earth elements) Column: 11 × 1.2 cm ID operated at the same rate as the other two columns; the resin is pretreated with 20 mℓ of eluent (a) a) 0.24 *M* α-HIBA solution (60 mℓ) of pH 5 (adsorption of Y and light rare earths; into the effluent pass Lu, Tm, Yb, Ho, and Er) b) 0.36 *M* α-HIBA solution (15 mℓ) of pH 5 (elution of Y) Arsenazo III was used for the spectrophotometric determination of Y Spectrophotometric procedures were also used to determine P, V, Zr, and Nb, while all the other elements mentioned above were determined by atomic absorption spectrophotometry; before the first ion exchange separation step, the sample (international standard rock sample) (100 mg) is decomposed with $HClO_4$-HF	254

Table 88
DETERMINATION OF RARE EARTHS AND OTHER ELEMENTS IN ENVIRONMENTAL MATERIALS AFTER ANION EXCHANGE SEPARATION IN CONCENTRATED HCl

Environmental material	Ion exchange resin, separation conditions, and remarks	Ref.
Long-range fallout debris	I) Isolation of Ce, Zr, and Nb (Dowex® 1-X2; chloride form) a) 12 *M* HCl (adsorption of Zr and Nb; elution of Ce and other rare earths) b) 6.5 *M* HCl (elution of Zr) c) 4.5 *M* HCl - 0.06 *M* HF (elution of Nb) From the eluates, the elements are isolated by precipitation (Zr as mandelate, Nb as hydroxide, and Ce as hydroxide and finally as iodate) and determined radiometrically; before this separation, the sample of fallout debris is subjected to carbonate fusion in the presence of carriers; the melt is leached with water and the insoluble portion (containing Ce, Y, other rare earths, Zr, Nb, Sr, and Ba) is treated with HCl; this treatment gives a fraction (which is subjected to the separation [I] outlined above) and a fraction (containing Y, other rare earths, Zr, Nb, Sr, and Ba) from which the elements are isolated by means of the two-column procedure outlined below (following HNO_3 separations) II) Isolation of Y, other rare earths, Nb, and Zr A) First column operation (Dowex® 1-X2; chloride form) The same separation principle as described above under (I) is used; for final purification of Y and other rare earths contained in the 12 *M* HCl eluate, the following separation is carried out B) Second column operation (Dowex® 50-X8; H^+ form) With 4.25% lactic acid solutions of pH 3.5 and 3.7 first the Y and then the other rare earths, respectively, are eluted from the cation exchanger In the eluates the elements were determined radiometrically	278
Deposits and airborne particulate matter	Strongly basic anion exchange resin The neutron-irradiated sample is treated with HF-HNO_3-H_2SO_4 and the solution in 12 *M* HCl is passed through a column of the resin; Sc and Cr in the effluent are separated by extracting the Sc with TTA-benzene; Co, Fe, Sb, and Zn are separated by successive elution from the ion exchange column The elements were determined radiometrically	560

Table 89
DETERMINATION OF RARE EARTHS IN NATURAL WATERS AFTER ANION EXCHANGE SEPARATION IN HCl SYSTEMS

Water	Ion exchange resin, separation conditions, and remarks	Ref.
Seawater	Dowex® 1 (chloride form) From 8 *M* HCl, Fe(III), U(VI), and other elements are adsorbed on a column of this resin, while ^{144}Ce passes into the effluent in which it is determined radiometrically following precipitation as hydroxide, fluoride, and again as hydroxide Before this ion exchange separation ^{144}Ce is isolated from the seawater (200-ℓ samples) by coprecipitation on iron hydroxide in the presence of Ce-carrier; the hydroxide precipitation is repeated	561
	Strongly basic anion exchange resin (chloride form) From 10 *M* HCl, Fe(III) is adsorbed on a column of the resin and ^{144}Ce, which passes into the effluent, is isolated by precipitation as a hydroxide; following extraction of Ce(IV) into isobutylketone from 9 *M* HNO_3 in the presence of bromate, the Ce is back extracted with water containing H_2O_2 to reduce it back to the trivalent state; subsequently, Ce is precipitated as oxalate and determined radiometrically; before the ion exchange separation, the Ce is isolated from the seawater (100-ℓ sample) by coprecipitation with ferric hydroxide	562
	Deacidite® FF-1P (SRA 71) (100-200 mesh; chloride form) Column: 12 × 1.5 cm conditioned with conc HCl (several 20-mℓ portions) From 12 *M* HCl (20 mℓ) trivalent Fe is adsorbed on the resin and Sc is eluted with conc HCl (150 mℓ); from the eluate Sc is isolated by extraction with oxine in butanol and finally determined by atomic absorption spectrophotometry Before the ion exchange separation, the Sc was isolated from the water sample (20 ℓ) by coprecipitation with ferric hydroxide; the recovery is 99—100%	563
Rainwater	Dowex® 1 (100-200 mesh; chloride form) From 12 *M* HCl (5 mℓ), Fe(III) and ^{95}Zr are adsorbed on a column of the resin while ^{147}Pm and other rare earths pass into the effluent; following washing of the resin bed with 12 *M* HCl (10 mℓ), further purification of the lanthanides is accomplished by precipitation and solvent extraction; ^{147}Pm is finally precipitated with Nd-oxalate as a carrier, converted to the oxide, and determined radiometrically; before the ion exchange separation, the Pm is isolated from the water sample by coprecipitation with the hydroxides of Nd and Ce (added as carriers)	564

Table 90
DETERMINATION OF RARE EARTH ELEMENTS IN SEAWATER AFTER ION EXCHANGE SEPARATIONS USING TWO- OR THREE-COLUMN METHODS

Ion exchange resin	Separation conditions and remarks	Ref.
Dowex® 1 (chloride form) and Dowex® 50 (H^+ form)	A) First column operation (anion exchange resin) To separate Fe(III) and U from Ce, a solution saturated with HCl gas is passed through a column of this resin under which condition Fe and U are retained by the exchanger while Ce and other lanthanides pass into the effluent; from the latter the rare earths are isolated by coprecipitation with ferric hydroxide and the precipitate is subjected to neutron irradiation; the irradiated sample is dissolved in acid in the presence of carriers and the lanthanides are separated as a group using a series of fluoride and hydroxide precipitations; then individual lanthanides are isolated by using the following procedure B) Second column operation (cation exchange resin) a) 0.1 *M* HCl (adsorption of lanthanides) b) α-HIBA solution (continuous gradient elution of the rare earths) Following precipitation with oxine the rare earths were determined radiometrically Before the first column operation, Ce and other lanthanides are isolated from the water sample (10 ℓ) by coprecipitation with ferric hydroxide in the presence of ^{88}Y tracer A very similar analytical scheme has been used in connection with the determination of stable Ce and other rare earths in marine sediments[395] For the first column operation Dowex® 1-X10 and 12 *M* HCl were utilized	169, 255
Dowex® 1 (chloride form) and Dowex® 50-X12 (particles of settling rate: 0.5—1.5 cm/min; H^+ form)	A) First column operation (anion exchange resin) From a solution saturated with HCl gas, Fe and U are adsorbed on this resin, while the rare earths pass into the effluent from which they are isolated by repeated coprecipitation on ferric hydroxide; the latter is irradiated with neutrons, dissolved in HCl, known amounts of all the lanthanides and barium are added, and the rare earths are separated and purified as a group using a series of fluoride and hydroxide precipitations; the final hydroxide precipitate is dissolved in dilute HCl and the individual lanthanides are separated by means of the following procedure B) Second column operation (cation exchange resin) To the dilute HCl solution containing the rare earths, 80 mg of the resin are added to adsorb the elements and then the resin is transferred to the top of an ion exchange resin column (25 cm × 3.6 mm ID) heated to 70—80°C The lanthanides are eluted using α-HIBA and continuous gradient elution (variation of pH and of concentration of eluent); from each of the fractions obtained, the rare earths are precipitated with 8-hydroxyquinoline and then determined radiometrically; prior to the anion exchange separation step, the acidified sample (10 ℓ) is equilibrated for 1—4-month periods after addition of Fe carrier and ^{88}Y and ^{139}Ce tracers and then the rare earths are coprecipitated with ferric hydroxide	192

Table 90 (continued)
DETERMINATION OF RARE EARTH ELEMENTS IN SEAWATER AFTER ION EXCHANGE SEPARATIONS USING TWO- OR THREE-COLUMN METHODS

Ion exchange resin	Separation conditions and remarks	Ref.
Dowex® 1-X8 (50-100 mesh; chloride and nitrate forms) and Dowex® 50-X12 (colloidal mesh; NH_4^+ form)	A) First column operation (anion exchange resin in the chloride form) (removal of U and Pa) Column: 12 × 0.5 cm conditioned with 12 *M* HCl From 12 *M* HCl (2 mℓ), U(VI) and Pa(V) are retained by the resin, while the rare earths and other elements such as Th pass into the effluent; after washing the resin bed with 12 *M* HCl (4 × 2 mℓ) the rare earths are isolated from the combined effluent and washings by hydroxide precipitation and Ra-isotopes, ^{210}Pb, and some Th are removed by a $BaSO_4$ scavenging operation Prior to the following ion exchange separation step (B), the RE are again precipitated as hydroxides and the precipitate is dissolved in 7.5 *M* HNO_3 B) Second column operation (anion exchange resin in the nitrate form) (removal of Th) Column: 7 × 1.0 cm conditioned with 7.5 *M* HNO_3 From 7.5 *M* HNO_3 solution (2 mℓ) of the rare earths, Th is retained by the resin (flow rate not exceeding 0.2 mℓ/min); from the effluent and washings (the column is washed with 7.5 *M* HNO_3 until 25 mℓ of the effluent are obtained) containing the rare earths, the ^{144}Ce is isolated by several precipitations (first as Ce[IV]-iodate) and determined radiometrically; from the supernate of the ceric iodate precipitation, ^{147}Pm and other rare earths are precipitated as hydroxides and the individual elements are separated by means of the following separation step C) Third column operation (cation exchange resin) Column: 35 × 1 cm conditioned with the lactate eluent The hydroxide precipitate is dissolved in a minimum amount of dilute HCl, the solution is equilibrated with about 1 g of the resin (100-200 mesh), which is then transferred to a column containing the same resin (colloidal mesh); subsequently, the rare earths are eluted at room temperature with 0.75 *M* lactate solution adjusted to pH 3.32 with ammonia solution using a flow rate of 0.15—0.20 mℓ/min To the Pm fraction Nd-carrier is added, and after precipitation as oxalate the ^{147}Pm is determined radiometrically Prior to the first column operation the ^{144}Ce and ^{147}Pm are isolated from the water sample (100-200 ℓ) by coprecipitation on ferric hydroxide in the presence of rare earth carriers	202

Table 91
DETERMINATION OF RARE EARTH ELEMENTS IN BIOLOGICAL MATERIALS AFTER ANION EXCHANGE SEPARATION IN HCl SYSTEMS[a]

Material	Ion exchange resin, separation conditions, and remarks	Ref.
Biological tissues (e.g., marine organisms)	A) First column operation (Dowex® 1-X8; 100-200 mesh; chloride form) Column of 0.6 cm ID containing 1.7 mℓ of the resin and operated at a flow rate of 2.5—3 mℓ/min From 8 *M* HCl (2 mℓ), Fe(III) and other elements are adsorbed on the resin, while Eu passes into the effluent and washings (~10 mℓ 8 *M* HCl) from which it is isolated by means of the following procedure B) Second column operation (Dowex® 50-X8; 100-200 mesh; H^+ form) Column of 0.6 cm ID containing 2.5 mℓ of the resin a) Very dilute HCl solution (~3—4 mℓ ~ 2 *M* HCl) (adsorption of Eu and other elements) b) 2 *M* HCl (9 mℓ) followed by 1.75 *M* HNO_3 (15 mℓ) (elution of Ta activity and of other elements) (flow rate: ~0.5 mℓ/min) c) 6 *M* HNO_3 (12 mℓ) (elution of the rare earths and Cr, Zr, and Hf) The Eu content of this eluate was determined radiometrically; before the anion exchange separation step, the sample (about 100 mg) of dry or ashed tissue is irradiated with neutrons and the tissue ash is dissolved in HCl-H_2O_2; the dry tissue is wet ashed using HNO_3-$HClO_4$-H_2O_2; the chemical yield of Eu was 98%, the sensitivity of the method was 0.4 ng, and, for the determination of 70 ppb of Eu in dry tissue, the coefficient of variation was ≃9% (ten results)	565
Opium	A) First column operation (Dowex® 1-X2; 200-400 mesh; chloride form) Column: 5 × ~0.7 cm From 8 *M* HCl, Fe(III) plus other elemnts are adsorbed on this resin and the rare earths are isolated from the effluent utilizing a hydroxide precipitation; subsequently, some individual rare earth elements are separated by using the following cation exchange procedure B) Second column operation (Dowex® 50-X12; particles with a settling rate between 2.5 and 5.0 cm/min) Column: 100—120 × 2 mm operated at 87°C a) 0.05 *M* HCl (few drops) (adsorption of the rare earths) b) 0.3—0.4 *M* ammonium lactate solutions of pH 4.15 (concentration gradient elution of the rare earths) The rare earth elements were determined radiometrically; before the anion exchange separation, the sample is ashed, irradiated with neutrons, decomposed with HCl, and the rare earths are coprecipitated with La-oxalate; this precipitation is followed by hydroxide and fluoride precipitations	181
Modern corals	Dowex® 1-X8 (chloride form) From 11.9 *M* LiCl - 0.4 *M* HCl more than 99% of the Sc is retained on a column of this resin, while lanthanides (La, Ce, Nd, Sm, Eu, Tb, Yb, and Lu) pass into the effluent from which they are isolated by precipitations as hydroxides and oxalates The rare earths were determined radiometrically; prior to the separation on the anion exchanger, the sample was subjected to neutron irradiation, dissolved in conc HCl containing La and Sc carrieres, and a hydroxide precipitation was carried out	427

Table 91 (continued)
DETERMINATION OF RARE EARTH ELEMENTS IN BIOLOGICAL MATERIALS AFTER ANION EXCHANGE SEPARATION IN HCl SYSTEMS[a]

Material	Ion exchange resin, separation conditions, and remarks	Ref.
Blood plasma	Dowex® 1-X8 (50-100 mesh; chloride form) Column: 25 × 1 cm conditioned with 0.6 *M* HCl and then washed with 0.003 *M* HCl until the effluent shows a pH of ~2.5 The following eluents are used after addition of the sample solution (0.1—0.2 mℓ) to the top of the column a) 0.003 *M* HCl (88 mℓ, of which the first 8 mℓ can be discarded) (adsorption of phosphate and elution of the lanthanides) (flow rate: 2.5 mℓ/min) b) 0.6 *M* HCl (250 mℓ) (elution of phosphate) In the effluent, the rare earths were determined spectrophotometrically using arsenazo I; before the separation, the sample (1 mℓ) is ashed and its solution in dilute nitric acid is heated to hydrolyze pyrophosphates and polyphosphates	566
Rice plants	Diaion® SA-100 (100-200 mesh; chloride form) Column: 6 × 1 cm From 11 *M* HCl (5—10 mℓ) interfering elements are retained by the resin, while the rare earth elements pass into the effluent from which they are isolated by TTA-benzene extraction; focusing chromatography is then used for their fractionation Before the ion exchange separation, the sample (20 g) is ashed, the ash irradiated with neutrons, and then the ash dissolved in aqua regia	567

[a] See also Table 13 in the chapter on Zinc, Volume IV.

Table 92
DETERMINATION OF RARE EARTH ELEMENTS IN U MATERIALS AFTER ANION EXCHANGE SEPARATION USING HCl MEDIA IN ONE-COLUMN PROCEDURES

Material	Ion exchange resin, separation conditions, and remarks	Ref.
High purity U	Dowex® 1-X8 (200-400 mesh; chloride form) Column: 6 × 0.4 cm ID From 8 *M* HCl, residual U is adsorbed on this resin, while the rare earth elements (Ce, Nd, Sm, Eu, and Gd) pass into the effluent in which they are determined by mass spectrometric analysis; the first three column volumes contain essentially all the rare earths; before this anion exchange separation step, the sample of U oxide (~5 g) is dissolved in conc HNO_3 (in the presence of tracers) and the bulk of U is removed by ether extraction from 8 *M* HNO_3 With this isotope dilution method, trace concentrations of the rare earths down to the ppb range can be measured with good sensitivity and accuracy; errors of <2% for Ce, Nd, Sm, and Gd at the 1-ppm level and 6% for Eu at the level of 0.67 parts per 10^9 have been observed	568
U	Dowex® 1-X8 (20-50 mesh; chloride form) From 8.7 *M* HCl, the U is adsorbed on a column of this resin, while the rare earths pass into the effluent in which they are determined spectrographically; this method permits the separation of microgram quantities of rare earth elements from gram amounts of U	569
^{233}U dioxide	Deacidite® FF (100-200 mesh; chloride form) Column of 1.3 cm ID containing 33 mℓ of the resin conditioned with HCl and operated at a flow rate of 1 mℓ/min a) ~12 *M* HCl (~10 mℓ) (adsorption of U and elution of the rare earths) b) 8 *M* HCl (130 mℓ) (elution of residual rare earths and other impurities) c) 0.1 *M* HCl (80 mℓ) (elution of U) The rare earths were determined by emission spectrography The method was used for the determination of ppm quantities of the rare earths and other impurities contained in samples of ^{233}U dioxide	570
Irradiated U	Dowex® 1-X8 (100-200 mesh; chloride form) Column containing 2 mℓ of the resin and operated at a flow rate of 10 drops/min From 12 *M* HCl (~10 mℓ) residual U and other elements (fission products) are adsorbed on this resin, while radioactive Ce passes into the effluent from which it is isolated by solvent extraction prior to its radiometric determination; before the column operation, the sample is dissolved in conc HCl and the bulk of U and fission products are extracted into a 0.1 *M* solution of trioctylamine oxide in xylene	571
U oxide-Y oxide mixture	Bio-Rad® AG1-X2 (50-100 mesh; chloride form) (volume of resin bed ≃8 mℓ) The U is adsorbed from 0—2 *M* HCl nearly saturated with NH_4Cl (K_d U ≃24—38), while Y passes into the effluent in which it is determined by titration with EDTA; since organic matter is not leached from the resin (as is the case when strong HCl solutions are used) this determination can be carried out without the need to heat the effluent to fumes with $HClO_4$ The precision for 27 determinations was ± 0.3%	573

Table 92 (continued)
DETERMINATION OF RARE EARTH ELEMENTS IN U MATERIALS AFTER ANION EXCHANGE SEPARATION USING HCl MEDIA IN ONE-COLUMN PROCEDURES

Material	Ion exchange resin, separation conditions, and remarks	Ref.
Neutron-irradiated solution containing microgram amounts of U	Dowex® 2-X8 (50-100 mesh) and Dowex® 1-X8 (200-400 mesh) Column: 5 × 0.9 cm ID conditioned with 12 *M* HCl and operated at a flow rate of 0.25 mℓ/min The sample in conc HCl (~100 mℓ) (containing Ce-carrier) is passed through the column containing these two resins to adsorb Ce fission products (U, Pa, and Zr are adsorbed on both resins, while Ru and Mo are best retained on Dowex® 1 which forms the upper layer of this composite column; on the lower layer [Dowex® 2] of the resin bed, Nb is best adsorbed), while Ce passes into the effluent and washings (15 mℓ conc HCl) unadsorbed; after precipitation as the iodate, the Ce is determined by γ-ray spectrometry; the method is applicable to fission product mixtures, e.g., from U in raw cores; recoveries of Ce range from 95—98%	572

Table 93
DETERMINATION OF RARE EARTH ELEMENTS IN U MATERIALS AFTER ION EXCHANGE SEPARATIONS USING TWO-COLUMN PROCEDURES[a]

Material	Ion exchange resin, separation conditions, and remarks	Ref.
U	A) First column operation From 9 *M* HCl (650 mℓ), the U is adsorbed on a column (4.5 cm ID containing 530 mℓ of Zerolit® FF (SRA-69) X7-9; 14-52 mesh; chloride form) (flow rate: 500 mℓ/hr), while the rare earths pass into the effluent and washings (400 mℓ of 9 *M* HCl) to which Al-nitrate is added and a hydroxide precipitation is carried out; the precipitate is dissolved in conc HCl (25 mℓ) and the solution passed through a Dowex® 1-X8 (100-200 mesh; chloride form) column (15 × 1 cm; containing 12 mℓ of the resin) to remove residual U; the rare earths and Al pass into the effluent and washings (25 mℓ 9 *M* HCl) and following precipitation and heating at 1100°C the sample is irradiated with neutrons; then the sample is dissolved in conc HNO_3, La- and Zr-carriers are added, and the rare earths are purified by a cycle of fluoride and hydroxide precipitations; subsequently, the individual elements are separated by using the following procedure B) Second column operation (Dowex® 50-X8; grain size corresponding to a sedimentation rate of 0.2—0.4 cm/min) (NH_4^+ form) Column: 9 cm × 2 mm a) 0.05 *M* HCl (adsorption of rare earths) b) Ammonium α-HIBA solutions (0.075, 0.1, and 1.8 *M*) of pH 4.75 (concentration gradient elution of the rare earths at a flow rate of ≃0.04 mℓ/min); the lanthanides pass into the effluent in the following order: Lu, Yb, Tm, Er, Ho, Y, Dy, Tb, Gd, Eu, Sm, Pm, Nd, Pr, Ce, and La In the fractions thus obtained the rare earths were determined radiometrically; prior to the anion exchange separation steps, the sample (50 g) is dissolved in HNO_3 and nitrate is removed by repeated treatment with HCl; the method was used to determine traces of the lanthanides in gram amounts of U	182, 183
Nuclear reactor fuel	A) First column operation (Dowex® 2-X8; 100-200 mesh; chloride form) From 10 *M* HCl, U is adsorbed on this resin, while other elements including the rare earths pass into the effluent from which the latter are isolated and further purified by hydroxide and fluoride precipitations; finally, the rare earths are separated into the individual elements by using the following cation exchange separation technique B) Second column operation (Dowex® 50-X8; 200-400 mesh; NH_4^+ form) For this separation, concentration gradient elution at pH 4.35 and 80°C utilizing 0.13—0.5 *M* α-HIBA as the eluent is employed; in the various fractions obtained, the rare earths were determined radiometrically; before the first column operation, the sample is irradiated with neutrons and then dissolved in 10 *M* HCl in the presence of carriers for each rare earth element	184
Uranyl chloride	A) First column operation (Dowex® 1-X8; chloride form) From 90% methanol - 10% 12 *M* HCl the U (1 g uranyl chloride) is adsorbed on a column of this resin, while the rare earths pass into the effluent; further purification of these elements is achieved by means of the following cation exchange separation procedure B) Second column operation (Dowex® 50-X8; H^+ form) a) Effluent from (A) (adsorption of rare earths) b) 4 *M* HCl (elution of rare earths) In the eluate, the lanthanides were determined spectrographically	433

[a] See also Tables 28 and 30 in the chapter on Actinides, Volume II.

Table 94
ION EXCHANGE SEPARATIONS OF HEAVIER RARE EARTHS FROM NEUTRON-IRRADIATED URANIUM TARGETS

Ion exchange resin	Separation conditions and remarks	Ref.
Dowex® 1-X8 (50-100 mesh; chloride and nitrate forms) and Dowex® 50-X8; 200-400 mesh; NH_4^+ form)	A) First column operation (chloride form of anion exchanger) (separation of the rare earths from the bulk of U and fission products) From the sample solution 8 *M* in HCl, U is adsorbed on a column (25 × 0.8 cm) of this resin, while the rare earths pass into the effluent from which they are isolated and purified by precipitations as hydroxides and fluorides; the final hydroxide precipitate is dissolved in ~8 *M* HNO_3 and from this solution ^{239}Np (an activation product) is removed using the following separation B) Second column operation (nitrate form of anion exchanger) (separation of the rare earths from Np) Following reduction of Np to the tetravalent state with ferrous sulfamate in the presence of $NH_2OH \cdot HCl$ as a holdback reductant, this actinide is adsorbed on a column (20 × 0.8 cm) of the resin, while the rare earths pass into the effluent (the decontamination factor achieved in one column operation is $\simeq 10^3$; seven such operations remove all ^{239}Np) Major contaminants of this effluent are Ba, Sr, Ce, and Y; Ba and Sr are removed as nitrates and Ce as the iodate; decontamination from the ^{93}Y—^{90}Y activity (a high fission yield product) is achieved by the third column operation described below C) Third column operation (cation exchange resin in a column (50 × 1 cm) operated at 80°C) (separation of heavier rare earths from Y) For this purpose a 0.5% solution of nitrilotriacetic acid (NTA) of pH 5 (adjusted with ammonia solution) or a 0.2 *M* α-HIBA solution of pH 4.65 can be used as eluent; the heavier rare earths are eluted before the Y D) Fourth column operation (cation exchange resin in a column of 50 × 1 cm) (separation of individual rare earths) This separation is effected at 80°C by elution of the rare earths with 0.2 *M* α-HIBA of pH 4.65 using a flow rate of 4—5 drops/min From the eluate fractions the heavier rare earths (Lu, Yb, Tm, Er, Ho, Dy, and Tb) are isolated by precipitation as oxalates and determined radiometrically; before the first column operation, the sample (1—2 g) of natural U is irradiated with neutrons, dissolved in 8 *M* HCl, and milligram amounts of carriers for the heavy rare earths and other elements are added The method is particularly suitable for the separation of low yield (10^{-5}—10^{-7}%), highly asymmetric rare earth fission products, viz., ^{179}Lu, ^{177}Lu, ^{175}Yb, ^{173}Tm, ^{172}Er, ^{171}Er, ^{167}Ho, ^{161}Tb, and ^{160}Tb in the neutron-induced fission of natural and depleted U targets	256

Table 95
ION EXCHANGE SEPARATIONS OF RARE EARTH ELEMENTS FROM FISSION PRODUCT MIXTURES USING TWO-COLUMN PROCEDURES

Ion exchange resin	Separation conditions and remarks	Ref.
Dowex® 2 (200-400 mesh; chloride form) and Dowex® 50W-X4 (200-400 mesh; NH_4^+ form)	A) First column operation (anion exchange resin) Column of 0.3 cm ID containing resin to give a column height of 7 cm and operated at a flow rate of 1 drop/12—15 sec; from 10 *M* HCl, accompanying fission products are adsorbed, while the rare earths pass into the effluent and washings (10 *M* HCl) (12 mℓ), which are evaporated to dryness and the residue taken up in 1 *M* α-HIBA for further separation B) Second column operation (cation exchange resin conditioned with eluent [a]) Column: 17.5 ± 0.2 in. × 0.3 cm a) 1 *M* α-HIBA of pH 1.9 (adsorption of the rare earths) b) 1 *M* α-HIBA (pH-gradient elution of the rare earths by increasing the pH from 2.79—3.40) (at room temperature, using a flow rate of 1 mℓ/10 min; 10-min fractions are collected); this separation which takes 35 hr can be speeded up by using a forced flow rate of 1 mℓ/min With the separation step (B), results are also obtained more rapidly (in 14 hr) by eluting at a higher pH gradient (pH 3.83—8.34) and with a more dilute eluent, i.e., 0.25 *M* α-HIBA[574] Before the anion exchange separation step, the rare earths are extracted into HDEHP-toluene to separate them from other fission products; this process, as well as the first column operation, is repeated to achieve further decontaminations The method can be used for the carrier-free separation of the individual rare earth nuclides from fission product mixtures	172
Dowex® 1-X8 (chloride form) and Dowex® 50-X8	A) First column operation (anion exchange resin) From conc HCl, U and Np are adsorbed on this resin while Y passes into the effluent; subsequently, the Y is subjected to fluoride and hydroxide precipitation steps, and the final hydroxide precipitate is dissolved in HCl and Y is separated from other rare earths by means of the following procedure B) Second column operation (cation exchange resin) Separation from other elements is achieved by elution of ^{94}Y with 0.2 *M* α-HIBA (pH 4.20) which is then determined radiometrically; before the anion exchange separation step the Y is precipitated as the fluoride and then as the hydroxide; chemical yields were found to range from 45—55% and the decontamination factor was $\simeq 10^7$	257
Dowex® 2-X8 (chloride form) and Bio-Rad® AG 50W-X8 (200-400 mesh)	A) First column operation (anion exchange resin column) From 8 *M* HCl, Mo, Pd, Cd, and Zn are adsorbed on this resin while the rare earths pass into the effluent; subsequently, they are separated from one another by means of the following technique B) Second column operation (cation exchange resin) Column: 60 × 0.8 cm operated at 87°C and using a flow rate of 23 and 34 mℓ/hr a) Solution of rare earths (pH $\simeq$1) (adsorption of the rare earths) b) Boiling water (as a rinse) c) 0.26—0.6 *M* ammonium lactate solution of pH 5 (concentration gradient elution of the rare earths)	185

Table 95 (continued)
ION EXCHANGE SEPARATIONS OF RARE EARTH ELEMENTS FROM FISSION PRODUCT MIXTURES USING TWO-COLUMN PROCEDURES

Ion exchange resin	Separation conditions and remarks	Ref.
	In place of eluent (c) the elution can also be effected with 0.20—0.45 *M* α-HIBA solutions of pH 3.2, 3.6, and 4.0 (on a column of Dowex® 50W-X4 at room temperature) The use of lactate as eluent is suitable for fission product samples, but improved separation of metals of the Y group is obtained with simultaneous gradients of pH and of α-HIBA concentrations; before the anion exchange separation, the rare earths are isolated as a group by using a series of precipitation steps and a separation from main impurities (U, Np, Pu, Zr, and Nb) is achieved by extraction into HDEHP-heptane	
Dowex® 2 (100-200 mesh; chloride form) and Dowex® 50 (settling rate in water of 0.6—1 in./min; NH_4^+ form)	A) First column operation (anion exchange resin) Column: 10 × 0.6 cm From conc HCl (5 mℓ containing 2 drops of HNO_3) accompanying activities are adsorbed on this resin, while the rare earths pass into the effluent and washings (5 mℓ) from which they are isolated by a hydroxide precipitation (this column operation is repeated twice); the precipitate is dissolved in a minimum of HCl or $HClO_4$ and the individual rare earths are separated using the following cation exchange separation step B) Second column operation (cation exchange resin) Column: 60 cm × 10 mm; operated at 87°C a) Dilute acid solution of the rare earths (adsorption of the rare earths) b) Water (as a rinse) c) 0.87 *M* lactic acid at pH 3.0 (elution of the rare earths until Ho comes off; the rare earths elute in reverse order to their atomic number; Y falls between Ho and Dy) d) 1.25 *M* lactic acid at pH 3.0 (elution until Nd comes off) e) 1.25 *M* lactic acid at pH 3.3 (elution until the La comes off) A flow rate of 10—25 mℓ/hr is used and fractions are taken every 5—15 min From the appropriate fractions the rare earth activities are isolated by hydroxide and oxalate precipitations and determined radiometrically Prior to the anion exchange separation step, rare earth-carriers, ^{147}Pm tracer, and Ba-holdback carrier are added to the solution and a hydroxide precipitation is carried out; to prepare solid-free solutions of rare earth activities as required for 4 π-counting, the fractions obtained after the cation exchange step (B) are acidified with HCl to pH 2; the lactic acid from the buffer is then extracted with diethyl ether and the solution is passed through a column (60 × 0.2 cm) of Dowex® 50 resin; after washing the resin bed with water to remove dissolved solids, the rare earth element is eluted with 6 *M* HCl	270, 271
Dowex 2® (100-200 mesh; chloride form) and Dowex® 50 (200-400 mesh of which the fraction with a settling rate (in water) of 15 cm/2—15 min was used; NH_4^+ form)	A) First column operation (anion exchange resin conditioned with 12 *M* HCl) Column: 30 × 0.6 cm From conc HCl, Zr and accompanying radioactive impurities are adsorbed on a column of this resin, while the rare earths pass into the effluent and washings (one col-	186

Table 95 (continued)
ION EXCHANGE SEPARATIONS OF RARE EARTH ELEMENTS FROM FISSION PRODUCT MIXTURES USING TWO-COLUMN PROCEDURES

Ion exchange resin	Separation conditions and remarks	Ref.
	umn volume of 12 *M* HCl) from which they are isolated by a hydroxide precipitation after addition of Zr-carrier; subsequently, the precipitate is dissolved in 6 *M* HCl and the anion exchange separation, as well as the hydroxide precipitation steps, is repeated B) Second column operation (cation exchange resin) Column: 60 cm × 7 mm; operated at 95°C (separation of individual rare earths) a) 0.1 *M* HCl (adsorption of the rare earths) b) Boiling water (as a rinse) c) 0.2—0.6 *M* ammonium lactate solution of pH 5 (concentration gradient elution of Y, Eu, Nd, Pr, and Ce at a rate of 0.5—1 mℓ/min) d) 0.24 *M* ammonium lactate solution of pH 5 (elution of Y, Er, and Sm at constant concentration of lactate) The rare earth elements in the 3-mℓ fractions collected are precipitated as oxalates, which are ignited to the oxides, and the rare earths are determined radiometrically Before the anion exchange seperation step, the rare earths are isolated from the fission product mixture by a cycle of fluoride and hydroxide precipitations in the presence of added rare earths and Zr as carriers	

Table 96
DETERMINATION OF Sc IN VARIOUS METALS AFTER ANION EXCHANGE SEPARATION IN HCl MEDIA[a]

Sample material	Ion exchange resin, separation conditions, and remarks	Ref.
Be metal and Be acetate	Dowex® 1-X8 (100-200 mesh; chloride form) Column: 15 cm × 1 cm^2 conditioned with 9 *M* HCl and operated at a flow rate of 0.5 mℓ/min a) 9 *M* HCl (adsorption of Fe[III], Co, and Zn; into the effluent pass Be, Sc, and Cr[III]) b) 8 *M* HCl (100 mℓ) (elution of residual Be, Sc, and Cr) c) 4 *M* HCl (80 mℓ) (elution of Co) d) 0.4 *M* HCl (80 mℓ) (elution of Fe) e) 0.005 *M* HCl (80 mℓ) (elution of Zn) In the eluates, ^{46}Sc, ^{51}Cr, ^{60}Co, ^{59}Fe, and ^{65}Zn were determined radiometrically; in the case of Sc, this element was first separated from Be and Cr by TBP-extraction Before the anion exchange separation, the sample is subjected to neutron irradiation and dissolved in HCl after the addition of carriers	575
High-purity Ga	Dowex® 1-X8 (200-400 mesh; chloride form) Column: 10 × 0.8 cm a) 6 *M* HCl (2 mℓ) (adsorption of Ga, Fe[III], Sn, and Zn; elution of Sc) b) 2 *M* HCl (elution of Ga and Fe) c) 2 *M* HCl - 3 *M* HF (elution of Sn) d) 0.005 *M* HCl (elution of Zn) Radiometric measurements based on substoichiometric methods were used for the determination of Sc and Zn; Ti was determined spectrophotometrically using the oxine method Before the ion exchange separation, the sample is irradiated with neutrons, dissolved in HCl in the presence of carriers, Pt, Au, and Hg are precipitated as the metals, and most of the Ga is removed by extracting it into diethyl ether from 6 *M* HCl	576
Standard Al sample	Dowex® 1-X8 (100-200 mesh; chloride form) Column: 8 cm × 0.8 cm^2 operated at a flow rate of 0.8 mℓ/cm^2 min From 8 *M* HCl (few milliliters), Fe, Co, and Zn are adsorbed on a column of this resin, while Sc passes into the effluent from which it is isolated by TBP-extraction A similar procedure has been used in connection with determination of Sc and other trace metals in Al;[578] the sample (40—50 mg) is irradiated with neutrons, dissolved in conc HCl (containing carriers), and the solution is passed through a column of hydrated Sb_2O_5 to remove ^{24}Na; then the anion exchange separation is carried out in 8 *M* HCl which elutes Sc, Cr, As, Hf, La, Sm, and Eu, while Fe, Co, Cu, Zn, Ga, Mo, and Cd remain on the resin; Sc and the other trace impurities are determined radiometrically; recovery of most of the elements was 96—99% except for As (88%)	577

[a] See also Table 17 in the chapter on Copper, Volume III; Table 5 in the chapter on Iron, Volume V; Table 16 in the chapter on Copper, Volume III; and Table 10 in the chapter on Gallium, Volume VI.

Table 97
ANION EXCHANGE SEPARATIONS OF Y IN HCl MEDIA AS APPLIED TO SAMPLES OF STEEL AND Sr CARBONATE TARGETS

Material	Ion exchange resin, separation conditions, and remarks	Ref.
Steel	Lewatit® MP 5080 (70-150 mesh; chloride form) Column of 1 cm ID filled with the resin to a height of 20 cm; the resin is conditioned with 8 *M* HCl From 8 *M* HCl (10 mℓ) the Fe is adsorbed on this resin, while Y passes into the effluent and washings (30—40 mℓ) to which graphite powder is added; following evaporation, the Y is determined by X-ray fluorescence spectroscopy Before the ion exchange separation the sample (1 g) is dissolved in $HCl\text{-}HNO_3$	269
Irradiated ^{87}Sr carbonate target	A) Isolation of carrier-free ^{87}Y (Dowex® 1-X8; chloride form) From 6 *M* HCl, Fe(III) is adsorbed on a column of this resin, while ^{87}Y passes into the effluent which is evaporated to dryness; the residue is taken up in 0.01 *M* ammonium carbonate (10—20 mℓ) and this solution is used to prepare the ^{87}Y generator (see below under [B]) B) Preparation of the ^{87}Y generator (Bio-Rad® AG1-X10; 100-200 mesh; carbonate form) The carbonate solution mentioned above under (A) is passed through a column of this resin on which ^{87}Y is adsorbed, while ^{87m}Sr passes into the effluent; this Sr-daughter activity is eluted (''milked''), as needed, by passing 0.01 *M* ammonium carbonate solution through the column (after allowing time for grow-in of a suitable amount of ^{87m}Sr) The first column operation is performed after dissolution of a cyclotron-irradiated ^{87}Sr carbonate target [^{87}Sr (p,n) $\rightarrow$ ^{87}Y] and its Ni container in HCl and separation of Sr from Ni by coprecipitation with ferric hydroxide; this generator was used for medical purposes	579

Table 98
DETERMINATION OF Ce IN Fe AND ALLOYS AFTER ANION EXCHANGE SEPARATIONS USING ONE- OR TWO-COLUMN PROCEDURES

Material	Ion exchange resin, separation conditions, and remarks	Ref.
Cast iron	Dowex® 2-X10 (100-200 mesh; chloride form) Column: 9 × 3.5 cm operated at a flow rate of 5—10 mℓ/min a) ≃6 *M* HCl (20 mℓ) (corresponding to 200 mg of the original sample) (adsorption of Fe[III] and elution of Ce and Mn) b) 9 *M* HCl (100 mℓ) (elution of residual Ce) Ce was determined spectrophotometrically using the oxine method; interference by Mn in the eluate is serious and an initial precipitation as Mn ferrocyanide is neccessary; before the ion exchange separation the sample (1 g) is dissolved in 6 *M* HCl-H_2O_2; the accuracy and reproducibility are high, and the absolute error is not greater than ±0.003%	580
Fe and alloys	Dowex® 1-X4 (chloride form) a) 8 *M* HCl (adsorption of Fe; Ce passes into the effluent) b) Dilute HCl, e.g., 0.1 *M* (elution of Fe) The Ce in the percolate is titrated with EDTA solution using arsenazo I as indicator	581
Refractory alloys	Anionite AV-17 A) First column operation (on chloride form of the resin) From 10 *M* HCl, Fe, Cu, Ti, and Zr (partly) are adsorbed on this resin, while Ce together with Ni, Al, and Mn passes into the effluent from which it is isolated by the following anion exchange separation step B) Second column operation (on nitrate form of the resin) a) 98% acetic acid - 5 *M* HNO_3 (9:1) (adsorption of Ce; into the effluent pass Ni, Al, and Mn) b) 1 *M* HNO_3 (elution of Ce) Arsenazo I was used for the spectrophotometric determination of Ce Before the separations the sample was dissolved in HCl-HNO_3; separation (B) has also been employed to separate U, Th, and the rare earth elements[524]	582

Table 99
RADIOMETRIC DETERMINATION OF RARE EARTH ELEMENTS IN INDUSTRIAL MATERIALS AFTER ION EXCHANGE SEPARATIONS USING ONE- OR TWO-COLUMN PROCEDURES[a]

Material	Ion exchange resin, separation conditions, and remarks	Ref.
Steels	Dowex® 1-X8 (chloride form) Column of 8 cm height Removal of Fe is effected by adsorbing it on this resin from 9 *M* HCl (~10 mℓ) following dissolution of the neutron-irradiated sample (about 200 mg) in the same acid in presence of H_2O_2 (1 mℓ 30% H_2O_2) and La-carrier (10 mg); from the effluent and washings (3 × 5 mℓ 9 *M* HCl), the rare earths are isolated by precipitation as fluorides and then they are determined radiometrically; this separation is necessary for the assay of Nd; other rare earths such as La, Ce, and Pr can be determined directly by γ-spectrometry; the estimated error is ±5% for rare earth metal concentrations of 50—900 ppm	583
Organic filter media and high-purity quartz	A) First column operation (Dowex® 1-X10; 100-200 mesh; chloride form) (1-in. column) From ≃12 *M* HCl (5 mℓ), Fe and other elements are adsorbed on this resin, while the rare earths pass into the effluent in which they are precipitated as the hydroxides B) Second column operation (Dowex® 50-X4; −400 mesh) (separation of individual rare earths) 6 *M* HCl (6 drops) is used to load the rare earths onto the column (60 cm long) and then they are separated employing pH-gradient elution with 0.5 *M* α-HIBA at pH 2.9—3.5 From the various fractions obtained, the rare earths (Tm, Er, Ho, and Tb) are precipitated as oxalates and determined radiometrically Before the anion exchange separation the sample is irradiated with neutrons, decomposed with HF-HNO_3 in the presence of carriers, and the rare earths are preconcentrated and purified by a series of fluoride and hydroxide precipitations	173

[a] See also Table 2 in the chapter on Tungsten, Volume IV; Table 8 in the chapter on Gold, Volume III; Table 26 in the chapter on Copper, Volume III; and Tables 29 and 31 in the chapter on Zinc, Volume IV.

Table 100
DETERMINATION OF Gd IN HIGH-PURITY Eu AFTER ION EXCHANGE SEPARATIONS USING TWO- OR THREE-COLUMN PROCEDURES

Ion exchange resin	Separation conditions and remarks	Ref.
Diaion® SA-100 anion exchanger (100-200 mesh; chloride form) and Diaion® SK-1 cation exchanger (100-200 mesh)	I) Method (a) A) First column operation (anion exchanger) The neutron-irradiated Eu sample (37 mg) is dissolved in conc HNO_3 and nitrate is removed by evaporation with HCl; subsequently, the solution of the residue in 6 *M* HCl (20 mℓ) is passed through a column (15 × 1 cm) of this resin to remove interferences; the effluent and washings (15 mℓ 6 *M* HCl) containing the rare earths are evaporated and then these elements are separated by using the following separation step B) Second column operation (cation exchange resin of 100-200 mesh) Column: 37.5 cm × 10 mm a) 0.1 *M* HCl (adsorption of the rare earths) b) 0.33 *M* α-HIBA solution (230 mℓ) of pH 3.77 (elution of Eu and Gd) In the fractions containing the Gd impurity, this element (activity due to ^{159}Gd) was determined radiometrically II) Method (b) A) First column operation (same resin as above under [B]) Column: 38 cm × 10 mm This separation serves to separate the Gd impurity from the Eu matrix (unirradiated sample) in the presence of radioactive Eu tracer a) 0.2 *M* HCl (adsorption of the rare earths) b) Same α-HIBA eluent as used above under (B) (elution of the rare earths) B) Second column operation (same resin as above under [B]) Column: 7 cm × 8 mm a) Dilute HCl solution (prepared by the addition of 1 mℓ of 6 *M* HCl to the combined Gd fractions obtained by the separation step [A]) b) 1 *M* HCl (30 mℓ) (removal of α-HIBA) c) 6 *M* HCl (30 mℓ) (elution of Gd) After evaporation of the eluate, the Gd is irradiated with neutrons, and to facilitate the determination of ^{159}Gd activity, the rare earths are separated as a group by using the following ion exchange step C) Third column operation (cation exchange resin) Column: 20 cm × 8 mm a) 0.1 *M* HCl (adsorption of rare earths and other elements) b) Water (10 mℓ) (removal of HCl) c) 0.25 *M* citric acid solution (60 mℓ) of pH 2.80 (elution of the rare earths and other elements) The fractions containing the rare earths are combined and Gd is separated from Eu by cation exchange as described for method (a)	258

Table 101
ANION EXCHANGE SEPARATION OF Sc FROM Y, LANTHANIDES, AND OTHER ELEMENTS IN HCl MEDIA

Elements separated	Ion exchange resin, separation conditions, and remarks	Ref.
Sc from accompanying elements	Dowex® 1-X8 (chloride form) (200-400 mesh) From 12 *M* HCl (1 mℓ), Sc is somewhat stronger retained by a column (23 cm × 0.50 cm^2) of this resin than Y, lanthanides, and other elements which pass into the effluent unadsorbed; Sc is eluted in a subsequent fraction of this eluent and then determined spectrophotometrically; this method has been used for the separation of microgram amounts of Sc from elements interfering with its determination; a similar technique has been described for the separation of Sc from Y on a 40-cm × 0.4-cm^2 column of the resin	424
Sc from Y and lanthanides	Dowex® 1-X8 (50-100 mesh; chloride form) Column of 0.7 cm ID containing 5 mℓ of the resin a) 13.3 *M* HCl (~25 mℓ) (adsorption of Sc; into the effluent pass Y and the lanthanides) b) 6 *M* HCl (~30 mℓ) (elution of Sc) The method was used to separate milligram amounts of the elements	371
	Dowex® 1-X8 (100-200 mesh; chloride form) Column of 1.05 cm ID filled with the resin (6 g) to a height of ~13 cm, conditioned with eluent (a) (100 mℓ) a) 90% Glacial acetic acid - 10% 3 *M* HCl (90—160 mℓ) (adsorption of Sc; elution of Y and lanthanides — the light rare earths are eluted before Y and the heavy rare earths) (flow rate: 0.2 mℓ/min) b) 1 *M* HCl (elution of Sc) Eluent (a) is passed after adsorption of the elements from 5 mℓ of a solution of the same composition	425
Sc from lanthanides, Al, Th, and Fe	Diaion® SA-100 (100-200 mesh) A) First column operation (on chloride form of the resin) (separation of Sc and Th from lanthanides, Al, and Fe) Column: 15 × 1 cm pretreated with 13 *M* HCl a) 13 *M* HCl (20 mℓ) (adsorption of Sc, Th, and Fe[III]; into the effluent pass the lanthanides and Al) b) 6 *M* HCl (10 mℓ) (elution of Sc and Th) B) Second column operation (on nitrate form of the resin) Column: 17 × 1 cm conditioned with 8 *M* HNO_3 From 8 *M* HNO_3 (10 mℓ), Th is retained by the resin, while Sc passes into the effluent where it is determined spectrophotometrically using the quercetin method	382

Table 102
ANION EXCHANGE SEPARATION OF Sc FROM OTHER ELEMENTS IN HCl SYSTEMS[a]

Elements separated	Ion exchange resin, separation conditions, and remarks	Ref.
Sc from Cu	Dowex® 1-X8 (chloride form) Column: 20 × 2 cm From 6 *M* HCl (containing ≯200 mg Cu), the Cu is adsorbed on this resin (flow rate: 15—20 mℓ/min), while Sc passes into the effluent and washings (75 mℓ 6 *M* HCl) in which it is determined spectrophotometrically using the xylenol orange method	584
Sc from U(VI)	Dowex® 1-X8 (50-100 mesh; chloride form) Column of 1.1 cm ID containing the resin to a height of 10 cm, pretreated with 8 *M* HCl and operated at 2 mℓ/min a) 8 *M* HCl (10 mℓ) (adsorption of U and elution of Sc) b) 8 *M* HCl (4 × 10 mℓ) (elution of residual Sc) c) Water (elution of U) Arsenazo I was used for the spectrophotometric determination of Sc	542
Sc from Y	Dowex® 1-X8 (chloride form) a) Absolute ethanol containing the chlorides of Y and Sc (adsorption of Y and elution of Sc) b) 8:2:2 Mixture of ethanol, water, and HCl (elution of Y) The elements were determined by EDTA titration	585

[a] See also Table 16 in the chapter on Copper, Volume III.

Table 103
DETERMINATION OF RARE EARTH ELEMENTS IN GEOLOGICAL[a] AND INDUSTRIAL MATERIALS AFTER ANION EXCHANGE SEPARATION IN DILUTE SULFURIC ACID MEDIA

Material	Ion exchange resin, separation conditions, and remarks	Ref.
Silicate rocks, ores, and meteorites	Dowex® 1-X8 (100-200 mesh; sulfate form) Column of 1 cm ID containing the resin to a height of 18 cm a) 0.025 *M* H_2SO_4 - 0.07 *M* ammonium sulfate (20 mℓ) (adsorption of Sc, U[VI], Th, Zr, Mo[VI], Nb[V], and Ta[V]; elution of accompanying elements) b) Same eluent as (a) (60 mℓ) (elution of foreign activities at a flow rate of 0.5—1 mℓ/min) c) 2 *M* ammonium sulfate solution (60 mℓ) (elution of Sc) (at the same flow rate) From the eluate, ^{46}Sc is isolated by precipitation as a hydroxide and determined radiometrically; before the anion exchange separation, the sample (0.1—0.4 g) is irradiated with neutrons and then decomposed by fusion with Na_2O_2 in the presence of Sc-carrier; afterwards, Fe is removed by precipitating Sc-hydroxide in the presence of thioglycolic acid; the chemical yield of the method averages 85%	586
U	Dowex® 1-X8 (50-100 mesh; sulfate form) Column of 2.5 cm ID containing the resin to a height of 25 cm and operated at a flow rate of 5 mℓ/min From 0.5% sulfuric acid solution (25 mℓ) containing ammonium thiocyanate (1 g) the U (as a sulfate and/or thiocyanate complex) and Fe(III) (as a thiocyanate complex) are retained by the resin, while the rare earths pass into the effluent and washings (350 mℓ of 0.5% H_2SO_4) in which they are determined spectrophotometrically using the arsenazo I method With this method milligram amounts of the rare earths can be separated from ≃1 g of U In the absence of thiocyanate the Fe is coeluted with the rare earth elements	587

[a] See also Table 98 in the chapter on Actinides, Volume II.

Table 104
SEPARATION OF Sc FROM Y, LANTHANIDES, AND OTHER METAL IONS USING ANION EXCHANGE IN DILUTE SULFURIC ACID MEDIA

Elements separated	Ion exchange resin, separation conditions, and remarks	Ref.
Y, lanthanides, Th, Zr, U, and other elements	Dowex® 1-X8 (100-200 mesh; sulfate form) Column of 0.9 cm ID containing the resin to a height of 21 cm and pretreated with eluent (a) a) ≈0.1 *M* $(NH_4)_2SO_4$ - 0.025 *M* H_2SO_4 (10—20 mℓ) (adsorption of Sc, U[VI], Th, Zr, Mo[VI], and Ta[V]; into the effluent pass Y, lanthanides, and most other elements, e.g., Al, Be, Cd, Co, Cu, Mg, Ga, Ge, Mn, V[IV], and Zn) b) 0.1 *M* $(NH_4)_2SO_4$ - 0.025 *M* H_2SO_4 (60 mℓ) (elution of residual nonadsorbed elements) c) 1 *M* HCl (50 mℓ) or 0.5 *M* H_2SO_4 (60 mℓ) (elution of Sc) When Mo(VI) is accompanied by Sc(III) the latter is eluted first with ~200 mℓ of eluent (a), then Mo(VI) with 0.5 *M* NaOH - 0.5 *M* NaCl (90 mℓ); if Th(IV) and U(VI) are present, the column is washed first with eluent (a) to remove lanthanides and the other nonadsorbable ions; the Sc is recovered in the 100—200-mℓ fraction of this eluent, while Th and U remain adsorbed on the resin; Th is eluted with 4 *M* HCl (40 mℓ) and then U with 1 *M* $HClO_4$ (30 mℓ) If the sample solution contains lanthanides or Y, Sc, Th, Zr, and U, the following eluents are used a) 0.1 *M* $(NH_4)_2SO_4$ - 0.025 *M* H_2SO_4 (200 mℓ) (chromatographic elution of lanthanides or Y and Sc in this order) b) 1 *M* $(NH_4)_2SO_4$ - 0.025 *M* H_2SO_4 (60 mℓ) (elution of Th) c) 4 *M* HCl (50 mℓ) (elution of Zr) d) 1 *M* $HClO_4$ (30 mℓ) (elution of U)	588
Y, lanthanides, Al, Be, Mg, Co, Mn, Cu, Ni, Zn, Cd, Hg, Fe(II), V(IV), and alkali metals	Bio-Rad® AG1-X8 (100-200 mesh; sulfate form) Column: 50 × 2.0 cm containing 20 g of the resin conditioned with 0.01 *N* $(NH_4)_2SO_4$ of pH 3 a) 0.05 *N* H_2SO_4 (250 mℓ) adjusted to pH 3 with dilute ammonia (adsorption of Sc and other elements forming anionic complexes with sulfate; elution of Y, lanthanides, and all other elements listed in the first column) (flow rate: 3.0 ± 0.2 mℓ/min) b) 0.5 *N* H_2SO_4 (250 mℓ) (elution of Sc) (U[VI] and Zr[IV] are further retained by the resin, while Fe[III], Th, Hf, and In are coeluted with the Sc); titration with EDTA using xylenol orange as the indicator was used for the determination of Sc The method has been employed for the separation of milligram amounts of the elements	589
Y, lanthanides, Al, Be, Cu, Cd, Co, and other elements	Dowex® 1-X8 (100-200 mesh; sulfate form) Column of 0.9 cm ID containing the resin to a height of 21 cm and conditioned with eluent (a) and operated at a flow rate of 0.5—1.5 mℓ/min a) 0.1 *M* ammonium sulfate - 0.025 *M* H_2SO_4 (10 mℓ) (adsorption of Sc; into the effluent and washings [60 mℓ of eluent {a}] pass the other elements listed in the first column) b) 1 *M* HCl (50 mℓ) (elution of Sc) For the spectrophotometric determination of Sc, Eriochrome brilliant violet B was used	590

Table 105
SEPARATION OF RARE EARTH ELEMENTS BY ANION EXCHANGE IN THIOCYANATE MEDIA[a]

Elements separated	Ion exchange resin, separation conditions, and remarks	Ref.
Sc from La, Lu, Sm, Eu, Tb, Ca, Al, Cu, Hg, Zn, Cd, and In	Dowex® 1-X8 (100-200 mesh; thiocyanate form) Column of 1.1 cm ID containing the resin (10 g) to a height of ≃18 cm and conditioned with eluent (a) a) 2 *M* NH_4SCN - 0.5 *M* HCl (minimum volume) (adsorption of Sc and other elements; elution of lanthanides, Al, Ca, etc.) b) 3 *M* HCl (elution of Sc) (flow rate: 0.17 mℓ/min) c) Water (elution of Zn) The recovery of Sc is 99%	437
Sc from lanthanides and Th; lanthanides from Fe	Amberlyst® A-21 (macroreticular weak base anion exchange resin with $-N(CH_3)_2$ as the functional group) (thiocyanate form) A) Separation of Sc from lanthanides and Th a) 0.6 *M* KSCN - 0.5 *M* HCl (adsorption of Sc; elution of lanthanides and Th) b) 0.03 *M* KSCN - 0.5 *M* HCl (elution of Sc) B) Separation of lanthanides from Fe a) 0.5 *M* NH_4SCN - 0.5 *M* HCl (adsorption of Fe[III]; La, Er, and Lu pass into the effluent) b) 25:2:2 Mixture of acetone, water, and HCl (elution of trivalent Fe) In the eluates, the elements were determined by EDTA titrations	439
Y or Tb from In	Dowex® 1-X8 (20-50 mesh; thiocyanate form) Column: 3 × 1 cm ID operated at a flow rate of 10 mℓ/hr a) 0.1 *M* NH_4SCN (16 mℓ) (adsorption of In and elution of Y or Tb) b) 0.05 *M* EDTA (26 mℓ) (elution of In) The method was used to separate microgram amounts of the elements	442

[a] See also Tables 72, 110, and 111 in the chapter on Actinides, Volume II.

Table 106
SEPARATION OF Sc FROM V AND Ti BY ANION EXCHANGE IN HF MEDIA

Ion exchange resin	Separation conditions and remarks	Ref.
Dowex® 1-X8 (fluoride form)	0.5—2.5 *M* HF through which a stream of SO_2 is passed to reduce V to V(IV) (adsorption of Sc and Ti; elution of V[IV]) Before the separation, the sample (Ti target) is dissolved in HF This method has been used for the preparation of carrier-free radioisotopes of Sc and V from cyclotron targets	444

Table 107
MUTUAL SEPARATIONS OF RARE EARTH ELEMENTS BY ANION EXCHANGE IN PHOSPHORIC ACID MEDIA

Elements separated	Ion exchange resin, separation conditions, and remarks	Ref.
La + Nd, Tb, Yb, Lu, and Sc	Dowex® 1-X4 ($H_2PO_4^-$ form) a) 0.75 *M* H_3PO_4 (adsorption of Lu and Sc; eluted are Nd + La, Tm, and Yb in this order) b) 2 *M* H_3PO_4 (elution of Lu and Sc; Lu is eluted first) By means of these phosphoric acid media the light rare earths cannot be separated	446

Table 108
SEPARATION OF RARE EARTH ELEMENTS BY ANION EXCHANGE IN CARBONATE AND BICARBONATE MEDIA

Elements separated	Ion exchange resin, separation conditions, and remarks	Ref.
^{90}Y from ^{90}Sr and ^{140}La from ^{140}Ba	Dowex® 1-X8 (50-100 mesh; carbonate form) Column 7 cm × 0.4 cm^2 a) 0.08 *M* ammonium carbonate solution (50 mℓ) (adsorption of ^{90}Y and ^{140}La; into the effluent pass ^{90}Sr and ^{140}Ba) b) 0.01 *M* ammonium carbonate or water (50 mℓ in each case) (elution of residual Sr and Ba) (flow rate: 1.5 mℓ/min) c) 1 *M* HCl (50 mℓ) (elution of ^{90}Y and ^{140}La) This method has been used to obtain carrier-free ^{90}Y or ^{140}La; moreover, it is possible to separate ^{90}Y repeatedly from the carbonate solution by this technique	459
^{90}Y from ^{90}Sr	Dowex® 1-X8 (bicarbonate form) a) 0.2 *M* $KHCO_3$ (adsorption of ^{90}Y and elution of ^{90}Sr) b) Water (as a wash) c) 1 *M* HCl (elution of ^{90}Y)	460
Lanthanides, Th, and U	Dowex® 1-X8 (100-200 mesh; carbonate form) Column: 12 cm × 0.6 cm^2 operated at a flow rate of ≃1 mℓ/min a) 0.5 *M* K_2CO_3 solution (20 mℓ) (adsorption of lanthanides, U, and Th) b) 0.7 *M* K_2CO_3 solution (100 mℓ) or 0.4 *M* $(NH_4)_2CO_3$ solution (elution of lanthanides and Th) c) 1 *M* KCl solution (40 mℓ) (elution of U) The separated elements were determined by EDTA titrations The insolubility of La in carbonate solutions limits the determination of this element to micro amounts	461

Table 109
PRECIPITATION-SEPARATION OF ^{140}La FROM ^{140}Ba ON A STRONGLY BASIC RESIN IN THE HYDROXIDE FORM

Ion exchange resin	Separation conditions and remarks	Ref.
Dowex® 1-X8(100-200 mesh; hydroxide form)	a) <0.01 *M* in HCl (5—10 mℓ) (precipitation of ^{140}La-hydroxide in the resin bed; ^{140}Ba passes into the effluent) b) Water (100 mℓ) (elution of residual Ba) c) 6 *M* HNO_3 (10 mℓ) (elution of ^{140}La) The method was used for the carrier-free preparation of ^{140}Ba	462

Table 110
DETERMINATION OF RARE EARTH IMPURITIES IN HIGH-PURITY RARE EARTH OXIDES AFTER ANION EXCHANGE SEPARATION USING EDTA MEDIA

Material	Ion exchange resin, separation conditions, and remarks	Ref.
High-purity Y-oxide	I) Determination of Dy The sample (5—10 mg) is irradiated with neutrons, dissolved in HCl, and after addition of sufficient 0.1 *M* EDTA to form a 1:1 complex with Y, the solution (2—3 drops) is passed through a column (10 cm × 0.15 cm^2) of the anionite AV-17 (grain size: 15—25 μm); with 0.1 *M* EDTA (pH 4.42) the Y is eluted before the Dy which is determined radiometrically; flow rate: 0.6 mℓ/cm^2 min II) Determination of Eu, Sm, and Gd Following neutron-irradiation, the sample (100 mg) is dissolved in HCl and Y is precipitated as the hydroxide which is dissolved in hot water (10 mℓ) in the presence of a stoichiometric amount of EDTA; then the rare earths are separated using the following procedures A) First column operation (anionite AV-17; EDTA form) Column: 20 × 2.5 cm operated at a flow rate of 1.5 mℓ/cm^2 min a) EDTA solution of rare earths (see above) (adsorption of Eu, Sm, and Gd) b) 0.06 *M* EDTA (~400 mℓ) (elution of Y) c) 0.3 *M* EDTA (~500 mℓ) (elution of Eu, Sm, and Gd) To this eluate (c), HCl is added to precipitate free EDTA, and the rare earth metals are coprecipitated with ferric hydroxide (after destruction of remaining EDTA with ammonium persulfate and H_2O_2) B) Second column operation (anionite AV-17; chloride form) (removal of Fe) From strong HCl solution, the Fe(III) is adsorbed on a microcolumn of this resin, while the rare earths pass into the effluent from which they are isolated individually using the following procedure C) Third column operation (cationite KU-2; grain size: 10—15 μm; NH_4^+ form) On a microcolumn (10 × 0.1 cm) of this resin, Gd, Eu, and Sm are eluted in this order using 0.17 *M* ammonium α-HIBA as the eluent; the rare earths were determined radiometrically	591
Specpure Er-oxide	Amberlite® IRA-400 [EDTA form = (H_2Y^{2-})] With an EDTA solution of pH 4.5—4.7, the rare earth elements are eluted at 40—60°C on a column of this resin in the following order: Lu, Er, Ho, and Dy The elements were determined radiometrically; before this separation, the sample (20 mg) is subjected to neutron irradiation and then dissolved in EDTA solution This method was used for the determination of parts per million quantities of Lu, Ho, and Dy in Er-oxide; a similar technique has been utilized for the analysis of La-oxide and misch-metal[593]	592

Table 111
MUTUAL SEPARATION OF RARE EARTHS USING ANION EXCHANGE IN EDTA AND DCTA MEDIA

Elements separated	Ion exchange resin, separation conditions, and remarks	Ref.
Gd, Tm, and Tb	Dowex® 1-X4 (100-200 mesh; EDTA form) Column: 14 cm × 0.64 cm^2 (operated at 25°C) With 0.1 *M* EDTA of pH 4.8, the rare earth elements are eluted in the order Tm, Tb, and Gd using a flow rate of 0.51 mℓ/cm^2 min; complete separation was obtained in 93 fractions; if the rare earths are eluted with 0.01 *M* EDTA of pH 4.95 at 1.41 mℓ/cm^2 min the separation requires the collection of 300 fractions (5.5 mℓeach)	594
Y, Sc, Sm, Eu, Pm, Tm, Dy, La, Pr, and Cs	Dowex® 1-X4 (grain size: ≤0.07 mm; DCTA form) (DCTA = *trans*-1,2-diaminocyclohexane-NNN′N′-tetra acetic acid) Column: 9.9 cm × 0.0314 cm^2 a) 0.0011 *M* DCTA (disodium salt) of pH 4.8 (elution of Cs, Y, Sc, and Sm in this order at 25°C using a flow rate of 1.1 × 10^{-2} mℓ/cm^2 sec) Under the same conditions Cs, Tm, Dy, La, and Pr are eluted following this order b) 0.0025 *M* DCTA (disodium salt) of pH 4.8 (elution of Cs, Y, Eu, and Pm in this order at 30°C employing a flow rate of 9 × 10^{-3} mℓ/cm^2 sec)	595

Table 112
SEPARATION OF Sc FROM V AND Ti BY ANION EXCHANGE IN OXALIC ACID SOLUTIONS

Anion exchange resin	Separation conditions and remarks	Ref.
Dowex® 1 (formate form)	Column: 10 cm × 0.209 cm^2 a) 0.1 *M* oxalic acid (adsorption of Sc, V, and Ti) (milligram amounts of the elements) b) 0.1 *M* oxalic acid - 0.1 *M* HCl (~45 column volumes) (elution of Sc) c) 0.1 *M* oxalic acid - 0.4 *M* HCl (~25 column volumes) (elution of V[IV]) d) 0.1 *M* HCl (10 column volumes) (elution of Ti) Formate is weakly held by the resin and therefore does not interfere when passing eluents (a)—(d)	474
Dowex® 1-X8 (100-200 mesh; chloride form)	From 0.1 *M* oxalic acid - 0.1 *M* HCl, V is adsorbed on a column (25 × 1 cm) of this resin, while ^{48}Sc passes into the effluent and washings (2 column volumes of the eluent) in which it is determined radiometrically following decomposition of oxalic acid with H_2O_2 and by heating Before the ion exchange separation, the sample (V pentoxide or ammonium metavanadate) is subjected to neutron irradiation, dissolved in conc HCl, and Sc is extracted into diethyl ether from a thiocyanate medium to separate it from the bulk of V The method was used for the preparation of carrier-free ^{48}Sc	596

Table 113
SEPARATION OF RARE EARTH ELEMENTS BY ANION EXCHANGE IN OXALIC ACID SOLUTION

Elements separated	Ion exchange resin, separation conditions, and remarks	Ref.
Mixed fission products containing Cs, Sr, Y, Ce, Pm, Zr, Nb, and Ru	Dowex® 1-X7.5 (50-100 mesh; oxalate form) Column: 10 × 0.6 cm operated at a flow rate of 0.5 mℓ/min a) 0.5% NH_4Cl solution (~20 mℓ) of pH 6 (adsorption of all elements listed in the first column except ^{137}Cs and ^{90}Sr which pass into the effluent) b) Water (~20 mℓ) (as a rinse) c) 0.2 *M* HCl (~20 mℓ) (elution of ^{90}Y, ^{144}Ce, and ^{147}Pm) d) 0.5 *M* HCl (~20 mℓ) (elution of ^{95}Zr) e) 1 *M* HCl (~20 mℓ) (elution of ^{95}Nb) f) 7 *M* HNO_3 (~30 mℓ) (elution of ^{106}Ru)	597
^{90}Y from ^{90}Sr	Amberlite® IRA-400 (oxalate form) Column: 7.5 cm × 1.13 cm^2 operated at a flow rate of 30—40 drops/min; the resin is conditioned with 5% oxalic acid solution a) 0.5% NH_4Cl solution (20 mℓ) of pH 2—3 (adsorption of ^{90}Y + carrier; into the effluent passes ^{90}Sr + carrier) b) Water (40 mℓ) (elution of residual Sr) c) 0.5 *M* HCl (50 mℓ) (elution of ^{90}Y + carrier) The elements were determined radiometrically	598
Ce from Lu	Dowex® 1-X8 (100-200 mesh; oxalate form) Column volume: 0.92 mℓ a) 0.5 *M* oxalic acid (adsorption of Lu and elution of Ce) (15 fractions of 1.35 mℓ each) b) 0.75 *M* oxalic acid (elution of Lu) (~12 fractions of 1.35 mℓ each)	473

Table 114
SEPARATION OF Sc FROM OTHER ELEMENTS BY ANION EXCHANGE IN SYSTEMS CONTAINING MALONIC, ASCORBIC, AND TARTARIC ACID

Elements separated	Ion exchange resin, separation conditions, and remarks	Ref.
Sc from alkali and alkaline earth metals, Tl(I), Hg(II), Fe(II), Co, Ni, Mn, Zn, Cd, Pd, Sb, Cu, Cr, V, Al, Zr, In, Fe(III), Th, and Pb	Dowex® 2-X8 (50-100 mesh; malonate form) Column: 18 × 1.4 cm operated at a flow rate of 1 mℓ/min A) Separation of Sc from alkali and alkaline earth metals, Tl, Hg, and Fe(II) a) Malonate solution of pH 5 (adsorption of Sc and elution of accompanying elements) b) 1 *M* HCl (elution of Sc) B) Separation of Sc from Co, Ni, Mn, Zn, Cd, and Pd a) Malonate solution of pH 5 (adsorption of Sc and accompanying elements) b) Water (elution of accompanying elements) c) 1 *M* HCl (elution of Sc) C) Separation of Sc from Sb, Cu, Cr, V, Fe(III), Al, Zr, and In a) Malonate solution of pH 5 (adsorption of Sc and other elements) b) 0.5 *M* NH_4Cl (elution of Sb) c) 1 *M* NaCl (elution of Cu, Cr, and V) d) 1 *M* NH_4Cl (elution of Fe[III]) e) 1.5 *M* NH_4Cl (elution of Al) f) 1 *M* ammonium acetate (elution of Zr) g) 0.1 *M* HCl (elution of In) D) Separation of Sc from Th a) Malonate solution of pH 5 (adsorption of Sc and Th) b) 0.2 *M* HNO_3 (elution of Th) c) 0.5 *M* HNO_3 (elution of Sc) E) Separation of Sc from Pb a) Malonate solution of pH 5 (adsorption of Sc and Pb) b) 1 *M* HNO_3 (elution of Sc) c) 6 *M* HNO_3 (elution of Pb)	475
Sc from alkali and alkaline earth elements, Cr, Mn, Fe(II), Co, Ni, Pd, Zn, Cd, Al, Sb, Pb, Y, Ti, Zr, V, In, and lanthanides + Y	Dowex® 2-X8 (50-100 mesh; ascorbate form) Column: 18 × 1.4 cm operated at a flow rate of 1 mℓ/min A) Separation of Sc from alkali and alkaline earth elements a) Ascorbate solution of pH 6.5 (adsorption of Sc and elution of the other elements) b) Water (as a rinse) c) 2 *M* HCl (elution of Sc) B) Separation of Sc from Cr, Mn, Fe(II), Co, Ni, Pd, Zn, Cd, Al, Sb, and Pb a) Ascorbate solution of pH 6.5 (adsorption of Sc and the other elements) b) Water (elution of the other elements) c) 2 *M* HCl (elution of Sc) C) Separation of Sc from Y, Ti, Zr, V, and In a) Ascorbate solution of pH 6.5 (adsorption of Sc and the other elements) b) 0.1 *M* HCl (elution of Y) c) 0.2 *M* HCl (elution of Ti and In) d)1 *M* NH_4Br (elution of V) e) 1 *M* ammonium acetate (elution of Zr) f) 2 *M* HCl (elution of Sc) D) Separation of Sc from lanthanides (La, Ce, Pr, Nd, Sm, and Gd) a) Ascorbate solution of pH 6.5 (or malonate solution of pH 5) (adsorption of Sc and lanthanides)	475

Table 114 (continued)
SEPARATION OF Sc FROM OTHER ELEMENTS BY ANION EXCHANGE IN SYSTEMS CONTAINING MALONIC, ASCORBIC, AND TARTARIC ACID

Elements separated	Ion exchange resin, separation conditions, and remarks	Ref.
	b) 0.05 *M* HCl (elution of lanthanides) c) 2 *M* HCl (elution of Sc) Titrimetric procedures were used for the determination of Sc and the other elements (both in the malonate and ascorbate systems described above) The malonate and ascorbate systems outlined were employed to separate milligram amounts of the elements	
Sc from V(IV), Zr, and Hf	Anionite AV-16 (tartrate form) a) Sorption solution of pH 2 (adsorption of Sc, V, Zr, and Hf as anionic tartrate complexes) b) 5% NaOH solution (elution of V[IV], Zr, and Hf) c) 1—4 *M* HCl (elution of Sc)	476

Table 115
SEPARATION OF LANTHANIDES BY ANION EXCHANGE USING CITRIC ACID MEDIA

Elements separated	Ion exchange resin, separation conditions, and remarks	Ref.
Nd from Sm	Amberlite® IRA-400 (−80 + 200 mesh BBS; citrate form) Column: 10 × 2.2 cm To separate these two elements from each other, their solution (pH 8) containing nitrilotriacetic acid or Na triphosphate is equilibrated with a batch of the resin which is then transferred to the top of the resin column; subsequently, Nd is eluted first using 0.05% citric acid - 0.1% NaCl solution of pH 2.5; flow rate: ≈0.5 mℓ/min	477
Pm from Eu	Dowex® 1 (250-500 mesh; citrate form) Column: 14.9 cm × 0.08 cm^2 operated at a flow rate of 1.5 mℓ/hr With 0.0125 *M* citric acid solution (70 mℓ) of pH 2.1, ^{147}Pm is eluted before ^{154}Eu; the first ~45 mℓ of eluent contains Pm	478

Table 116
SEPARATION OF Sc FROM Ce AND Y USING ANION EXCHANGE IN A XYLENOL ORANGE MEDIUM

Ion exchange resin	Separation conditions and remarks	Ref.
Amberlite® IRA-400 or Hitachi® 2632 (16—20 μm) (xylenol orange form)	Column: 14 × 0.8 cm ID equilibrated with eluent (a) and operated at a flow rate of 0.5 mℓ/min a) 2 × 10^{-4} *M* xylenol orange solution of pH 2.5 (adsorption of Sc; elution of Ce and Y) b) 0.25 *M* HCl (elution of Sc) The xylenol orange form of the resin is prepared by treating the chloride form with a large excess of xylenol orange solution followed by washing with deionized water to remove excess reagent The method was used for the separation of microgram amounts of the elements	479

Table 117
DETERMINATION OF RARE EARTH IMPURITIES IN RARE EARTH OXIDES AFTER SEPARATION BY ANION EXCHANGE IN METHANOL MEDIA CONTAINING α-HIBA OR LACTIC ACID

Rare earth oxide	Ion exchange resin, separation conditions, and remarks	Ref.
Er- and Y- oxides	Dowex® AG1-X4 (100-200 mesh; isobutyrate or lactate form) Column: 90 × 20 mm ID To separate rare earth impurities (La → Tb) from the Er matrix 60% methanol - 0.02 *M* α-HIBA is used as the eluent. Er is retained more strongly and the eluted impurities (parts per million amounts) are determined spectrographically In case of the Y matrix, the eluent is 50% methanol - 0.02 *M* α-HIBA which elutes the impurities (La → Eu) A similar separation on the same resin (lactate form), but using 0.1 *M* lactic acid in 27% methanol, has been employed to separate trace amounts of the lighter rare earths (up to Gd), Sr, and Ba from an Er sample	480, 481

Table 118
APPLICATIONS OF CHELATING RESINS TO THE ANALYTICAL CHEMISTRY OF THE RARE EARTH ELEMENTS[a]

Ion exchange resin	Applications, experimental conditions, and remarks	Ref.
Chelex® 100 (NH_4^+ form) and Dowex® A-1	Determination of rare earths and other elements in riverwater A) First column operation (preirradiation separation on a column: 12 cm × 6.5 mm) a) Sample solution (500 mℓ) neutralized with ammonia (adsorption of rare earths and heavy metals) b) 0.1 *M* ammonium acetate followed by water (elution of Na) Subsequently, the dried resin (1 g) is irradiated with neutrons and transferred to a column (15 cm × 8 mm) of Dowex® A-1 and the following solutions are used as eluents B) Second column operation (radiochemical separation) a) 0.1 *M* KNO_3 (removal of ^{24}Na and other radionuclides) Afterwards the column (1 g resin) is heated by means of a water jacket and eluent (b) is passed b) Hot 1 *M* Na carbonate (elution of the rare earths) (at room temperature, the recovery is less then 60% for ^{153}Sm) The following eluent (c) is then percolated through the resin bed at room temperature c) 2 *M* HNO_3 (elution of the heavy metals such as Mn, Cu, and Zn) The γ-ray spectrum of each fraction was measured; almost quantitative recovery was achieved for La, Sm, Eu, Dy, Mn, Cu, and Zn Prior to the first column operation the water sample is acidified to pH 1 with HNO_3 and filtered through a membrane filter	497
Resin I 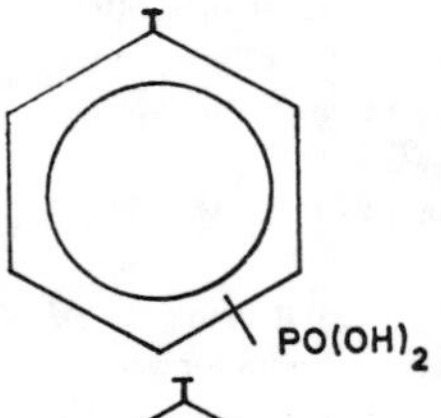 (structure: polystyrene-bound phenyl–$PO(OH)_2$) or resin II 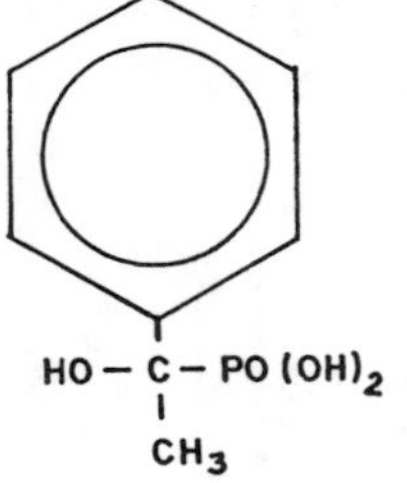 (structure: polystyrene-bound phenyl–$C(OH)(CH_3)-PO(OH)_2$)	Separation of Sc from Y and lanthanides These ion exchange resins were prepared from macroreticular styrene-DVB copolymer containing 4% of DVB From 1 *M* HNO_3, Sc is retained on a column containing resin I or II, while the other rare earths pass into the effluent The method can be used to separate milligram amounts of the elements	499
Dowex® A-1	Microchemical detection of Ce(III) On a spot test plate 50 mg of the resin are added to 4—5 drops of the test solution followed by 1 drop of each of 20% ammonia solution and 20% H_2O_2; in the presence of Ce a yellow color is formed which is due to the peroxy compound of Ce (probably CeO_3^{3+}); the dilution limit is 1:3 × 10^5	599
Resin containing PMBP (4-benzoyl-3-methyl-1-phenylpyrazolin-5-one (120-200 mesh)	Column: 6 × 0.5 cm containing 0.5 g of the resin and operated at a flow rate of 0.5 mℓ/min a) Sample solution (<10 mℓ or, for a 2-g sample, < 25 mℓ) (adsorption of rare earths; Ca passes into the effluent) b) Water (3 mℓ) (elution of residual Ca) c) 0.1 *M* HCl (3 mℓ) (elution of rare earths) The rare earths are determined spectrographically Before the separation the sample (0.5 g or up to 2 g) of nuclear-grade Ca is dissolved in 6 *M* HCl and the pH of the	600

Table 118 (continued)
APPLICATIONS OF CHELATING RESINS TO THE ANALYTICAL CHEMISTRY OF THE RARE EARTH ELEMENTS[a]

Ion exchange resin	Applications, experimental conditions, and remarks	Ref.
	solution is adjusted to ~5.5 with ammonia solution (after addition of 1 μg of Tm as internal standard and 0.1 g of ascorbic acid); finally, the sample solution (a) is obtained by adding 5% sulfosalicylic acid solution (5 mℓ) of pH 5.5; with this method down to 2.5 ppb of Yb and down to 0.1 ppm of Ce can be detected; sensitivities for the other metals are between these values	

[a] See also Table 160 in the chapter on Actinides, Volume II; Table 77 in the chapter on Copper, Volume III; and Table 19 in the chapter on Lead, Volume, VI.

REFERENCES

1. **Strelow, F. W. E.,** Accurate determination of rare earth metals and scandium in South African carbonatite by use of ion exchange chromatography for separation, *Anal. Chem.*, 38, 127, 1966.
2. **Strelow, F. W. E.,** Separation of tervalent rare-earth metals plus scandium from aluminum, gallium, indium, thallium, iron, titanium, uranium and other elements by cation exchange chromatography, *Anal. Chim. Acta,* 34, 387, 1966.
3. **Yokotsuka, S., Akutsu, E., and Ueno, K.,** Ion exchange behavior of inorganic ions in ion exchange resins of Diaion SKN-1 and SAN-1. Distribution ratios of ions in the system of Diaion-hydrochloric acid, *J. Nucl. Sci. Technol.*, 8(11), 622, 1971.
4. **Tohyama, I. and Otozai, K.,** Ion exchange in HCl, $NH_2OH.HCl$ and $N_2H_4.2HCl$ solutions, *Fresenius' Z. Anal. Chem.*, 286, 198, 1977.
5. **Šimek, M.,** Investigation of the ion-exchange characteristics of calcium, scandium and iron(III) in concentrated mineral acids, *Collect. Czech. Chem. Commun.*, 39(7), 1693, 1974.
6. **Nelson, F., Murase, T., and Kraus, K. A.,** Ion-exchange procedures. I. Cation exchange in concentrated HCl and $HClO_4$-solutions, *J. Chromatogr.*, 13, 503, 1964.
7. **Južnič, K.,** On the behavior of cerium(III) in hydrochloric acid, *Mikrochim. Acta,* 2(5 and 6), 345, 1983.
8. **Strelow, F. W. E. and Baxter, C.,** Separation of tervalent rare earth metals and scandium from aluminum, iron(III), titanium(IV), and other elements by cation exchange chromatography in hydrochloric acid-ethanol medium, *Talanta,* 16, 1145, 1969.
9. **Mehta, V. P. and Khopkar, S. M.,** Cation-exchange chromatographic separation of scandium from other elements on Dowex 50W-X8, *Sep. Sci. Technol.*, 13(10), 933, 1978.
10. **Hettel, H. J. and Fassel, V. A.,** Quantative separation of small amounts of rare earth elements from thorium, uranium and zirconium by ion exchange, Report ISC-851, U.S. Atomic Energy Commission, December 1956.
11. **Laktionova, N. V., Karyakin, A. V., Ryabukhin, V. A., Stroganova, N. S., and Ageeva, L. V.,** Spectrographic determination of rare earth metals in rocks with preliminary chemical concentration from a small sample, *Zh. Anal. Khim.*, 29(8), 1549, 1974.
12. **Ryabukhin, V. A., Stroganova, N. S., Gatinskaya, N. G., and Ermakov, A. N.,** Cation-exchange concentration of traces of rare earth metals from natural materials, *Zh. Anal. Khim.*, 28(11), 2166, 1973.
13. **Voldet, P. and Haerdi, W.,** Determination of europium and dysprosium in rocks by neutron-activation and high-resolution X-ray spectrometry, *Anal. Chim. Acta,* 87, 227, 1976.
14. **Hughson, M. R. and Sen Gupta, J. G.,** A thorian intermediate member of the britholite-apatite series: physical and chemical studies, *Am. Mineral.*, 49(7-8), 937, 1964.
15. **Grant, C. L.,** Spectrochemical determination of yttrium in biological materials, *Anal. Chem.*, 33, 401, 1961.
16. **Miro, M., de Padovani, O., Ramos, E., Román de Vega, V., and Lowman, F. G.,** Determination of stable scandium in plants, animals, sediments, sands, soils, rocks and minerals by neutron activation analysis, *Anal. Chim. Acta,* 35, 54, 1966.
17. **Hamaguchi, H., Ishida, K., Hikawa, I., and Kuroda, R.,** Separation of rare earth metals by anion-exchange chromatography in salt solution:magnesium nitrate medium, *Anal. Chem.*, 37, 1283, 1965.
18. **Walsh, J. N., Buckley, F., and Barker, J.,** Simultaneous determination of rare earth elements in rocks using inductively coupled plasma-source spectrometry, *Chem. Geol.*, 33(1-2), 141, 1981.
19. **Jackson, P. F. S. and Strelow, F. W. E.,** Rare earth content of the six international geological reference materials of South African Origin, *Chem Geol.*, 15(4), 303, 1975.

19a. **Hou, Q. L., Hughes, T. G., Haukka, M., and Hannaker, P.,** Determination of rare earth elements in geological materials by thin-film X-ray fluorescence and inductively coupled plasma atomic-emission spectrometry, *Talanta,* 32, 495, 1985.

19b. **Toyoda, K. and Haraguchi, H.,** Determination of rare earth elements in geological standard rock samples by inductively coupled plasma atomic emission spectrometry, *Chem. Lett.*, No. 7, 981, 1985.

20. **Strelow, F. W. E. and Jackson, P. F. S.,** Determination of trace and ultra-trace quantities of rare earth metals by ion exchange chromatography-mass spectrography, *Anal. Chem.*, 46, 1481, 1974.
21. **Schnetzler, C. C., Thomas, H. H., and Philpotts, J. A.,** Determination of rare earth metals in rocks and minerals by a mass-spectrometric stable-isotope-dilution technique, *Anal. Chem.*, 39, 1888, 1967.
22. **Masuda, A.,** Lanthanides in basalts of Japan with three distinct types, *Geochem. J.*, 1, 11, 1966.
23. **Meloni, S., Oddone, M., Cecchi, A., and Poli, G.,** Destructive neutron-activation analysis of rare earths in geological samples: comparison between two methods, *J. Radioanal. Chem.*, 71, 429, 1982.
24. **Loveless, A. J., Yanagita, S., Mabuchi, H., Ozima, M., and Russell, R. D.,** Isotopic ratios of gadolinium, samarium and europium in "Abee" enstatite chondrite, *Geochim. Cosmochim. Acta,* 33(6), 685, 1972.
25. **Mazzucotelli, A., Frache, R., Dadone, A., Baffi, F., and Cescon, P.,** Spectrophotometric determination of trace amounts of yttrium in silicates after cation-exchange separation with DL-2-hydroxybutyric acid, *Anal. Chim. Acta,* 99, 365, 1978.

26. **Jones, E. A., Watson, A. E., and Dixon, K.,** Separation and determination of trace and minor amounts of individual rare-earth metals in silicates, Report NIM-1428, National Institute for Metallurgy, Johannesburg, South Africa, 1972.
27. **Brenner, I. B., Watson, A. E., Steele, T. W., Jones, E. A., and Goncalves, M.,** Application of an argon-nitrogen inductively coupled radio-frequency plasma (ICP) to the analysis of geological and related materials for their rare earth contents, *Spectrochim. Acta Part B,* 36(8), 785, 1981.
28. **Volfovsky Spirn, R.,** Rare Earth Distribution in the Marine Environment, Ph.D. thesis, Massachusetts Institute of Technology, Cambridge, 1965.
29. **Mehta, V. P. and Khopkar, S. M.,** Cation-exchange studies of yttrium(III) and its separation from various other elements, *Chromatographia,* 11(9), 536, 1978.
30. **Jones, E. A.,** Separation of rare earths from solutions of phosphoric acid, Report NIM-1904, National Institute for Metallurgy, Randburg, South Africa, 1977.
31. **Cerrai, E., Pelati, L., and Triluzi, C.,** Radiochemical studies for radioactivity determination in marine plankton, Report CISE-95, 1963.
32. **Ward, M. and Foreman, J. K.,** Determination of rhodium-103, neodymium-143, samarium-149 and -151, and gadolinium-155 in irradiated uranium, *Talanta,* 10, 779, 1963.
33. **Zhou, J. and Wei, G.,** Spectrophotometric determination of traces of gadolinium, samarium, europium and dysposium in aluminum and aluminum-alloys, *Hua Hsueh Tung Pao,* No. 4, 24, 1981.
34. **Mazzucotelli, A. and Frache, R.,** Electrothermal atomisation for atomic-absorption determination of some rare earths in silicate rocks and minerals, *Mikrochim. Acta,* 2(3-4), 323, 1981.
35. **Huber-Schausberger, I.,** Emission spectrographic determination of trace elements in fluorspar, *Mikrochim. Acta,* No. 2, 240, 1970.
36. **Croudace, I. W.,** Use of pre-irradiation group separations with neutron-activation analysis for determination of rare earths in silicate rocks, *J. Radioanal. Chem.,* 59(2), 323, 1980.
37. **Kallmann, S., Oberthin, H. K., and Hibbits, J. O.,** Ion-exchange separation and spectrographic determination of some rare earths in beryllium, uranium, zirconium and titanium metals, alloys and oxides. Collection with calcium and magnesium fluoride and ion-exchange separation, *Anal. Chem.,* 32, 1278, 1960.
38. **Yu Zu Gen and Xie Ruyu,** The simultaneous determination of trace amounts of rare earths in rocks by optical emission spectrometry, private communication, 1985.
39. **Strelow, F. W. E., Victor, A. H., and Weinert, C. H. S. W.,** Ion exchange chromatography applied to the separation and accurate determination of some trace elements in rocks, *Geostand. Newsl.,* 2(1), 49, 1978.
40. **Edge, R. A. and Ahrens, L. H.,** Determination of scandium, yttrium, neodymium, cerium and lanthanum in silicate rocks by combined cation-exchange-spectrochemical method, *Anal. Chim. Acta,* 26, 355, 1962.
41. **Eby, G. N.,** Determination of rare earth, yttrium and scandium abundances in rocks and minerals by an ion-exchange-X-ray fluorescence procedure, *Anal. Chem.,* 44, 2137, 1972.
42. **Bolton, A., Hwang, J., and Van der Voet, A.,** Determination of scandium, yttrium and selected rare earth elements in geological materials by inductively coupled plasma optical emission spectrometry, *Spectrochim. Acta Part B,* 38(1-2), 165, 1983.
43. **Krähenbühl, U., Rolli, H. P., and von Gunten, H. R.,** Determination of rare earth metals in standard rocks and in lunar samples by activation analysis, *Helv. Chim. Acta,* 55(2), 697, 1972.
44. **Massart, D. L. and Hoste, J.,** Activation analysis for rare earth metals. I. Separation from accompanying elements, *Anal. Chim. Acta,* 42, 7, 1968.
45. **Massart, D. L. and Hoste, J.,** Activation analysis for rare earth metals. II. Determination of lutetium in gadolinite, *Anal. Chim. Acta,* 42, 15, 1968.
46. **Massart, D. L. and Hoste, J.,** Activation analysis for rare earth metals. III. Determination of rare-earth metals in minerals by the single-comparator technique, *Anal. Chim. Acta,* 42, 21, 1968.
47. **Murugaiyan, P., Verbeek, A. A., Hughes, T. C., and Webster, R. K.,** Separation scheme for the determination of alkali metals (K, Rb, Cs), alkaline earths (Sr, Ba) and rare earths (Ce, Nd, Sm, Eu) in silicate materials by isotopic dilution analysis, *Talanta,* 15, 1119, 1968.
48. **Shimizu, T.,** Spectrophotometric determination of scandium with bromopyrogallol red, *Talanta,* 14, 473, 1967.
49. **Shimizu, T.,** Determination of traces of scandium in silicate rocks by a combined ion-exchange-spectrophotometric method, *Anal. Chim. Acta,* 37, 75, 1967.
50. **Shimizu, T.,** Scandium content of igneous rocks and some oceanic sediments, *Bull. Chem. Soc. Jpn.,* 42, 1561, 1969.
51. **Shimizu, T. and Momo, E.,** Sensitive spectrophotometric determination of scandium with 4-(2-thiazolylazo)-resorcinol, *Anal. Chim. Acta,* 52, 146, 1970.
52. **Mathers, W. G. and Hoelke, C. W. K.,** Semi-automatic machine for group separation of radio-elements by cation exchange, *Anal. Chem.,* 35, 2064, 1963.
53. **Lin, H.-C. and Ting, G.,** Separation of zirconium-95 and niobium-95 from irradiated uranium by silica gel adsorption, Report INER-0186, Institute of Nuclear Energy, Research, Lung-Tan, Taiwan, 1975.

54. **Lin, H.-C. and Ting, G.,** Separation of zirconium-95 and niobium-95 from irradiated uranium by silica gel adsorption, *Radiochim. Acta,* 23(1), 1, 1976.
55. **Carpenter, J. H. and Grant, V. E.,** Concentration and state of cerium in coastal waters, *J. Mar. Res.,* 25, 228, 1967.
56. **Chen, C., Liu, Q., Xu, S., Xie, S., and Fu, H.,** Simple forced-flow apparatus for liquid chromatography separations, *Fenxi Huaxue,* 10(9), 572, 1982.
57. **Henriet, D. and de Gelis, P.,** Amperometric determination of cerium in steel and cast iron, *Chim. Anal.,* 50(10), 519, 1968.
58. **Belyavskaya, T. A., Alimarin, I. P., and Brȳkina, G. D.,** Ion-exchange chromatographic separation of scandium from manganese, *Vestn. Mosk. Univ. Ser. Khim,* No. 1, 53, 1967.
59. **Gerwien, U. and Oberhauser, R.,** Photometric determination of cerium in cast iron, steel and heat-conducting alloys after separation on a cation exchanger, *Materialpruefung,* 20(6), 233, 1978.
60. **Hakkila, E. A., Hurley, R. G., and Waterbury, G. R.,** Cation-exchange separation-X-ray spectrometric determination of scandium in plutonium, *Anal. Chem.,* 41, 665, 1969.
61. **Orlandini, K. A. and Korkisch, J.,** Cation-exchange separation of scandium from rare-earth elements, *Sep. Sci.,* 3(3), 255, 1968.
62. **Sherma, J. and Van Lenten, F. J.,** Ion exchange paper chromatography of metal ions with mixed aqueous-organic solvents containing mineral acid and a selective extractant, *Sep. Sci.,* 6(2), 199, 1971.
63. **Webster, R. K., Hughes, T. C., and Gibbons, D.,** Determination of cerium in steel by stable-isotopic dilution, Report AERE-M 2607, U.K. Atomic Energy Authority, 1973.
64. **Lee, C., Yin, Y. C., and Chung, K. S.,** Determination of individual lanthanide elements by neutron activation using new comparator technique, *J. Korean Nucl. Soc.,* 4(2), 83, 1972.
65. **Nevoral, V.,** Determination of trace contents of rare earth elements in mineral waters, *Collect. Czech. Chem. Commun.,* 43(9), 2274, 1978.
66. **Ma, T. C.,** Complete analysis of single minerals in yttrium-containing rocks, *Fen Hsi Hua Hsueh,* 8(3), 246, 1980.
67. **Yang, Y.-S.,** Ion-exchange separation of titanium and lanthanum, *Acta Chim. Sinica,* 28(4), 259, 1962.
68. **Mazzucotelli, A., Minoia, C., and Vannucci, R.,** Inductively coupled plasma atomic-emission spectroscopy of lanthanoids in silicates, *Rend. Soc. Ital. Mineral. Petrol.,* 38(2), 781, 1983.
69. **Gao, X., Zhang, M., and Hu, T.,** Studies on electro-analytical chemistry of rare earth elements. XI. Application of the adsorptive complex wave of the rare earth metal-catechol violet complex, *Fenxi Huaxue,* 11(10), 762, 1983.
70. **Bishop, J. G. and Hughes, T. C.,** Radiochemical neutron-activation analysis determination of rare earth elements in geological materials with ppb sensitivity, *J. Radioanal. Nucl. Chem.,* 84(2), 213, 1984.
71. **Haberer, K. and Stürzer, U.,** Rapid determination of radioactive cerium in water samples, *Gas Wasserfach. Wasser Abwasser,* 113(3), 122, 1972.
72. **Cheng, J. and Xu, S.,** Cation exchange behavior of copper, rare earth metals, iron and nickel in aqueous thiourea; rapid separation and determination of trace rare earth metals in copper, *Fenxi Huaxue,* 12(2), 122, 1984.
73. **Towell, D. G.,** Rare-Earth Distributions in Some Rocks and Associated Minerals of the Batholith of Southern California, Ph.D. thesis, Massachusetts Institute of Technology, Cambridge, 1963.
74. **Towell, D. G., Volfovsky Spirn, R., and Winchester, J. W.,** Rare-earth abundances in the standard granite, G-1, and the standard diabase, W-1, *Geochim. Cosmochim. Acta,* 29, 569, 1965.
75. **Shilling, J. G.,** Rare Earth Fractionation in Hawaiian Volcanic Rocks, Ph.D. thesis, Massachusetts Institute of Technology, Cambridge, 1966; *Science,* 153, 867, 1966.
76. **Kimura, T. and Hamada, T.,** Simple and rapid method for collection of radionuclides in rain-water, *Radioisotopes,* 27(6), 348, 1978.
77. **Stewart, D. C.,** Rapid separation of trace amounts of rare-earth elements of the yttrium group, *Anal. Chem.,* 27, 1279, 1955.
78. **Korkisch, J.,** Ion exchange in mixed and non-aqueous media, in *Progress in Nuclear Energy,* Series 9, Vol. 6, Pergamon Press, 1966.
79. **Korkisch, J. and Ahluwalia, S. S.,** Cation exchange of several elements in hydrochloric acid-organic solvent media, *Talanta,* 14, 155, 1967.
80. **Starobinets, G. L., Pishchulina, T. G., and Ganopol'skii, V. I.,** Exchange of certain rare earth metal ions in mixed aqueous methanolic media on cationite KU-2X8, *Izv. Akad. Nauk. Be. SSR Ser. Khim. Nauk,* No. 4, 115, 1968.
81. **Peterson, S. F., Tera, F., and Morrison, G. H.,** Isotopic ion exchange separation of radioisotopes using an acetone-water-hydrochloric acid system, *J. Radioanal. Chem.,* 2, 115, 1969.
82. **Bogatyrev, I. O., Zaborenko, K. B., and Tsibanov, V. V.,** Use of a mathematical model in choosing conditions for cation-exchange separation of micro amounts of rare earth metals in aqueous acetone solutions of mineral acids, *Zh. Fiz. Khim.,* 54(4), 934, 1980.
83. **Strelow, F. W. E., Van Zyl, C. R., and Bothma, C. J. C.,** Distribution coefficients and the cation exchange behavior of elements in hydrochloric acid-ethanol mixtures, *Anal. Chim. Acta,* 45, 81, 1969.

84. **Grigorescu-Sabău, C. and Spiridon, St.,** Ion exchange equilibria in the presence of organic solvents. II. Influence of organic solvents on the determination constant of scandium, *Stud. Cercet. Chim. Acad. Rep. Pop. Rom.*, 10(2), 235, 1962.
85. **Spiridon, St. and Grigorescu-Sabău, C.,** Ion-exchange equilibria in the presence of organic solvents. III. Influence of the nature of mineral acid on the distribution constant of scandium, *Stud. Cercet. Fiz. Acad. Rep. Pop. Rom.*, 14(1), 33, 1963.
86. **Grigorescu-Sabău, C.,** Ion exchange equilibria in the presence of organic solvents. I. Influence of organic solvents on the distribution constant of praseodymium, *J. Inorg. Nucl. Chem.*, 24, 195, 1962.
87. **Strelow, F. W. E.,** Partly non-aqueous media for accurate chemical analysis by ion exchange, *Ion Exch. Membr.*, 2, 37, 1974.
88. **Molnar, F., Horvath, A., and Khalkin, V. A.,** Micro-chromatographic separation of lanthanoids, *Magy. Kem. Foly.*, 80(10-11), 474, 1974.
89. **Lishchenko, T. V. and Korobko, G. P.,** Sorption of rare earth metals from nitrate medium, *Nauchn. Tr. Nauchno Issled. Proektn. Inst. Redkomet. Promsti.*, 45, 35, 1972.
90. **Korkisch, J., Feik, F., and Ahluwalia, S. S.,** Cation-exchange behavior of several elements in nitric acid-organic solvent media, *Talanta,* 14, 1069, 1967.
91. **Korkisch, J.,** Combined ion exchange-solvent extraction (CIESE): a novel separation technique for inorganic ions, *Sep. Sci.*, 1, 159, 1966.
92. **Boswell, C. R., Pierce, T. B., and Barclay, A. N.,** Determination of gadolinium by ion exchange and neutron absorption Report AERE-R6331, U.K. Atomic Energy Authority, 1974.
93. **Latimer, J. N., Bush, W. E., Higgins, L. J., and Shay, R. S.,** Handbook of Analytical Procedures, Report RMO-3008, Analytical Laboratory, U.S. Atomic Energy Commission, Grand Junction Office, February 1970.
94. **Strelow, F. W. E.,** Quantitative separation of lanthanoids and scandium from barium, strontium and other elements by cation-exchange chromatography in nitric acid, *Anal. Chim. Acta,* 120, 249, 1980.
95. **Crock, J. G. and Lichte, F. E.,** Determination of rare-earth elements in geological materials by inductively coupled plasma-atomic emission spectrometry, *Anal. Chem.*, 54, 1329, 1982.
96. **Crock, J. G. and Lichte, F. E.,** Group separation of rare earth elements and yttrium from geological materials by cation exchange chromatography, *Chem. Geol.*, 45(1-2), 149, 1984.
97. **Barlow, I. C. and Van Valkenburg, R. K.,** Thin-film analysis of gadolinium-iron oxide and gadolinium-iron by a combined ion exchange-X-ray fluorescence method, *Mikrochim. Acta,* No. 4, 827, 1968.
98. **Fritz, J. S. and Garralda, B. B.,** Cation-exchange separation of bivalent metal ions from rare-earth metals, *Talanta,* 10, 91, 1963.
99. **Craft, T. F., Erb, T. L., and Hill, D. F.,** Chemical analysis of radioactive materials in water, *J. Am. Water Works Assoc.*, 56(10), 1357, 1964.
100. **Sinegribova, O. A., Bokova, T. A., Semenov, G. I., and Yagodin, G. A.,** Adsorption of rare-earth metals from aqueous-alcoholic solution, *Tr. Mosk. Khim. Tekhnol. Inst.*, No. 65, 24, 1970.
101. **Šimek, M.,** Sorption of scandium(III) and iron(III) ions by cation exchanger from solutions of concentrated acids and organic solvents, *Chem. Zvesti,* 28(3), 343, 1974.
102. **Kostÿgov, A. S. and Saunkin, O. F.,** Radioactive indication method of determining traces of praseodymium in compounds of cerium, *Zavod. Lab.*, 33(6), 678, 1967.
103. **Okabe, Y.,** Separation of quadrivalent cerium from tervalent lanthanons by means of cation exchange, *Bull. Electrotech. Lab. Tokyo,* 19, 801, 1955.
104. **Strelow, F. W. E. and Bothma, C. J. C.,** Separation of scandium from yttrium, lanthanum and the rare-earth metals by cation exchange chromatography, *Anal. Chem.*, 36, 1217, 1964.
105. **Alimarin, I. P., Miklishanskii, A. Z., and Yakovlev, Yu. V.,** Neutron-activation analysis for rare earth impurities in metallic uranium, *J. Radioanal. Chem.*, 4(1), 45, 1970.
106. **Anikina, L. I., Bagreev, V. V., Dobrolyubskaya, T. S., Zolotov, Yu. A., Karyakin, A. V., Miklishanskii, A. Z., Nikitina, N. G., Palei, P. N., and Yakovlev, Yu. V.,** Luminescence determination of gadolinium, europium and samarium as impurities in metallic uranium, *Zh. Anal. Khim.*, 24(7), 1014, 1969.
107. **Braier, H. A.,** Separation of traces of rare-earth elements in uranium by ion exchange, 2nd U.N. Int. Conf. on the Peaceful Uses of Atomic Energy, Report A/CONF.15/P/1568 (Rev. 1), Geneva, 1958.
108. **Chang, C. C., Shih, J. S., and Chang, C. T.,** Spectrophotographic determination of trace rare-earth elements in high-purity thorium, *Fresenius' Z. Anal. Chem.*, 270, 187, 1974.
109. **Cogan, E.,** Separation and determination of rare-earth metals in zirconium-rare earth alloys, Report Invest. No. 6430, U.S. Bureau of Mines, 1964.
110. **Kuroda, R., Nakagomi, Y., and Ishida, K.,** Separation of scandium by cation exchange in acid ammonium sulfate media, *J. Chromatogr.*, 22(1), 143, 1966.
111. **Guan, J., Li, D., and Gao, B.,** Spectrographic determination of microgram quantities of gadolinium, dysprosium, samarium, and europium separated chemically from uranium hexafluoride, *Fenxi Huaxue,* 12(1), 40, 1984.

112. **Golovatŷi, R. N. and Oshchapovskii, V. V.,** Chromatographic separation of cerium from manganese and some other metals, *Ukr. Khim. Zh.*, 31(3), 310, 1965.
113. **Afanas'ev, Yu. A., Ryabinin, A. I., and Karuna, A. G.,** Sorption of cerium(III) and yttrium by a mixture of cationite KU-2 and lead sulfate, *Radiokhimiya*, 15(3), 435, 1973.
114. **Isaeva, L. B., Belyavskaya, T. A., and Brykina, G. D.,** Sorption of scandium by ionites from aqueous and aqueous acetone solutions of oxalic or tartaric acid mixed with sulfuric acid, *Vestn. Mosk. Univ. Ser. Khim.*, 12(5), 586, 1971.
115. **Carter, J. A. and Dean, J. A.,** Determination of traces of certain rare-earth elements in uranium compounds by ion exchange and spectrography, *Appl. Spectrosc.*, 14(2), 50, 1960.
116. **Hettel, H. J. and Fassel, V. A.,** Determination of traces of certain rare earths in zirconium, *Anal. Chem.*, 27, 1311, 1955.
117. **Nikitin, M. K.,** Study of ion-exchange in hydrofluoric acid solutions. Group separation of elements by means of ion-exchange resins, *Dokl. Akad. Nauk SSSR*, 148(3), 595, 1963.
118. **Caletka, R. and Krivan, V.,** Cation-exchange of 43 elements from hydrofluoric acid solution, *Talanta*, 30, 543, 1983.
119. **Hamaguchi, H., Kuroda, R., and Onuma, N.,** Ion-exchange study of thiocyanatochloro complexes of rare-earth elements, *Talanta*, 10, 120, 1963.
120. **Hamaguchi, H., Kuroda, R., Aoki, K., Sugisita, R., and Onuma, N.,** Cation-exchange separation of scandium, *Talanta*, 10, 153, 1963.
121. **Pietrzyk, D. J. and Kiser, D. L.,** Cation-exchange separations using ammonium thiocyanate-organic solvent-water eluants, *Anal. Chem.*, 37, 233, 1965.
122. **Chuveleva, E. A., Nazarov, P. P., and Chmutov, K. V.,** Application of distribution chromatography to the separation of the rare-earth elements, *Zh. Fiz. Khim.*, 36(5), 1022, 1962.
123. **Jian, L.-G., Gao, Y., and Liu, X.-Q.,** Separation of lanthanoids by the ion exchange-organic extractant elution method, *Heh Hua Hsueh Yu Fang She Hua Hsueh*, 3(1), 55, 1981.
124. **Small, H.,** Gel-liquid extraction. II. The separation of metallic species using ion exchange resins in conjunction with water-immiscible organic extractants, *J. Inorg. Nucl. Chem.*, 19, 160, 1961.
125. **Smith, H. L. and Hoffmann, D. C.,** Ion exchange separations of the lanthanides and actinides by elution with ammonium α-hydroxyisobutyrate, *J. Inorg. Nucl. Chem.*, 3(3-4), 243, 1956.
126. **Strelow, F. W. E., Hanekom, M. D., Victor, A. H., and Eloff, C.,** Distribution coefficients and cation-exchange behavior of elements in hydrobromic acid-acetone media, *Anal. Chim. Acta*, 76, 377, 1975.
127. **Nelson, F. and Michelson, D. C.,** Ion-exchange procedures. IX. Cation exchange in HBr solutions, *J. Chromatogr.*, 25, 414, 1966.
128. **Korkisch, J. and Klakl, E.,** Cation-exchange behavior of several elements in hydrobromic acid-organic solvent media, *Talanta*, 16, 377, 1969.
129. **Nelson, F. and Kraus, K. A.,** Ion-exchange procedures. X. Cation exchange in concentrated $HClO_4$-HCl solutions, *J. Chromatogr.*, 178, 163, 1979.
130. **Whitney, D. C. and Diamond, R. M.,** Ion-exchange studies in concentrated solutions. III. The nature of the resin sulfonate-cation bond in strong acid solutions, *J. Phys. Chem.*, 68, 1886, 1964.
131. **Alexa, J.,** Ion exchange chromatography of rare-earth metals in aqueous organic media. II. Effects of aqueous alcohol solutions on sorption in non-complex-forming media, *Collect. Czech. Chem. Commun.*, 30(7), 2351, 1965.
132. **Martell, A. E. and Calvin, M.,** *Chemistry of Metal Chelate Compounds*, Prentice-Hall, Englewood Cliffs, N.J., 1952, chap. 9.
133. **Mikler, J.,** Separation of rare earth metals by ion exchange. III. Reversal in the sequence of elution of europium and yttrium from Dowex 50W-X8 with ammonium acetate, *Mh. Chem.*, 96(3), 1007, 1965.
134. **Bleyl, H. J. and Münzel, H.,** Rapid radiochemical separations. IV. Separation of rare earth metals on cation exchange columns, *Radiochim. Acta*, 9, 149, 1968.
135. **Campbell, D. O.,** Rapid rare-earth-metal separation by pressurised ion exchange chromatography, *J. Inorg. Nucl. Chem.*, 35(11), 3911, 1973.
136. **Schoebrechts, F., Merciny, E., and Duykaerts, G.,** High performance liquid chromatographic separation of trivalent lanthanides on a cation exchanger using EDTA. I. Synthesis and general properties of polystyrene-divinylbenzene sulfonate pellicular ion exchangers, *J. Chromatogr.*, 174, 351, 1979.
137. **Schoebrechts, F., Merciny, E., and Duyckaerts, G.,** Contribution of high performance liquid chromatography to separation of tervalent lanthanoids on cation-exchange resins in presence of EDTA. II. Effect of the diameter of grains of homogeneous resins and of the case of pellicular resins on chromatographic resolution, *J. Chromatogr.*, 179(1), 63, 1979.
138. **Hirose, A., Iwasaki, Y., Iwato, I., Ueda, K., and Ishii, D.,** Post-column colorimetric detection with xylenol orange in micro h.p.l.c. of rare earth metals, *HRC CC J. High Resolut. Chromatogr. Commun.*, 4(10), 530, 1981.
139. **Niu, C., Zhang, S., Zhang, J., and Yu, D.,** Determination of terbium in a mixture of light lanthanoids by high speed ion-exchange chromatography, *Fenxi Huaxue*, 11(7), 521, 1983.

140. **Elchuk, S. and Cassidy, R. M.,** Separation of lanthanoids on high-efficiency bonded phases and conventional ion-exchange resins, *Anal. Chem.*, 51, 1434, 1979.
141. **Wang, W.-N., Chen, Y.-J., and Wu, M.-T.,** Complementary analytical methods for cyanide, sulfide, certain transition metals and lanthanides in ion chromatography, *Analyst (London)*, 109, 281, 1984.
142. **Qaim, S. M., Ollig, H., and Blessing, G.,** Separation of lanthanoids by preparative high-pressure liquid chromatography, *Radiochim. Acta*, 26(1), 59, 1979.
143. **Lengyel, T.,** Investigations on ion-exchange equilibria by a radioactive-tracer method. X. Application of the gradient-elution technique in extractive ion-exchange separation, *Acta Chim. Hung.*, 52(4), 391, 1967.
144. **Haskin, L. A., Wildeman, T. R., and Haskin, M. A.,** An accurate procedure for the determination of the rare earths by neutron activation, *J. Radioanal. Chem.*, 1, 337, 1968.
145. **Chen, Y.-M. and Sung, N.-K. Y.,** Mutual separation of rare earths in monazite with cation-exchange elution chromatography, *J. Chin. Chem. Soc. Taipei*, 22(2), 113, 1975.
146. **Marsh, S. F.,** Separation of lanthanide fission products from nuclear fuels by extraction chromatography and cation exchange for isotope-dilution mass spectrometric analysis, *Anal. Chem.*, 39, 641, 1967.
147. **Wolfsberg, K.,** Determination of rare earths in fission products by ion exchange at room temperature, *Anal. Chem.*, 34, 518, 1962.
148. **Ruf, H., von Baeckmann, A., and Ganter, E.,** Determination of neodymium-148 in rare-earth fission products by neutron activation, *Mikrochim. Acta*, No. 5, 1029, 1970.
149. **Ishii, D., Hirose, A., and Iwasaki, Y.,** Studies of micro high-performance liquid chromatography. VII. Cation-exchange separation of sixteen rare-earth metals by microscale high performance liquid chromatography, *J. Radioanal. Chem.*, 46(1), 41, 1978.
150. **Zhao, G.-W., Luo, S.-Q., Yao, S.-R., and Yu, S.-K.,** Separation of sixteen rare earth metals by high-performance liquid chromatography, *Chung-Kuo K'o Hsueh Chi Shu Ta Hsueh Hsueh Pao*, 11(12), 128, 1981.
151. **Zhao, G., Luo, S., Yao, S., and Yu, S.,** Separation of sixteen rare earth metals by high-performance liquid chromatography, *Hua Hsueh Tung Pao*, No. 8, 17, 1981.
152. **Zhao, G., Luo, S., Yao, S., and Yu, S.,** Separation of sixteen rare earth metals by high-performance liquid chromatography, *Jpn. Analyst*, 31(2), 63, 1982.
153. **Story, J. N. and Fritz, J. S.,** Forced-flow chromatography of the lanthanoids with continuous in-stream detection, *Talanta*, 21, 892, 1974.
154. **Massart, D. L. and Hoste, J.,** Separation of lutetium, ytterbium and thulium on cation-exchange resin by ammonium α-hydroxyisobutyrate, *Anal. Chim. Acta*, 28, 378, 1963.
155. **Fujii, I.,** Determination of individual rare earth elements in rare earth minerals by neutron activation. I. Determination of individual rare-earth elements in monazite and allanite, *J. At. Energy Soc. Jpn.*, 3(1), 9, 1961.
156. **Fujii, I.,** Determination of individual rare-earth elements in rare-earth-minerals by neutron-activation. II. Determination of individual rare-earth elements in gadolinite, xenotime, fergusonite, euxenite and yttrialite, *J. At. Energy Soc. Jpn.*, 3(3), 186, 1961.
157. **Lesbats, A.,** Systematic analysis of high-purity magnesium, *Ann. Chim.*, 4(1), 29, 1969.
158. **Gaittet, J.,** Systematic neutron-irradiation analysis of very-high purity aluminum and iron, *Ann. Chim. Paris*, 5(9-10), 1219, 1960.
159. **Nervik, W. E.,** An improved method for operating ion-exchange resin columns in separating the rare earth elements, *J. Phys. Chem.*, 59(8), 690, 1955.
160. **Almásy, A.,** Determination of rare-earth metals in their mixtures, *Acta Chim. Acad. Sci. Hung.*, 17(1), 55, 1958.
161. **Yamabe, T. and Hayashi, T.,** Elution behavior of rare-earth elements on a mixed ion-exchange column, *J. Chromatogr.*, 76(1), 213, 1973.
162. **Inczédy, J., Nemeshegyi, G., and Erdey, L.,** Separation and determination of rare earth-metals by ion-exchange chromatography. I. Photometric determination, *Acta Chim. Hung.*, 43(1), 1, 1965.
163. **Yamabe, T.,** Ion-exchange separations on mixed columns, *J. Chromatogr.*, 83, 59, 1973.
164. **Inczédy, J., Nemeshegyi, G., and Erdey, L.,** Separation and determination of rare earth metals by ion-exchange chromatography. II. Chromatographic separation of rare earth metals combined with chemical detection, *Acta Chim. Hung.*, 43(1), 9, 1965.
165. **Takata, J. and Arikawa, Y.,** Application of controlled potential coulometry to automated monitoring of liquid chromatography. VII. Cation-exchange chromatography of rare earth elements, *Jpn. Analyst*, 24(12), 762, 1975.
166. **Hayashi, T. and Yamabe, T.,** Elution behavior of the rare earth elements on single and mixed ion-exchange columns, *J. Chromatogr.*, 87(1), 227, 1973.
167. **Brücher, E. and Szarvas, P.,** Separation of rare earth fission products by elution with lactic acid-ethylenediaminetetraacetic acid, *Acta. Chim. Hung.*, 52(1), 31, 1967.
168. **Sugihara, T. T., James, H. I., Troianello, E. J., and Bowen, V. T.,** Radiochemical separation of fission products from large volumes of seawater. Strontium, cesium, cerium, and promethium, *Anal. Chem.*, 31, 44, 1959.

169. **Høgdahl, O. T.,** Analytical procedures involved in the measurement of radioactive and stable cerium in the marine environment, in *Reference Methods for Marine Radioactivity Studies,* International Atomic Energy Agency, Vienna, 1970, 187.
170. **Mosen, A. W., Schmitt, R. A., and Vasilevskis, J.,** Determination of the rare earth elements, lanthanum through lutetium, in chrondritic, achondritic and iron meteorites by neutron activation analysis, *Anal. Chim. Acta,* 25, 10, 1961.
171. **Schmitt, R. A., Smith, R. H., Lasch, J. E., Mosen, A. W., Olehy, D. A., and Vasilevskis, J.,** Radiochemical procedures for the determination of the rare earth elements, scandium and yttrium, *Geochim. Cosmochim. Acta,* 27(6), 616, 1963.
172. **Wish, L. and Foti, S. C.,** Carrier-free separation of the individual rare earth radionuclides from fission-product mixtures, *J. Chromatogr.,* 20(3), 585, 1965.
173. **Gleit, C. E., Benson, P. A., and Holland, W. D.,** Determination of silver, cadmium, indium, tantalum, rhenium, gold and rare earths at low concentrations by neutron activation and radiochemical analysis, *Anal. Chem.,* 36, 2067, 1964.
174. **Hwang, J.-M., Shih, J.-S., Yeh, Y.-C., and Wu, S.-C.,** Determination of rare earths in monazite sand and rare earth impurities in high-purity rare earth oxides by high-performance liquid chromatography, *Analyst (London),* 106, 869, 1981.
175. **Dogadkin, N. N., Kuchinskaya, O. I., Tustanovskii, V. I., and Yakovlev, Yu. V.,** Determination of rare earth alloying additives in special steels by neutron-activation analysis, *J. Radioanal. Chem.,* 29(2), 251, 1976.
176. **Zeligman, M. M.,** Ion-exchange separation of fission-product rare-earth metals with 2-hydroxyisobutyric acid, *Anal. Chem.,* 37, 524, 1965.
177. **Kubota, M.,** Separation of rare earths by an ion-exchange method. Determination of impurities in high-purity rare earth oxides, *J. Radioanal. Chem.,* 23(1-2), 73, 1974.
178. **Massart, D. L. and Hoste, J.,** Activation analysis for rare-earth metals. IV. Determination of traces of rare-earth impurities in gadolinium trioxide, *Anal. Chim. Acta,* 42, 166, 1968.
179. **Sisson, D. H., Mode, V. A., and Campbell, D. O.,** High-speed separation of the rare-earth metals by ion exchange. II, *J. Chromatogr.,* 66(1), 129, 1972.
180. **Freiling, E. C. and Bunney, L. R.,** Ion exchange as a separation method. VII. Near optimum conditions for the separation of fission product rare earths with lactic acid eluant at 87°C, *J. Am. Chem. Soc.,* 76, 1021, 1954.
181. **Pappas, A. C. and Alstad, J.,** Determination of trace elements in opium by means of activation analysis. I. Methods of analysis and studies referring to the determination of the origin of opium, *Radiochim. Acta,* 1, 109, 1963.
182. **Vobecký, M., Maštalka, A., and Mareček, J.,** Determination of lanthanides in uranium by activation analysis, *Collect. Czech. Chem. Commun.,* 31(8), 3309, 1966.
183. **Vobecký, M. and Maštalka, A.,** Radiochemical isolations. II. Chromatographic isolation of rare-earth elements, *Collect. Czech. Chem. Commun.,* 28(3), 709, 1963.
184. **Bächmann, K. and Lieser, K. H.,** Activation analysis for rare-earth metals in nuclear-reactor fuel, *Fresenius' Z. Anal. Chem.,* 250, 172, 1970.
185. **Gusmini, S. and Dubuquoy, C.,** Total elementary separation of rare-earth metals on cation-exchange resins, Report CEA, R-4102, Centre d'Etudes de Bruyeres-le-Chatel, France, Commissariat a l'Energie Atomique, 1971.
186. **Clanet, F.,** Radiochemical separation of the lanthanides in fission products, *J. Chromatogr.,* 5(4), 356, 1961.
187. **Rengan, K. and Meinke, W. W.,** Rapid radiochemical separation and activation analysis of rare earth elements, *Anal. Chem.,* 36, 157, 1964.
188. **Rengan, K. and Meinke, W. W.,** Rapid radiochemical separation and activation analysis of rare earth elements, Report TID-19.094, U.S. Atomic Energy Commission, 1963.
189. **Baker, J. D., Gehrke, R. J., Greenwood, R. C., and Meikrantz, D. H.,** Advanced system for separation of rare earth fission products, *J. Radioanal. Chem.,* 74(1-2), 117, 1982.
190. **Dubuquoy, C. and Metzger, G.,** High-performance liquid chromatographic separation of rare-earth metals, *Analusis,* 5(7), 314, 1977.
191. **Hubicka, H. and Hubicki, Z.,** Ion exchange separation of yttrium from light medium lanthanoids, *Zesz. Nauk. Politech. Slack. Chem.,* 639(93), 123, 1980.
192. **Høgdahl, O. T., Melsom, S., and Bowen, V. T.,** Neutron activation analysis of lanthanide elements in seawater, *Adv. Chem. Ser.,* 73, 308, 1968.
193. **Massart, D. L. and Bossaert, W.,** Prediction of elution maxima in the gradient elution of rare earth-metals, *J. Chromatogr.,* 32(1), 195, 1968.
194. **Wu, N., Zhang, X., and Zhao, G.,** High-performance ion-exchange chromatographic separation and spectrophotometric determination of individual rare-earth metals in cobalt-based alloys, *Fenxi Huaxue,* 12(2), 128, 1984.

195. **Hayakawa, T., Moriyasu, M., and Hashimoto, Y.,** High-performance liquid chromatography of rare earth-metals with post-column reaction with arsenazo III, *Jpn. Analyst,* 32(2), 136, 1983.
196. **Campbell, D. O. and Ketelle, B. H.,** Rapid chromatographic elution of tracer lanthanide using high pressure ion exchange, *Inorg. Nucl. Chem. Lett.,* 5, 533, 1969.
197. **Campbell, D. O. and Buxton, S. R.,** Rapid ion-exchange separations. Chromatographic lanthanide separations using a high-pressure ion-exchange method, *Ind. Eng. Chem. Process Design Dev.,* 9, 89, 1970.
198. **Yoshida, K. and Haraguchi, H.,** Determination of rare-earth elements by liquid chromatography-inductively coupled plasma atomic-emission spectrometry, *Anal. Chem.,* 56, 2580, 1984.
199. **Pinajian, J. J. and Raman, S.,** The preparation of of ^{145}Pm from neutron-irradiated ^{144}Sm, *J. Inorg. Nucl. Chem.,* 30, 3151, 1968.
200. **Krtil, J., Moravec, A., Mencl, J., and Alexa, J.,** Isolation of stable isotopes of nedymium from the solution of fission products of uranium by ion exchange for isotope-dilution mass-spectrometric analysis, *Radiochem. Radioanal. Lett.,* 3(2), 113, 1970.
201. **Bulovic, V., Krtil, J., Djordjevic, M., and Moravec, A.,** Determination of burn-up of nuclear fuel on the basis of mass-spectrometric measurement of the stable isotopes of neodymium. II. Procedures for isolating and measuring the neodymium isotopes, *Radiochem. Radioanal. Lett.,* 5(4-5), 229, 1970.
202. **Sugihara, T. T., James, H. I., Troianello, E. J. and Bowen, V. T.,** Radiochemical separation of fission products from large volumes of sea-water. Strontium, caesium, cerium and promethium, *Anal. Chem.,* 31, 44, 1959.
203. **Stewart, D. C.,** Controlling rare-earth separations by means of varying resin column operating conditions, *J. Inorg. Nucl. Chem.,* 4, 131, 1957.
204. **Schonfeld, E. and Stewart, D. C.,** Effect of temperature on rare earth separations on cation resin columns with lactate, *Radiochim. Acta,* 5, 176, 1966.
205. **Auboin, G. and Laverlochere, J.,** Improvement of chemical separations on ion-exchange resins: influence of particle size, *J. Radioanal. Chem.,* 1(2), 123, 1968.
206. **Alexa, J.,** Study of macro-porous cation exchange resins for their suitability for ion exchange chromatography, *Collect. Czech. Chem. Commun.,* 33(6), 1933, 1968.
207. **Alexa, J.,** Chromatographic separation of rare earths by means of the macroreticular cation exchange resin "Katex KP-1", *Collect. Czech. Chem. Commun.,* 29, 2851, 1964.
208. **Firsova, L. A., Chuveleva, E. A., Nazarov, P. P., and Glazunov, P. Ya.,** Influence of high absorbed doses of ionizing radiation on the chromatographic separation of rare earth metals. I. Influence of separated suppressor ions. Separation on KU-2 cationite (zinc form), *Zh. Fiz. Khim.,* 54(10), 2659, 1980.
209. **Firsova, L. A., Chuveleva, E. A., Nazarov, P. P., and Glazunov, P. Ya.,** Influence of high absorbed doses of ionizing emissions on the chromatographic separation of rare earth metals. II. Influence of separated suppressor ions. Separation on KU-2 cationite (hydrogen form), *Zh. Fiz. Khim.,* 54(10), 2662, 1980.
210. **Brunisholz, G. and Quinche, J. P.,** Analytical chemistry of the rare-earth elements. Simple chromatographic procedure for the quantitative analysis of mixtures, *Chimia,* 14(10), 343, 1960.
211. **Brunisholz, G. and Quinche, J. P.,** Chromatographic procedure for the determination of rare earth-metals, particularly cerium-earth metals, *Chimia,* 13(10), 331, 1959.
212. **Brunisholz, G., Collet, P., Klipfel, L., and Roulet, R.,** Simultaneous determination of rare earth metals by ion-exchange chromatography, *Rev. Chim. Miner.,* 10(1), 131, 1973.
213. **Moret, R. and Brunisholz, G.,** Improvement of the analysis of the rare earths by displacement chromatography on ion-exchangers, *Chimia,* 15, 313, 1961.
214. **Brunisholz, G. and Roulet, R.,** Separation of rare earth metals by ion-exchange chromatography. II. Separation and analysis of the pair europium-gadolinium, *Helv. Chim. Acta,* 53(1), 126, 1970.
215. **Brunisholz, G. and Roulet, R.,** Separation of rare-earth metals by ion exchange chromatography. III. Simultaneous determination of erbium, thulium, ytterbium and lutetium, *J. Chromatogr.,* 75(1), 101, 1973.
216. **Frank, B.,** Separation of lanthanoids by ion exchange with NTA as eluent and iron(III) as retaining or displacing ion, *Biul. Lubel. Tow. Nauk. Mat. Fiz. Chem.,* 21(1), 29, 1979.
217. **Quinche, J. P.,** Separation of lanthanoids by ion exchange chromatography with use of N-(2-hydroxyethyl)-iminodiacetic acid, *Helv. Chim. Acta,* 56(3), 1073, 1973.
218. **Quinche, J. P., Quinche-Sax, S., and Thonney, M.,** Simultaneous determination of lanthanum, cerium, praseodymium, neodymium, samarium and europium (on gadolinium) by displacement chromatography on cation exchange resin, *Chim. Anal.,* 52(4), 400, 1970.
219. **Kubota, M.,** Analysis of the elution curve in eluting lutetium from cation-exchanger with 2-hydroxy-2-methylpropionic acid, *Bull. Chem. Soc. Jpn.,* 48(12), 3603, 1975.
220. **Karol, P. J.,** Chromatographic separation of rare earth-metals by cation exchange. Dependence of lanthanoid peak position on eluent pH, *J. Radioanal. Chem.,* 24(1), 17, 1975.
221. **Seyb, K. E.,** Cation-exchange separation of rare earth metals from nuclear fission products, *Fresenius' Z. Anal. Chem.,* 226, 159, 1967.
222. **Ross, R. and Römer, J.,** Separation of terbium, gadolinium, and europium in milligram amounts. I. Separation by cation-exchange, *Z. Phys. Chem.,* 233(1-2), 17, 1966.

223. **Fenyo, J. C., Selegny, E., Gusmini, S., and Dubuquoy, C.,** Separation of lanthanides by chromatography on ion-exchange paper SA-2 in 2-hydroxyisobutyrate medium, *J. Chromatogr.*, 49(2), 269, 1970.
224. **Foti, S. C. and Wish, L.,** Carrier-free method separates rare earth nuclides, *Chem. Eng. News*, 44(39), 60, 1966.
225. **Wish, L.,** Quantitative radiochemical analysis by ion-exchange. Calcium, strontium and barium, *Anal. Chem.*, 33, 53, 1961.
226. **Lengyel, T.,** Investigations on ion exchange equilibria by a radioactive-tracer method. IX. Theory and use of extractive ion-exchange in the separation of lanthanides, *Acta Chim. Hung.*, 52(4), 375, 1967.
227. **Deelstra, H. and Verbeek, F.,** The determination of the stability constants of some lanthanide-α-hydroxy-isobutyrate complexes using cation-exchange methods, *Bull. Soc. Chim. Belg.*, 73, 597, 1964.
228. **Deelstra, H. and Verbeek, F.,** Separation of the lanthanum series and yttrium by cation-exchange elution with ammonium α-hydroxyisobutyrate and lactate, *J. Chromatogr.*, 17(3), 558, 1965.
229. **Alexa, J.,** Ion-exchange chromatography of rare-earth metals in aqueous organic media. VII. Influence of electrolytes, *Collect. Czech. Chem. Commun.*, 35(6), 1921, 1970.
230. **Alexa, J.,** Ion-exchange chromatography of rare earth-metals in aqueous-organic media. I. Effect of some mixed media on the separation of a mixture of europium with yttrium and promethium, *Collect. Czech. Chem. Commun.*, 30(7), 2344, 1965.
231. **Alexa, J.,** Ion-exchange chromatography of rare-earth metals in aqueous-organic media. III. Effect of aqueous alcohol solutions on sorption from complex-forming media, *Collect. Czech. Chem. Commun.*, 30(7), 2361, 1965.
232. **Alexa, J.,** Ion-exchange chromatography of rare earth metals in aqueous-organic media. IV. Effect of aqueous alcohol solutions on the separation factor, *Collect. Czech. Chem. Commun.*, 30(7), 2368, 1965.
233. **Alexa, J.,** Ion-exchange chromatography of rare earths in aqueous-organic media. VI. Stability constants of terbium with α-hydroxyisobutyric acid in the system methanol-water, *Collect. Czech. Chem. Commun.*, 33, 2731, 1968.
234. **Alexa, J.,** Ion-exchange chromatography of rare earths in aqueous organic media. VIII. The distribution of solvents in the ion exchanger phase in mixed media system, *Collect. Czech. Chem. Commun.*, 36, 1670, 1971.
235. **Walter, C. W. and Korkisch, J.,** Anion-exchange of metal ions in aqueous-organic solvent systems containing nitric acid. V. Theoretical considerations, *Mikrochim. Acta*, No. 1, 194, 1971.
236. **Maštalka, A., Vobecký, M., and Vojtěch, O.,** Radiochemical separations. III. Influence of medium on the chromatographic isolation of lanthanides, *Collect. Czech. Chem. Commun.*, 28, 743, 1963.
237. **Vojtěch, O. and Beranová, H.,** Effects of some polyhydric alcohols on ion-exchange and the separation of rare earth metals with the aid of α-hydroxyisobutyric acid, *Collect. Czech. Chem. Commun.*, 30(6), 1799, 1965.
238. **Karol, P. J.,** Rare-earth metal separations by cation-exchange column chromatography: comparison of 2-hydroxyisobutyric acid and 2-hydroxy-2-methyl-butyric acid as eluents, *J. Chromatogr.*, 79, 287, 1973.
239. **Nishi, T. and Fujiwara, I.,** Separation of the lanthanides by ion exchange with α-hydroxy-α-methylbutric acid, *J. At. Energy Soc. Jpn.*, 6(1), 15, 1964.
240. **Dubuquoy, C., Gusmini, S., Poupard, D., and Verry, M.,** Separation of lanthanoids by chromatography on ion-exchange paper SA-2, *J. Chromatogr.*, 57(3), 455, 1971.
241. **Wildeman, T. R. and Haskin, L.,** Rare earth elements in ocean sediments, *J. Geophys. Res.*, 70, 2905, 1965.
242. **Albert, P.,** Analysis of high-purity iron, *Mem. Sci. Rev. Metall.*, 65, 3, 1968.
243. **Lee, C., Yim, Y. C., Chung, K. S., and Bak, H.-I.,** Average fission neutron cross section for ^{93}Nb (n, α) ^{90}Y, ^{90}Zr(n, p), ^{90}Y, ^{93}Nb (n, α) ^{90m}Y and ^{90}Zr(n, p) ^{90m}Y reactions, *J. Korean Chem. Soc.*, 17(1), 20, 1973.
244. **Wakat, A. and Griffin, C.,** Rapid isolation of individual rare-earth metals from fission products and identification of neodymium-152, *Radiochem. Radioanal. Lett.*, 2(6), 351, 1969.
245. **Kubota, M.,** Determination of trace impurities in high-purity lutetium oxide by neutron activation with the aid of cation exchange separation, *J. Nucl. Sci. Technol.*, 11(8), 334, 1974.
246. **Ko, C.-P.,** Determination of neodymium in gadolinium, gallium garnet, *Fen Hsi Hua Hsueh*, 8(2), 196, 1980.
247. **Choppin, G. R. and Silva, R. J.,** Separation of the lanthanides by ion-exchange with α-hydroxyisobutyric acid, *J. Inorg. Nucl. Chem.*, 3(2), 153, 1956.
248. **Natsume, H., Umezewa, H., Suzuki, T., Ichikawa, F., Sato, T., Baba, S., and Amano, H.,** Systematic radiochemical analysis of fission products, *J. Radioanal. Chem.*, 7, 189, 1971.
249. **Baraniak, L. and Zahn, H.,** Neodymium-148 analysis for burn-up determination in thermal-reactor fuel elements, *Kernenergie*, 19(1), 5, 1976.
250. **Hermann, A., Baraniak, L., Hübener, S., Nebel, D., Niese, U., and Trebeljahr, A.,** Experiences with respect to the destructive determination of the burn-up of reactor fuel elements, *Isotopenpraxis*, 12(7-8), 277, 1976.

251. **Kulus, E., Molnar, F., and Szabo, E.,** Concentration and determination of rare earth impurities in an yttrium matrix, *J. Radioanal. Chem.*, 7(2), 347, 1971.
252. **Molnár, F., Horváth, A., and Khalkin, V. A.,** Anion-exchange behavior of light rare earth metals in aqueous methanol solutions containing neutral nitrates. I. Separation of carrier-free light rare-earth metals, *J. Chromatogr.*, 26(1), 215, 1967.
253. **Molnár, F., Horváth, A., and Khalkin, V. A.,** Anion-exchange behavior of light rare earth metals in aqueous methanol solutions containing neutral nitrates. II. Macro-micro separations, *J. Chromatogr.*, 26(1), 225, 1967.
254. **Frache, R., Mazzucotelli, A., Dadone, A., Baffi, F., and Cescon, P.,** Systematic analysis of major, minor and trace elements in silicates by ion-exchange separations, *Analusis*, 6(7), 294, 1978.
255. **Høgdahl, O. T., Melsom, S., and Bowen, V. T.,** Neutron activation analysis of lanthanide elements in seawater. Trace inorganics in water, *Adv. Chem. Ser.*, 73, 308, 1968.
256. **Bhargava, V. K., Rao, V. K., Marathe, S. G., Sahakundu, S. M., and Iyer, R. H.,** Separation of heavier rare earth metals from neutron-irradiated uranium targets, *J. Radioanal. Chem.*, 47(1-2), 5, 1978.
257. **James, W. D., Adams, D. E., Sigg, R. A., Beck, J. N., and Kuroda, P. K.,** Quick radiochemical separation of yttrium from mixed fission products, *Radiochem. Radioanal. Lett.*, 18(1), 7, 1974.
258. **Kuboto, M.,** Activation analysis of gadolinium impurity in high-purity europium oxide, *J. Nucl. Sci. Technol.*, 13(8), 449, 1976.
259. **Ryabukhin, V. A., Gatinskaya, N. G., and Ermakov, A. N.,** Ion-exchange pre-concentration of elements of the cerium sub-group from "pure" yttrium, *Zh. Anal. Khim.*, 32(5), 909, 1977.
260. **Petrow, H. G.,** Radiochemical determination of neodymium, praseodymium and cerium in fission products, *Anal. Chem.*, 26, 1514, 1954.
261. **Alberti, G. and Massucci, M. A.,** Chromatographic separation and identification of some individual rare-earth metals on paper by means of fluorescence, *J. Chromatogr.*, 11(3), 394, 1963.
262. **Dubuquoy, C., Gusmini, S., and Poupard, D.,** Separation of the lanthanoids by chromatography on ion-exchange paper SA-2. III, *J. Chromatogr.*, 70(1), 216, 1972.
263. **Dubuquoy, C., Gusmini, S., Poupard, D., and Verry, M.,** Separation of lanthanoids by ion-exchange chromatography on SA-2 paper: studies and applications, Report CEA R-4440, Centre d'Etudes de Bruyeres-le-Chatel, France, Commissariat a l'Energie Atomique, 1973.
264. **Wish, L., Freiling, E. C., and Bunney, L. R.,** Ion-exchange as a separation method. VIII. Relative elution positions of lanthanide and actinide elements with lactic acid eluant at 87°C, *J. Am. Chem. Soc.*, 76, 3444, 1954.
265. **Lieser, K. H. and Bächmann, K.,** Investigation of the separation of strontium-90-yttrium-90 and of barium-140-lanthanum-140 with Dowex 50 as ion exchanger and lactic acid as complexing agent, *Fresenius' Z. Anal. Chem.*, 225, 379, 1967.
266. **Almásy, A.,** Stepwise elution analysis for the separation of rare earth-metals. Preliminary communication, *Acta Chim. Acad. Sci. Hung.*, 10(1-3), 303, 1956.
267. **Karol, P. J.,** Rare-earth-metal separations by mixed ion-exchange columns, *J. Chromatogr.*, 89(2), 346, 1974.
268. **Cuninghame, J. G., Sizeland, M. L., Willis, H. H., Eakins, J., and Mercer, E. R.,** The rapid separation of rare earth fission products by cation exchange using lactic acid, *J. Inorg. Nucl. Chem.*, 1, 163, 1955.
269. **Ciba, J. and Jurczyk, J.,** Determination of yttrium in steels and in chromium-nickel alloys by wavelength-dispersive X-ray fluorescence; separation of the matrix, *Fresenius' Z. Anal. Chem.*, 313, 542, 1982.
270. **Bunney, L. R., Freiling, E. C., McIsaac, L. D., and Scadden, E. M.,** Radiochemical procedure for individual rare earths, *Nucleonics*, 15(2), 81, 1957.
271. **Pascual, J. and Freiling, E. C.,** Rare-earth solutions for 4 π-counting, *Nucleonics*, 15(5), 94, 1957.
272. **Senegačnik, M. and Paljk, Š.,** Fallout monitoring of waters. II. Radiochemical determination of cerium-144, caesium-137, strontium-89,90 and barium-140 by ion-exchange, *Fresenius' Z. Anal. Chem.*, 244, 375, 1969.
273. **McNaughton, G. S. and Woodward, R. N.,** Studies in radioactive fallout in New Zealand. I. The measurement of radiocaesium-strontium-barium, and cerium in rainwater, *N. Z. J. Sci.*, 4(3), 523, 1961.
274. **Pozdnyakov, A. A.,** Chromatographic separation of radio isotopes of elements of the yttrium group obtained by fission of ytterbium and hafnium by protons of high energy, *Zh. Anal. Khim.*, 11(5), 566, 1956.
275. **Beranová, H. and Petržila, V.,** Rapid separation of rare earth and trans-plutonium elements by ion exchange, *Collect. Czech. Chem. Commun.*, 29(2), 500, 1964.
276. **Rane, A. T. and Bhatki, K. S.,** Rapid radiochemical separations of strontium-90-yttrium-90 and calcium-45-scandium-46 on a cation exchange resin, *Anal. Chem.*, 38, 1598, 1966.
277. **Miyake, Y.,** A sequential procedure for the radiochemical analysis of marine material, unpublished working paper to IAEA Safety Series No. 11, Methods of Surveying and Monitoring Marine Radioactivity, 1964.
278. **Welford, G. A., Collins, W. R., Morse, R. S., and Sutton, D. C.,** The sequential analysis of long range fallout debris, *Talanta*, 5, 168, 1960.

279. **Haberer, K. and Stürzer, U.,** Rapid determination of radio-nuclides for the safeguarding of drinking water, *Fresenius' Z. Anal. Chem.*, 299, 177, 1979.
280. **Haberer, K. and Stürzer, U.,** Rapid determination of radioactive rare earths in water samples, *Gwf Wasser/Abwasser*, 114(5), 241, 1973.
281. **Draganić, I. G., Draganić, C. D., and Dizdar, Z. I.,** Application of ion-exchangers for the determination of traces of impurities in uranium, *Bull. Inst. Nucl. Sci. Boris Kidrich (Belgrade)*, 55, 37, 1954.
282. **Dolar, D. and Draganić, Z.,** Separation of traces of the rare earth elements from uranium by means of Amberlite IR-120 resin, *Bull. Inst. Nucl. Sci. Boris Kidrich (Belgrade)*, 2, 77, 1953.
283. **Blaedel, W. J., Olsen, E. D., and Buchanan, R. F.,** Sequential ion-exchange separation scheme for the identification of metallic radioelements, *Anal. Chem.*, 32, 1866, 1960.
284. **Shibate, M., Tanabe, I., Okuda, K., and Kadota, K.,** Separation of the cerium-group rare earth metals by ion exchange using a citrate-acetate mixed solution as eluting agent, *Bull. Chem. Soc. Jpn.*, 41(11), 2627, 1968.
285. **Nesmeyanov, A. N. and Volkov, A. A.,** Use of hydroxides of iron(III), aluminum(III) and zinc(II) as collectors for yttrium-90 in sea-water, *Radiokhimiya*, 24(2), 247, 1982.
286. **Khramov, V. P.,** The ion-exchange separation of the rare earth elements of cerium sub-group, *Tr. Sarat. Inst. Mekh. Selsk. Khoz.*, No. 19, 15, 1959 (1960).
287. **Spedding, F. H. and Powell, J. E.,** Separation of rare earths by ion-exchange. VIII. Quantitative theory of the mechanism involved in elution by dilute citrate solutions, *J. Am. Chem. Soc.*, 76, 2550, 1954.
288. **Spedding, F. H. and Powell, J. E.,** Separation of rare earths by ion exchange. VII. Quantitative data for the elution of neodymium, *J. Am. Chem. Soc.*, 76(9), 2545, 1954.
289. **Topp, N. E.,** Modern techniques for separating the rare earth elements, *J. Less Common Met.*, 7(6), 411, 1964.
290. **Harris, D. H. and Tompkins, E. R.,** Ion exchange as a separation method. II. Separation of several rare earths of the cerium-group (La, Ce, Pr and Nd), *J. Am. Chem. Soc.*, 69, 2792, 1947.
291. **Ketelle, B. H. and Boyd, G. E.,** Further studies on the ion-exchange separation of the rare earths, *J. Am. Chem. Soc.*, 73, 1862, 1951.
292. **Ketelle, B. H. and Boyd, G. E.,** The exchange adsorption of ions from aqueous solutions by organic zeolites. IV. The separation of the yttrium group rare earths, *J. Am. Chem. Soc.*, 69, 2800, 1947.
293. **Tompkins, E. R., Khym, J. X., and Cohn, W. E.,** Ion-exchange as a separation method. I. The separation of fission-produced radioisotopes, including individual rare earths, by complexing elution from Amberlite resin, *J. Am. Chem. Soc.*, 69, 2769, 1947.
294. **Almásy, A.,** Stepwise elution analysis for the separation of rare earth metals, *Acta Chim. Hung.*, 10(1 to 3), 303, 1956.
295. **Jász, A. and Lengyel, T.,** Investigations on ion exchange equilibria with radioactive tracer method. VIII. Use of equilibrium data for prediction of ion-exchange separation efficiency, *Acta Chim. Hung.*, 40(2), 167, 1964.
296. **Ohyoshi, A., Ohyoshi, E., Ono, H., and Yamakawa, S.,** A study of citrate complexes of several lanthanides, *J. Inorg. Nucl. Chem.*, 34, 1955, 1972.
297. **Brykina, G. D., Isaeva, L. B., and Belyavskaya, T. N.,** Sorption of scandium by ion-exchange resins from aqueous and aqueous-acetone solutions of tartaric, citric and hydrochloric acids, *Radiokhimiya*, 14(4), 515, 1972.
298. **Spedding, F. H. and Powell, J. E.,** Methods for separating rare earth elements in quantity as developed at Iowa State College, *J. Met.*, 1131, October 1954.
299. **Powell, J. E. and Spedding, F. H.,** The separation of rare earths by ion exchange, *Trans. Metall. Soc. AIME*, 215, 457, 1959.
300. **Topp, N. E. and Young, D. D.,** Separation of rare earth elements by ion-exchange chromatography. I. Use of NH_4^+-H^+ development columns, *J. Chromatogr.*, 14(3), 469, 1964.
301. **Topp, N. E.,** The use of complexing agents for rare earth separation by ion exchange techniques, *Chem. Ind.*, 17, 1320, 1956.
302. **Fouarge, J. and Fuger, J.,** Chromatographic separation of the alkali metals, the alkaline earths and the rare earths, UNESCO/NS/RIC/52, 1957.
303. **Dybczyński, R.,** Investigations of the effect of flow rate, loading and other factors on the effectiveness of radiochemical separations by ion-exchange chromatography, *J. Radioanal. Chem.*, 31(1), 115, 1976.
304. **Bunisholz, G. and Roulet, R.,** Separation of rare earth metals by ion-exchange chromatography. I. Buffered-eluent technique, *Helv. Chim. Acta*, 52(7), 1847, 1969.
305. **Brunisholz, G.,** Procedure for pH-control during the separation of rare earths by ion-exchange chromatography, *Chimia*, 12(6), 180, 1958.
306. **Brunisholz, G.,** Separation of rare earths by means of ethylene-diaminetetra-acetic acid. IV. Separation on ion-exchangers, *Helv. Chim. Acta*, 40(6), 2004, 1957.
307. **Misumi, S. and Taketatsu, T.,** Separation of the rare earth-elements of the cerium-group with cation-exchange resin and zinc-EDTA complex, *Jpn. Analyst*, 8(10), 673, 1959.

308. **Powell, J. E. and Burkholder, H. R.,** Improvement in separation of gadolinium and of europium from samarium in ion-exchange elution with EDTA, *J. Chromatogr.*, 29(1), 210, 1967.
309. **Chmutov, K. V., Nazarov, P. P., Maslova, G. B., and Sheptunov, V. N.,** Isolation of promethium by displacement chromatography due to complex formation, *J. Chromatogr.*, 59(2), 415, 1971.
310. **Powell, J. E., Burkholder, H. R., and Gonda, K.,** Experimental determination of elution requirements in displacement ion exchange, *J. Chromatogr.*, 54(2), 259, 1971.
311. **Winget, J. O. and Lindstrom, R. E.,** Amino acids as retaining agents in displacement chromatography of the rare earth elements, *Sep. Sci.*, 4(3), 209, 1969.
312. **Spedding, F. H., Powell, J. E., Daane, A. H., Hiller, M. A., and Adams, W. H.,** Methods for preparing pure scandium oxide, *J. Electrochem. Soc.*, 105(11), 683, 1958.
313. **Spedding, F. H., Powell, J. E., and Wheelwright, E. J.,** The separation of adjacent rare-earths with ethylenediaminetetraacetic acid by elution from an ion-exchange resin, *J. Am. Chem. Soc.*, 76, 612, 1954.
314. **Shiokawa, T., Kudo, H., and Omori, T.,** Chemical effects associated with beta-decay process. III. The chemical behavior of Pr-144 decayed from Ce(III)-144-EDTA complexes in the solid phase, *Bull. Chem. Soc. Jpn.*, 42(2), 436, 1969.
315. **Spedding, F. H., Powell, J. E., and Wheelwright, E. J.,** The use of copper as the retaining ion in the elution of rare earths with ammonium ethylenediamine tetraacetate solutions, *J. Am. Chem. Soc.*, 76, 2557, 1954.
316. **Powell, J. E., Burkholder, H. R., and James, D. B.,** Elution requirements for the resolution of ternary mixtures of rare-earth metals, *J. Chromatogr.*, 32(3), 559, 1968.
317. **James, D. B., Powell, J. E., and Burkholder, H. R.,** Displacement ion exchange separation of ternary rare-earth-metal mixtures with chelating eluents, *J. Chromatogr.*, 35(3), 423, 1968.
318. **Fritz, J. S. and Umbreit, G. R.,** Ion-exchange separation of metal-ions, *Anal. Chim. Acta*, 19, 509, 1958.
319. **Fuger, J.,** Separation of the rare earths by means of ethylene-diaminetetra-acetic acid, *Bull. Soc. Chim. Belg.*, 66(1-2), 151, 1957.
320. **Brücher, E. and Szarvas, P.,** Effect of the nature of the retaining ion on the displacement chromatographic separation of rare earth-metals with EDTA, *Magy. Kem. Foly.*, 75(2), 63, 1969.
321. Michigan Chemical Corporation, Apparatus for the Separation of Metal Ions by Ion Exchange, U.S. Patent 3,059,777; date appl. November 21, 1958.
322. **Rekhkolainen, G. I.,** X-ray-fluorescence determination of rare earth-metals in solution, *Zavod. Lab.*, 31(4), 442, 1965.
323. **Bredenfel'd, N. V. and Astakhov, K. V.,** Effect of acidity on the ratio of ionic species in rare-earth metal-EDTA-solution, *Nauchn. Tr. Nauchno Issled. Proektn. Inst. Redkomet. Promsti.*, 45, 50, 1972.
324. **Sheptunov, V. N., Maslova, G. B., Nazarov, P. P., and Chmutov, K. V.,** Separation of rare earth-metals by displacement chromatography in presence of complexans. IV. Effect of concentration of displacement ion, *Zh. Fiz. Khim.*, 50(8), 2100, 1976.
325. **Hagiwara, Z. and Noguchi, M.,** Separation factors for some lanthanide pairs obtained at 60° in the ion-exchange system involving EDTA, *Bull. Chem. Soc. Jpn.*, 43(2), 401, 1970.
326. **Taketatsu, T.,** Ion-exchange separation of cerium and thorium with lead-EDTA complex as eluting agent, *Bull. Chem. Soc. Jpn.*, 32(3), 291, 1959.
327. **Loriers, J. and Lenoir, C.,** Use of copper as a cation exchanger in the separation of rare earth-elements by ion-exchange using ethylenediaminetetra - acetic-acid, *Comptes Rendus*, 247(4), 468, 1958.
328. **Noddack, W. and Korneli, W.,** Influence of acidity of eluents on the separation of rare earth-metals by ion-exchange, *Helv. Chim. Acta*, 44(7), 1829, 1961.
329. **Weidmann, G. and Liebold, G.,** Influence of copper, nickel and cobalt charging of ion-exchange resin columns on the separation of a praseodymium-neodymium mixture by elution with Trilon A, *Angew. Chem.*, 69(23), 753, 1957.
330. **Martynenko, L. I. and Mitrofanova, N. D.,** Sorption of rare-earth nitrilotriacetates by cation-exchange resin, *Vestn. Mosk. Univ. Ser. Khim.*, No. 1, 18, 1969.
331. **Firsova, L. A., Chuveleva, E. A., Nazarov, P. P., and Glazunov, P. Ya.,** Influence of high absorbed doses of ionizing radiation on the chromatographic separation of rare earth-metals. IV. Use of sodium nitrilo-triacetate as eluent, *Zh. Fiz. Khim.*, 54(11), 2949, 1980.
332. **Kharitonov, O. V., Radevich, V. S., Chuveleva, E. A., Nazarov, P. P., and Chmutov, K. V.,** Influence of citric acid on the separation of lanthanoids with diethylenetriaminepentaacetic acid and nitrilotriacetic acid by displacement chromatography, *Zh. Fiz. Khim.*, 50(7), 1870, 1976.
333. **Das, N. R. and Bhattacharyya, S. N.,** Use of nitrilotriacetic acid and different retaining ions in the separation of rare earth metals on the tracer scale by ion exchange, *Int. J. Appl. Radiat. Isot.*, 28(8), 731, 1977.
334. **Noddack, W. and Oertel, G.,** Ion exchange equilibria between nitrilotriacetic acid - rare earth complexes and the ion exchanger Dowex 50 in the Cu^{++}-, Ni^{++-}-, and Na^{+}-form, *Z. Elektrochem. Ber. Bunsenges. Phys. Chem.*, 61(9), 1216, 1957.

335. **Holleck, L. and Hartinger, L.,** Quantitative separation of cerite rare-earth elements by ion exchange and the effect of the pH of the eluent, *Angew. Chem.*, 68(12), 411, 1956.
336. **Hartinger, L. and Holleck, L.,** Quantitative separation of small quantities of rare earth elements from uranium fission products, *Angew. Chem.*, 68(12), 412, 1956.
337. **Holleck, L. and Hartinger, L.,** The position of yttrium in the exchanger separation of yttrium earths, where nitrilotriacetic acid is the eluent, *Angew. Chem.*, 68(12), 412, 1956.
338. **Loriers, J. and Carminati, D.,** Use of nitrilotriacetic acid for separation of ceric earths by ion exchange, *Comptes Rendus*, 237, 1328, 1953.
339. **Hagiwara, Z. and Oki, H.,** Ion exchange of a mixture of erbium and thulium with N′-(2-hydroxyethyl)ethylenediamine-NNN′-triacetic acid, *Bull. Chem. Soc. Jpn.*, 42(11), 3177, 1969.
340. **Powell, J. E., Spedding, F. H., and James, D. B.,** The separation of rare earths. A project for high school chemistry students, *J. Chem. Educ.*, 37, 629, 1960.
341. **Merciny, E. and Duyckaerts, G.,** Separation of lanthanides and actinides by N-hydroxyethylenediamine-NN′N′-tri-acetic acid. I. Study of effect of salt concentration on the efficiency of the column, *J. Chromatogr.*, 22(1), 164, 1966.
342. **Brunisholz, G. and Roulet, R.,** Effect of temperature on the efficiency of chromatographic separation of some pairs of rare earth metals, *Chimia*, 21(5), 188, 1967.
343. **Merciny, E. and Duyckaerts, G.,** Separation of lanthanides and actinides on ion-exchangers with N-hydroxyethylenediamine NN′N′-triacetic acid. II. Influence of pH on the separation factor and efficiency of the column, *J. Chromatogr.*, 26(2), 471, 1967.
344. **Merciny, E.,** The separation of lanthanides and trivalent actinides on ion exchangers using hydroxyethylenediamine-triacetic acid, *Fresenius' Z. Anal. Chem.*, 236, 498, 1968.
345. **Merciny, E. and Duyckaerts, G.,** Separation of lanthanides and actinides on ion exchangers using N-(2-hydroxyethyl)ethylenediamine-NN′N′-triacetic acid. IV. Formation of a complex of the type$[MeY_2Hn]^{(3-n)}$ between the lanthanides and HEDTA, *J. Chromatogr.*, 35(4), 549, 1968.
346. **Powell, J. E. and Burkholder, H. R.,** Augmenting the separation of adjacent pairs of rare-earth metals in elutions with N-(2-hydroxyethyl)ethylenediamine NN′N′-triacetic acid (HEDTA), *J. Chromatogr.*, 36(1), 99, 1968.
347. **Winget, J. O. and Lindstrom, R. E.,** Amino acids as retaining agents for separation of rare-earth metals on ion-exchange resin, Report Invest. RI 7175, U.S. Bureau of Mines, 1968.
348. **Yoshimura, J. and Waki, H.,** A systematic ion exchange separation of metallic elements. I. Group separation of metallic elements by anion exchange with hydrochloric acid, *Bull. Chem. Soc. Jpn.*, 35(3), 416, 1962.
349. **Amer Amezaga, S.,** Elution of rare earth metals from a cation-exchanger with N-(hydroxyethyl)ethylenediaminetriacetic acid with use of zinc as retaining ion, *An. Quim.*, 67, 739, 1971.
350. **Lishchenko, T. V., Korobko, G. P., Churilina, N. V., and Rozhdestvenskaya, O. I.,** Use of ethylenediaminedi(isopropylphosphoric) and N′-(2-hydroxyethyl)ethylenediamine-NNN′-triacetic acids for the separation of rare earth elements, including yttrium, *Nauchn. Tr. Gos. Nauchno Issled. Proektn. Inst. Redkomet. Promsti. Giredmet*, 91, 48, 1979.
351. **Kogan, L. and Ratner, R.,** Separation of yttrium from the lanthanides by ion-exchange with diethylenetriaminepenta acetic acid as eluent. I. Separation of yttrium-erbium mixtures, *J. Chromatogr.*, 62(3), 449, 1971.
352. **Katayama, M., Masuda, Y., Tarutani, T., and Misumi, S.,** Ion-exchange of some complexes of metals with diethylenetriamine-NNN′N″N″-penta-acetic acid, *Jpn. Analyst*, 20(4), 451, 1971.
353. **Firsova, L. A., Chuveleva, E. A., Nazarov, P. P., and Glazunov, P. Ya.,** Influence of high absorbed doses of ionizing emissions on the chromatographic separation of rare earth metals. III. Application of sodium diethylenetriaminepenta-acetate as eluent, *Zh. Fiz. Khim.*, 54(10), 2664, 1980.
354. **Delle Site, A. and Kappelmann, F. A.,** Complexation-behavior of the lanthanoid-iminodiacetic acid system and its possible application to the separation of light rare earth elements from transplutonium elements by ion exchange, *Inorg. Nucl. Chem. Lett.*, 10(1), 81, 1974.
355. **Kolleganov, M. Yu., Nazarov, P. P., Martynenko, L. I., Mitrofanova, N. D., and Spitsyn, V. I.,** Chromatographic separation of europium and gadolinium mixtures by ethylenediaminedisuccinic acid, *Izv. Akad. Nauk SSSR Ser. Khim.*, No. 2, 249, 1985.
356. **Fritz, J. S. and Umbreit, G. R.,** Ion-exchange separation of metal ions, *Anal. Chim. Acta*, 19(6), 509, 1958.
357. **Higuchi, H., Tomura, K., and Hamaguchi, H.,** Determination of rare earth-metals in rock samples by neutron-activation analysis. Use of a lithium-drifted germanium detector in conjunction with separation of light lanthanoids from heavy lanthanoids by cation exchange, *J. Radioanal. Chem.*, 5(2), 207, 1970.
358. **Tobia, S. K.,** Separation of the lanthanides from Egyptian monazite, *J. Appl. Chem.*, 13(4), 189, 1963.
359. **Krawczyk, I.,** Separation of uranium and rare-earth-elements from ethylenediaminetetraacetic acid solutions on ion exchange resin, Report PAN-213 (VIII), Institute of Nuclear Research, Warsaw, February 1961.
360. **Gordon, L., Firsching, F. H., and Shaver, K. J.,** Concentration of microgram amounts of rare earths in thorium. Precipitation and ion-exchange procedures, *Anal. Chem.*, 28, 1476, 1956.

361. **Kustas, V. L., Lazebnaya, G. V., and Zagorskaya, M. K.,** Spectrographic determination of impurities in lanthanum oxide of high-purity after their concentration by a chromatographic method, *Zh. Anal. Khim.*, 18(1), 99, 1963.
362. **Duyckaerts, G. and Fuger, J.,** Contribution to the study of the separation of rare earth fission products using ethylenediaminetetra-acetic acid, *Anal. Chim. Acta,* 14, 243, 1956.
363. **Shakhtakhtinskaya, N. G. and Iskenderov, M. G.,** Chromatographic separation of erbium from aluminum, iron and titanium, followed by arsenate-iodometric analysis, *Azerb. Khim. Zh.*, No. 3, 99, 1975.
364. **Prášilová, J.,** Sub-stoichiometric determination of heavy rare earth metals by isotope dilution analysis, *Talanta,* 13, 1567, 1966.
365. **Macášek, F. and Čech, R.,** Chromatographic separation of radiochemically pure yttrium-90 from strontium-90, *Chem. Zvesti,* 19(2), 107, 1965.
366. **Lavrukhina, A. K., Kalicheva, I. S., and Kolesov, G. M.,** Determination of scandium in meteorites by neutron activation with use of substoichiometric separation and γ-spectrometry, *Geokhimiya,* No. 6, 651, 1967.
367. **Das, A. K., Pal, J. C., and Banerjee, S.,** Separation of rare-earth metals from other elements by selective cation exchange, *Anal. Chim. Acta,* 47, 162, 1969.
368. **Vladescu, L. and Voicu, D.,** Separation of cerium(IV) from uranium(VI), titanium(IV) and zirconium(IV) by ion-exchange chromatography, *Rev. Chim. (Bucharest),* 30(5), 474, 1979.
369. **Vladescu, L. and Voicu, D.,** Separation and concentration of cerium(IV) by means of cation exchangers, *Rev. Chim. (Bucharest),* 24(6), 881, 1979.
370. **Vladescu, L. and Voicu, D.,** Ion-exchange chromatographic separation of cerium(IV), cerium(III) and lanthanum(III), *Rev. Chim. (Bucharest),* 30(4), 373, 1979.
371. **Yoshimura, J., Takashima, Y., and Waki, H.,** Separation of scandium from lanthanide elements with ion-exchange resin, *J. Chem. Soc. Jpn. Pure Chem. Sect.*, 79(10), 1169, 1958.
372. **Mikler, J.,** Separation of rare earth metals by ion-exchange. V. Separation of dysprosium and yttrium, *Mh. Chem.*, 98(5), 1899, 1967.
373. **Subbotina, A. I.,** Use of acetic acid in the chromatographic separation of yttrium and cerium, *Tr. Khim. Khim. Tekhnol. (Gor'kii),* No. 1, 30, 1960.
374. **Eusebius, L. C. T., Mahan, A., Ghose, A. K., and Dey, A. K.,** Cation-exchange sorption of some metal ions from aqueous ammonium acetate medium; separation of cerium(IV) from cerium(III), lanthanum(III) and other metal ions, *Indian J. Chem. Sect. A,* 15(5), 438, 1977.
375. **Lu, G. and Yi, Y.,** Separation and determination of rare earth metals, calcium, strontium and barium in bastanaesite, *Fenxi Huaxue,* 11(3), 227, 1983.
376. **Ghosh, B. K., Mahan, A., Ghose, A. K., and Dey, A. K.,** Adsorption behavior of some metal ions on cation-exchange resin (Dowex 50W-X4) from ammonium-acetate-dimethylformamide media, *J. Indian. Chem. Soc.*, 57, 591, 1980.
377. **Mahan, A., Ghose, A. K., and Dey, A. K.,** Cation-exchange characteristics of some metal ions in nitric acid-ammonium acetate medium, *Sep. Sci.*, 6(6), 781, 1971.
378. **Petru, F. and Pesek, J.,** Chemistry of the rare earth elements. LXXI. Separation of scandium from yttrium and lanthanum by means of solutions of chloroacetic acid and sulfuric acid, *Collect. Czech. Chem. Commun.*, 36(11), 3752, 1971.
379. **Tsubota, H. and Kitano, Y.,** A rapid method for determining fission products contained in waters using an ion-exchanger, *Bull. Chem. Soc. Jpn.*, 33(6), 765, 1960.
380. **Tsubota, H.,** Formation constants of some metal formate complexes, and the use of formate buffer solution as an elutriant for cation exchange chromatography, *Bull. Chem. Soc. Jpn.*, 35(4), 640, 1962.
381. **Orlandini, K. A.,** Cation-exchange separation of scandium from rare earth-metals in oxalic acid media, *Inorg. Nucl. Chem. Lett.*, 5(4), 325, 1969.
382. **Hamaguchi, H., Kuroda, R., Sugisita, R., Onuma, N., and Shimizu, T.,** Colour reaction of scandium with quercetin and its application to the analysis of scandium, *Anal. Chim. Acta,* 28, 61, 1963.
383. **Suzuki, Y.,** Preparation of carrier-free Y-90 from Sr-90 with ion-exchange, *Int. J. Appl. Radiat. Isot.*, 15, 599, 1964.
384. **Belyavskaya, T. A., Brykina, G. D., and Isaeva, L. B.,** Adsorption of sandium from aqueous acetone solutions of hydrochloric and oxalic acids with ion-exchangers, *Vestn. Mosk. Univ. Ser. Khim.*, 12(4), 468, 1971.
385. **Heininger, C. and Lanzafame, F. M.,** High-speed separation of the light rare earth elements on centrifugally accelerated ion-exchange paper, *Anal. Chim. Acta,* 30, 148, 1964.
386. **Senegačnik, M., Paljk, S., and Južnič, K.,** Fall-out analysis of atmospheric water precipitations. II. Determination of cerium-144, caesium-137, strontium-89, strontium-90, and barium-140 by ion exchange, *Fresenius' Z. Anal. Chem.*, 233, 81, 1968.
387. **Das, N. R. and Bhattacharyya, S. N.,** Use of ascorbic acid as an eluent for cation-exchange separation of rare earth tracers, *Int. J. Appl. Radiat. Isot.*, 31(9), 575, 1980.

388. **Dadone, A., Baffi, F., Frache, R., Cosma, B., and Mazzucotelli, A.,** Organic acid solutions in the chromatography of inorganic ions. VIII. Cation-exchange of Mn(II), Cd(II), Co(II), Ni(II), Zn(II), Cu(II), Fe(III), Sc(III), Y(III), Eu(III), Dy(III), Ho(III), Yb(III), Ti(IV), and Nb(V) in malate media, *Chromatographia,* 14(1), 32, 1981.
389. **Kunbazarov, A., Sorochan, A. M., and Senyavin, M. M.,** Chromatographic separation of mixtures of cerium and yttrium with the aid of organophosphorus compounds, *Zh. Neorg. Khim.,* 16(3), 651, 1971.
390. U.K.A.E.A., Improvements in or Relating to the Separation of Rare Earth-Elements, British Patent 1,197,491; date appl. May 26, 1967.
391. **El'khilyali, A. E., Martynenko, L. I., and Spitsyn, V. I.,** Study of use of glycine complexes for separating a mixture of neodymium and praseodymium on a cationite, *Izv. Akad. Nauk SSSR Ser. Khim.,* No. 3, 517, 1970.
392. **Iyer, S. G. and Venkateswarlu, Ch.,** Exchange of metal ions on Dowex A-1(H^+), *Indian J. Chem.,* 14A(6), 437, 1976.
393. **Major, W. J., Lee, K., Wessman, B. A., and Peters, R.,** A rapid method for radiochemical analysis of strontium-90 and the rare earths from seawater by extraction techniques, *Trans. Am. Nucl. Soc.,* 10, 71, 1967.
394. **Hayes, D. W., Hood, D. W., and Slowey, J. F.,** Rare earth elements and trace elements content of particulate matter in seawater, in the chemistry and analysis of trace metals in seawater, Annual report, 64-27A, Department of Oceanography and Meteorology, Texas A & M University, 1964.
395. **Goldberg, E. D., Koide, M., Schmitt, R. A., and Smith, R. H.,** Rare earth distribution in the marine environment, *J. Geophys. Res.,* 68, 4209, 1963.
396. **Wang, W.-N., Chen, Y.-J., and Wu, M.-T.,** Complementary analytical methods for cyanide, sulfide, certain transition metals and lanthanides in ion chromatography, *Analyst (London),* 109, 281, 1984.
397. **Buchanan, R. F. and Faris, J. P.,** Adsorption of the elements from nitric acid by anion exchange, in *Radioisotopes in the Physical Sciences and Industry,* International Atomic Energy Agency, Vienna, 1962, 361.
398. **Faris, J. P. and Buchanan, R. F.,** Anion exchange characteristics of the elements in nitric acid and nitrate solutions and application in trace element analysis, Report ANL-6811, Argonne National Laboratory, Argonne, Ill., July 1964.
399. **Naumann, D. and Ross, R.,** The anion exchange behavior of cerium, thorium and uranium, *Kernenergie,* 3, 425, 1960.
400. **Roberts, F. P.,** Determination of small amounts of cerium activity in strontium-90 using anion exchange, Report HW-69737, U.S. Atomic Energy Commission, 1961.
401. **Borkar, M. D.,** Determination of cerium in marine sediments, *Water Air Soil Pollut.,* 13(2), 133, 1980.
402. **Fedorova, N. D., Fokina, L. S., and Makarov, M. K.,** Use of chromatographic separation in determination of cerium in high-temperature alloys, *Tr. Vses. Nauchno Issled. Inst. Stand. Obraztsov Spektr. Etalonov.,* No. 6, 98, 1970.
403. **Macdonald, D.,** Carrier-free separation of praseodymium-144 from cerium-144 by anion exchange, *Anal. Chem.,* 33, 1807, 1961.
404. **Marcus, Y. and Givon, M.,** Anion exchange of metal complexes. XIV. The effect of acidity on the sorption of lanthanides from lithium nitrate solutions, *J. Phys. Chem.,* 68, 2230, 1964.
405. **Marcus, Y. and Abrahamer, I.,** Anion-exchange of metal complexes. VII. The lanthanides-nitrate system, *J. Inorg. Nucl. Chem.,* 22, 141, 1961.
406. **Hamaguchi, H., Hikawa, I., and Kuroda, R.,** Anion-exchange behavior of scandium, rare earth metals, thorium and uranium in magnesium nitrate solution, *J. Chromatogr.,* 18(3), 556, 1965.
407. **Walter, C. W. and Korkisch, J.,** Anion exchange of metal ions in aqueous-organic solvent systems containing nitric acid. I. Methanol, *Mikrochim. Acta,* No. 1, 81, 1971.
408. **Walter, C. W. and Korkisch, J.,** Anion exchange of metal ions in aqueous-organic solvent systems containing nitric acid. II. Ethanol, isopropanol and methyl glycol, *Mikrochim. Acta,* No. 1, 137, 1971.
409. **Walter, C. W. and Korkisch, J.,** Anion exchange of metal ions in aqueous-organic solvent systems containing nitric acid. III. Tetrahydrofuran and acetone, *Mikrochim. Acta,* No. 1, 158, 1971.
410. **Walter, C. W. and Korkisch, J.,** Anion-exchange of metal ions in aqueous-organic solvent systems containing nitric acid. IV. Acetic acid and dimethyl sulfoxide, *Mikrochim. Acta,* No. 1, 181, 1971.
411. **Korkisch, J., Hazan, I., and Arrhenius, G.,** Ion-exchange in mixed solvents. Adsorption behavior of the rare earths and some other elements on a strong base anion exchange resin from nitric acid-alcohol media. Methods for separation and spectrophotometric determination, *Talanta,* 10, 865, 1963.
412. **Edge, R. A.,** The anion-exchange behavior of yttrium, neodymium and lanthanum in dilute nitric acid solutions containing ethanol, *J. Chromatogr.,* 5(6), 526, 1961.
413. **Morrow, R. J.,** Anion-exchange separation techniques with methanol-water solutions of hydrochloric and nitric acids, *Talanta,* 13, 1265, 1966.
414. **Van Acker, P.,** Separation of lanthanoids on anion exchange resin in anhydrous acetic acid medium, *Anal. Chim. Acta,* 113, 149, 1980.

415. **Edge, R. A.,** Adsorption of europium and cerium on a strong-base anion exchange resin from dilute nitric acid solutions containing aliphatic alcohols, *J. Chromatogr.*, 8(3), 419, 1962.
416. **Molnar, F. and Lebedev, N. A.,** Concentration of all rare earth impurities of cerium sub-group rare-earth matrices, *J. Radioanal. Chem.*, 2(1-2), 91, 1969.
417. **Molnar, F. and Lebedev, N. A.,** Concentration of rare earth metal impurities in analysis of elements of the cerium sub-group by anion exchange chromatography, *Zh. Anal. Khim.*, 24(8), 1152, 1969.
418. **Küppers, G. and Erdtmann, G.,** Determination of praseodymium in lanthanum compounds by neutron activation and ion exchange separation, *Fresenius' Z. Anal. Chem.*, 307, 369, 1981.
419. **Jones, E. A.,** Improved ion exchange separation of rare earth elements for spectrographic analysis, Report NIM-1956, National Institute for Metallurgy, Randburg, South Africa, 1978.
420. **Kraus, K. A., Nelson, F., and Smith, G. W.,** Anion-exchange studies. IX. Adsorbability of a number of metals in hydrochloric acid solutions, *J. Phys. Chem.*, 58, 11, 1954.
421. **Marcus, Y.,** Metal-chloride complexes studied by ion-exchange and solvent extraction methods. I. Non-transition-metal ions, lanthanides, actinides and d^0 transition metal ions, *Coord. Chem. Rev.*, 2, 195, 1967.
422. **Marcus, Y.,** Metal-chloride complexes studied by ion-exchange and solvent extraction methods. II. Transition-metal elements and the hexavalent actinides, *Coord. Chem. Rev.*, 2, 257, 1967.
423. **Laul, J. C., Case, D. R., Wechter, M., Schmidt-Bleek, F., and Lipschutz, M. E.,** An activation analysis technique for determining groups of trace elements in rocks and chondrites, *J. Radioanal. Chem.*, 4, 241, 1970.
424. **Macdonald, J. C. and Yoe, J. H.,** Spectrophotometric determination of scandium with anthrarufin-2,6-disulphonic acid (disodium salt), *Anal. Chim. Acta*, 28, 264, 1963.
425. **Kuroda, R. and Hikawa, I.,** Separation of scandium and rare earth by anion exchange in acetic acid-hydrochloric acid mixtures, *J. Chromatogr.*, 25(2), 408, 1966.
426. **Faris, J. P.,** Separation of metal ions by anion exchange in mixtures of hydrochloric acid and hydrofluoric acid, Report ANL-78-78, Argonne National Laboratory, Argonne, Ill., December 1978.
427. **Livingston, H. D. and Bowen, V. T.,** Activation analysis of lanthanides in modern corals-germanium detector procedures, U.S.A.E. Rep. NYO-2174-70, presented at Symp. Rare Earths, Am. Geophys. Union 49th meeting, Washington, D.C., 1968.
428. **Kuroda, R., Ishida, K., and Kiriyama, T.,** Adsorption behavior of a number of metals in hydrochloric acid on a weakly basic anion exchange resin, *Anal. Chem.*, 40, 1502, 1968.
429. **Edge, R. A.,** The adsorption of yttrium, neodymium and lanthanum on a strong base anion exchange resin from dilute hydrochloric acid solutions containing ethanol, *J. Chromatogr.*, 5, 539, 1961.
430. **Peters, J. M. and del Fiore, G.,** Distribution coefficients for 52 elements in hydrochloric acid-water-acetone mixture on Dowex 1-X8, *Radiochem. Radioanal. Lett.*, 21(1-2), 11, 1975.
431. **Korkisch, J. and Hazan, I.,** Anion exchange behavior of uranium, thorium, the rare earths and various other elements in hydrochloric acid-organic solvent media, *Talanta*, 11, 1157, 1964.
432. **Fritz, J. S. and Pietrzyk, D. J.,** Non-aqueous solvents in anion-exchange separations, *Talanta*, 8, 143, 1961.
433. **Wodkiewicz, L. and Dybczyński, R.,** Separation of trace amounts of rare earth elements from uranium in aqueous-non-aqueous media by ion-exchange chromatography, *Chem. Anal. (Warsaw)*, 19(1), 175, 1974.
434. **Strelow, F. W. E. and Bothma, C. J. C.,** Anion exchange and a selective scale for elements in sulfuric acid media with a strongly basic resin, *Anal. Chem.*, 39, 595, 1967.
435. **Hamaguchi, H., Ohuchi, A., Onuma, N., and Kuroda, R.,** Anion-exchange behavior of rare earth metals in potassium sulfate medium, *J. Chromatogr.*, 16(2), 396, 1964.
436. **Kuroda, R., Oguma, K., Kono, N., and Takahashi, Y.,** Adsorption behavior of various metals on a weakly basic anion-exchange resin in sulfuric acid media, *Anal. Chim. Acta*, 62, 343, 1972.
437. **Hamaguchi, H., Onuma, N., Kishi, M., and Kuroda, R.,** Anion-exchange behavior of scandium in chloride-thiocyanate media, *Talanta*, 11, 495, 1964.
438. **Turner, J. B., Philp, R. H., and Day, R. A.,** Adsorption of some metals on anion exchange resins from potassium thiocyanate solutions, *Anal. Chim. Acta*, 26, 94, 1962.
439. **Fritz, J. S. and Kaminski, E. E.,** Anion exchange separations of metal ions in thiocyanate media, *Talanta*, 18, 541, 1971.
440. **Surls, J. P. and Choppin, G. R.,** Ion exchange study of thiocyanate complexes of the actinides and lanthanides, *J. Inorg. Nucl. Chem.*, 4, 62, 1957.
441. **Hamaguchi, H., Ishida, K., and Kuroda, R.,** Anion exchange behavior of rare earth metals, thorium, protactinum and uranium in thiocyanate-chloride media, *Anal. Chim. Acta*, 33, 91, 1965.
442. **Singh, D. and Tandon, S. N.,** Anion exchange studies of metal thiocyanates in aqueous and mixed solvent systems, *Talanta*, 26, 163, 1979.
443. **Johnson, R. G. and Wandless, G. A.,** Determination of carrier yields for neutron activation analysis using energy dispersive X-ray spectrometry, *J. Radioanal. Nucl. Chem.*, 81(1), 21, 1984.
444. **Schindewolf, U. and Irvine, J. W.,** Preparation of carrier-free vanadium, scandium and arsenic activities from cyclotron targets by ion exchange, *Anal. Chem.*, 30, 906, 1958.

445. **Faris, J. P.,** Adsorption of the elements from hydrofluoric acid by anion exchange, *Anal. Chem.*, 32, 520, 1960.
446. **Palkowska-Motrenko, H. and Dybczyński, R.,** Anion-exchange separation of rare earth elements in orthophosphoric acid medium, *J. Radioanal. Chem.*, 59(1), 31, 1980.
447. **Marsh, S. F., Alarid, J. E., Hammond, C. F., McLeod, M. J., Roensch, F. R., and Rein, J. E.,** Anion exchange of 58 elements in hydrobromic acid and in hydroiodic acid, Report LA-7084 Los Alamos National Laboratory, Los Alamos, N.M., February 1978.
448. **Klakl, E. and Korkisch, J.,** Anion exchange behavior of several elements in hydrobromic acid-organic solvent media, *Talanta*, 16, 1177, 1969.
449. **Korkisch, J. and Hazan, I.,** Anion-exchange separations in hydrobromic acid-organic solvent media, *Anal. Chem.*, 37, 707, 1965.
450. **Wodkiewicz, L. and Dybczyński, R.,** Anion-exchange behavior of elements on a weakly basic anion exchanger in hydrobromic acid medium, *J. Chromatogr.*, 102, 277, 1974.
451. **Wodkiewicz, L. and Dybczyński, R.,** Anion exchange behavior of several elements in the system: AG3-X4A [Br^-]-HBr aq. Comparison with analogous systems involving other weakly and strongly basic anion exchange resins, *Chem. Anal. (Warsaw)*, 26, 419, 1981.
452. **Taketatsu, T.,** Dissolution and anion exchange behavior of the rare earth elements in potassium and ammonium carbonate solutions, *Bull. Chem. Soc. Jpn.*, 35(9), 1573, 1962.
453. **Taketatsu, T.,** Dissolution and anion exchange behavior of rare earth and other metallic elements in potassium hydrogen carbonate, potassium carbonate and ammonium carbonate solutions, *Bull. Chem. Soc. Jpn.*, 36(5), 549, 1963.
454. **Taketatsu, T.,** The solubility and anion exchange behavior of heavy lanthanide elements in a potassium carbonate solution, *Bull. Chem. Soc. Jpn.*, 37(6), 906, 1964.
455. **Taketatsu, T.,** Solubilities and anion-exchange behavior of rare earth metals in potassium carbonate solution, *Anal. Chim. Acta*, 32, 40, 1965.
456. **Misumi, S. and Taketatsu, T.,** Anion exchange studies of beryllium(II), cerium(IV), thorium(IV) and uranium(VI) carbonate complex ions, *Bull. Chem. Soc. Jpn.*, 32(8), 877, 1959.
457. **Saito, N. and Sekine, T.,** Anion-exchange behavior of yttrium, *Nature*, 180, 753, 1957.
458. **Eristavi, V. D. and Kashakashvili, L. L.,** Use of carbonate forms of anionites in analytical chemistry of scandium, yttrium and lanthanum, *Izv. Akad. Nauk Gruz. SSR Ser. Khim.*, 3(1), 38, 1977.
459. **Misumi, S. and Taketatsu, T.,** Separation of yttrium-90 from strontium-90 and lanthanum-140 from barium-140 with anion exchange resin of carbonate form, *J. Inorg. Nucl. Chem.*, 20, 127, 1961.
460. **Ogura, T., Tarutani, T., and Misumi, S.,** Separation of yttrium-90 from strontium-90 with an anion exchange resin of bicarbonate form, and studies on yttrium-bicarbonate complex ion, *Mem. Fac. Sci. Kyushu Univ. Ser. C*, 5(4), 129, 1964.
461. **Taketatsu, T.,** Anion-exchange separation of rare earth elements and thorium from uranium with carbonate solutions, *Talanta*, 10, 1077, 1963.
462. **Perkins, R. W.,** Filtration-precipitation separation of barium-140 from lanthanum-140, *Anal. Chem.*, 29, 152, 1957.
463. **Holcomb, H. P.,** Sequential analysis of radionuclides in heavy water moderator, *U.S.A.E.C. Res. Dev. Rep.*, DP-886, May 1964.
464. **Minczewski, J. and Dybczyński, R.,** Separation of rare earths on anion-exchange resin. II. Anion-exchange behavior of the rare earth complexes with EDTA, *J. Chromatogr.*, 7(1), 98, 1962.
465. **Minczewski, J. and Dybczyński, R.,** Anion-exchange behavior of the rare earth metal complexes with EDTA, *Chem. Anal. (Warsaw)*, 6(2), 275, 1961.
466. **Dybcziński, R.,** Separation of rare earth metals by means of anion exchange, *Chem. Anal. (Warsaw)*, 4(3), 531, 1959.
467. **Dybcziński, R.,** Separation of rare earth metals by means of anion exchange, *Chem. Anal. (Warsaw)*, 4(3), 531, 1959.
468. **Minczewski, J. and Dybczyński, R.,** Separation of rare earth metals on anion-exchange resins. III. Position of scandium in the separation scheme of rare earth complexes with ethylenediaminetetra-acetic acid, *J. Chromatogr.*, 7(4), 568, 1962.
469. **Dybczyński, R.,** Effect of cross-linking on the anion exchange separation of rare earth metal-EDTA complexes, *J. Chromatogr.*, 50(3), 487, 1970.
470. **Wodkiewicz, L. and Dybczyński, R.,** Effect of resin cross-linking on the anion exchange separation of rare earth complexes with DCTA, *J. Chromatogr.*, 68(1), 131, 1972.
471. **Dybczyński, R.,** Influence of temperature on tracer-level separation by ion-exchange chromatography, *J. Chromatogr.*, 31(1), 155, 1967.
472. **Dybczyński, R.,** Separation of rare earth on anion-exchange resins. IV. Influence of temperature on anion-exchange behavior of the rare earth-metal ethylenediaminetetra-acetates, *J. Chromatogr.*, 14(1), 79, 1964.
473. **De Corte, F., van den Winkel, P., Speecke, A., and Hoste, J.,** Distribution coefficients for twelve elements in oxalic acid medium on a strong basic anion-exchange resin, *Anal. Chim. Acta*, 42, 67, 1968.

474. **Walter, R. I.,** Anion-exchange studies of scandium(III) and vanadium(IV). Separation of scandium, titanium and vanadium, *J. Inorg. Nucl. Chem.*, 6(1), 58, 1958.
475. **Chakravorty, M. and Khopkar, S. M.,** Anion-exchange separation of scandium from yttrium, lanthanum, cerium and other elements in malonic and ascorbic acid media, *Chromatographia*, 10(2), 100, 1977.
476. **Brouchek, F. and Chirakadze, M. A.,** Chromatographic separation of scandium from vanadium(IV), zirconium and hafnium on AV-16 anionite modified with tartrate ions, *Soobshch. Akad. Nauk Gruz. SSR*, 73(2), 345, 1974.
477. **Kundra, S. K., Shah, S. A., Subbaraman, P. R., and Gupta, J.,** Anion-exchange fractionation of light lanthanides, *Indian J. Chem.*, 3(3), 108, 1965.
478. **Huffman, E. H. and Oswalt, R. L.,** A rare earth separation by anion exchange, *J. Am. Chem. Soc.*, 72, 3323, 1950.
479. **Chao, H.-E. and Suzuki, N.,** Adsorption behavior of scandium, yttrium, cerium and uranium from xylenol orange solutions on to anion exchange resins, *Anal. Chim. Acta*, 125, 139, 1981.
480. **Faris, J. P.,** Separation of the rare earth metals by anion exchange in water-solvent mixtures in the presence of 2-hydroxyisobutyric acid, *J. Chromatogr.*, 26(1), 232, 1967.
481. **Faris, J. P.,** Separation of the rare earth metals by anion exchange in water-solvent mixtures in the presence of lactic acid, *J. Chromatogr.*, 32(4), 795, 1968.
482. **van den Winkel, P., de Corte, F., and Hoste, J.,** Anion-exchange in acetic acid solutions, *Anal. Chim. Acta*, 56, 241, 1971.
483. **van den Winkel, P., de Corte, F., Speecke, A., and Hoste, J.,** Adsorption of some elements in acetic acid medium on Dowex 1-X8, *Anal. Chim. Acta*, 42, 340, 1968.
484. **Maslowska, J. and Pietek, W.,** Method for the separation of aluminum(III), iron(III) and cerium(III) complexes with 4-hydroxybenzoic acid on weakly alkaline anion exchangers, *Chromatographia*, 17(12), 693, 1983.
485. **Nakai, T., Yajima, S., Fujii, I., Kamemoto, Y., and Shiba, K.,** Radioactivation analysis of rare earth elements in metallic thorium and thorium oxide, *J. Chem. Soc. Jpn. Pure Chem. Sect.*, 82(2), 197, 1961.
486. **Samsahl, K.,** Neutron-activation analysis of biological material with high radiation levels, *Ab. Atomenergi Rep.*, AE-247, 1966.
487. **Budešinský, B.,** Determination of Carbon in Uranium Materials, International Atomic Energy Agency, Vienna, Reprint CN 16/64, 1963.
488. **Youh, C. C.,** Radiometric micro-determination of rare-earth metals in thorium oxide, *Hua Hsueh*, No. 1, 27, 1967.
489. **Křepelka, J. H., Vetejška, K., and Mazáček, J.,** Separation of iron from rare earths, *Collect. Czech. Chem. Commun.*, 24, 198, 1959.
490. **Ryabinin, A. I. and Sorokina, A. A.,** Separation of rare-earth metals by fractional precipitation on anion-exchange resins, *Zh. Prikl. Khim.*, 38(11), 2410, 1965.
491. **Tang, J.-H., Hsu, H.-C., and Yang, J.-K.,** Separation of rare earths and thorium by anion exchange resin in the citrate form, *Pei Ching, Ta Hsueh Pao Tzu Jan Ko Hsueh*, 11(1), 99, 1965.
492. **Shrobilgen, G. J. and Lang, C. E.,** Lanthanide distribution coefficients on Dowex chelating resin A-1, *J. Inorg. Nucl. Chem.*, 30, 3127, 1968.
493. **Winget, J. O.,** Separation of the lanthanide series and yttrium using phosphoric and iminodiacetic acid resins, Report Invest. RI-6510, U.S. Bureau of Mines, 1964.
494. **Hering, R.,** Ion exchange resins with complexing anchor groups. IV. Behavior of rare earth ions on iminodiacetic acid resin, *J. Inorg. Nucl. Chem.*, 24, 1399, 1962.
495. **Hering, R.,** Ion exchange resins with complexing anchor groups. X. Behavior of trivalent metal ions towards a iminodiacetic acid resin and solutions containing iminodiacetic acid, *Z. Chem.*, 3(6), 1, 1963.
496. **Hering, R.,** The causes for the separability of complex forming metal ions, especially of the rare earths, on an iminodiacetic acid chelating resin, by elution with iminodiacetic acid solutions, in Proc. Symp. on Ion Exchangers, Balatonszeplak, Hungary, 1963, 147.
497. **Hirose, A., Kobori, K., and Ishii, D.,** Determination of rare earth elements and heavy metals in river-water by preconcentration on Chelex 100 and neutron activation, *Anal. Chim. Acta*, 97, 303, 1978.
498. **Padmanabhan, P. K. and Venkateswarlu, Ch.,** Exchange of cations on Zeo-Karb 226 (H^+), *Indian J. Chem.*, 13, 264, 1975.
499. **Marhol, M., Beranová, H., and Cheng, K. L.,** Selective ion-exchangers containing phosphorus in their functional groups. I. Sorption and separation of some bivalent and trivalent ions, *J. Radioanal. Chem.*, 21, 177, 1974.
500. **Miklishanskii, A. Z., Leikin, Yu. A., Yakovlev, Yu. V., and Savel'ev, B. V.,** Sorption properties of ion exchangers containing phosphorus and their use in neutron activation analysis, *Zh. Anal. Khim.*, 29(7), 1284, 1974.
501. **Niu, F., Teng, R., and Ma, T.,** Separation of radioactive thullium from erbium in gram quantity with diethylhexylphosphoric acid-containing resin, *Fenxi Huaxue*, 11(5), 379, 1983.

502. **Hu, R., Lu, A., and Chai, X.,** Studies on the rare earth separation behavior of P507 extraction resins, *Huaxue Shiji,* 5(3), 131, 1983.
503. **Miyazaki, A. and Barnes, R. M.,** Complexation of some transition metals, rare earth elements, and thorium with poly(dithiocarbamate) chelating resin, *Anal. Chem.*, 53, 299, 1981.
504. **Zhang, L., Zhang, S., and Tang, S.,** Synthesis and application of chelating resins. II. Separation performance for microamounts of rare earth metals, *Fenxi Huaxue,* 13(3), 180, 1985.
505. **Nevoral, V.,** Enrichment of traces of rare earth metals in mineral waters by means of ion exchangers, *Fresenius' Z. Anal. Chem.*, 268(3), 189, 1974.
506. **Ehrlich, A. M.,** Rare Earths Abundances in Manganese Nodules, Ph.D. thesis, Massachusetts Institute of Technology, Cambridge, 1968.
507. **Tolk, A. and Raaphorst, J. G.,** Analysis of unirradiated uranium oxide nuclear fuel, in *Analytical Methods in the Nuclear Fuel Cycle,* International Atomic Energy Agency, Vienna, 1972, 175.
508. **Pesek, J. and Petru, F.,** Contributions to the chemistry of the rarer elements. LXXIX. Separation of scandium from yttrium, lanthanum, manganese(II), aluminum, gallium, cerium, indium, thallium, iron(III), iron(II), thorium and ytterbium and from other elements contained in monazite sand of Asiatic origin, *Collect. Czech. Chem. Commun.*, 37(8), 2528, 1972.
509. **Ripan, R. and Stanisav, C.,** Behavior of soluble metallic poly-phosphates towards ion exchangers. IV. Separation of cerium(IV) from cobalt(II) and copper(II) by cation-exchange, *Chim. Anal.*, 49(9), 469, 1967.
510. **Hamaguchi, H., Onuma, N., Kuroda, R., and Sugisita, R.,** Ultra-violet spectrophotometric determination of scandium with tiron, *Talanta,* 9, 563, 1962.
511. **Fan, B., Tang, Y., and Hu, Z.,** Separation and determination of heavy rare-earth metals by high-performance ion exchange chromatography, *Fenxi Huaxue,* 11(11), 842, 1983.
512. **Gan, W., Di, X., Li, J., Tian, L., Li, L., and Wu, S.,** Chromatopolarography of rare earth metals, *Fenxi Huaxue,* 13(1), 11, 1985.
513. **Cornish, F. W.,** A preliminary report on the determination of submicrogram quantities of individual rare earths by radioactivation using ion exchange separation, AERE Report C/R 1224, A.E.R.E., Harwell England, 1956.
514. **Yoshida, K., Fuwa, K., and Haraguchi, H.,** Determination of fifteen rare earth elements by liquid chromatography with inductively coupled plasma atomic-emission spectrometric detection, *Chem. Lett.*, No. 12, 1879, 1983.
515. **Roelandts, I., Duyckaerts, G., and Brunfelt, O. A.,** Anion-exchange isolation of rare-earth elements from apatite minerals in methanol-nitric acid medium, *Anal. Chim. Acta,* 73, 141, 1974.
516. **Roelandts, I.,** Determination of light rare-earth elements in apatite by X-ray fluorescence spectrometry after anion exchange extraction, *Anal. Chem.*, 53, 676, 1981.
517. **Brunfelt, A. O. and Steinnes, E.,** Determination of lutetium, ytterbium and terbium in rocks by neutron activation and mixed solvent anion exchange chromatography, *Analyst (London),* 94, 979, 1969.
518. **Desai, H. B., Iyer, R. K., and Das, M. S.,** Determination of scandium, yttrium, samarium and lanthanum in standard silicate rocks, G-1 and W-1, by neutron activation analysis, *Talanta,* 11, 1249, 1964.
519. **Brunfelt, A. O. and Steinnes, E.,** Cerium and europium contents of some standard rocks, *Chem. Geol.*, 2, 199, 1967.
520. **Hooker, P. J., O'Nions, R. K., and Pankhurst, R. J.,** Determination of rare earth metals in USGS standard rocks in mixed solvent ion exchange and mass-spectrometric isotope dilution, *Chem. Geol.*, 16, 189, 1975.
521. **Pankhurst, R. J.,** Determination of rare earth elements by mass-spectrometric isotope dilution, *Proc. Anal. Div. Chem. Soc.*, 13(5), 122, 1976.
522. **Van Puymbroeck, J. and Gijbels, R.,** Determination of rare earth elements in rocks by spark-source mass spectrometry and isotope dilution after ion exchange separation in mixed solvents, *Fresenius' Z. Anal. Chem.*, 309, 312, 1981.
523. **Van Puymbroeck, J. and Gijbels, R.,** Determination of rare earth elements by spark-source mass spectometry, *Bull. Soc. Chim. Belg.*, 87(11-12), 803, 1978.
524. **Korkisch, J. and Arrhenius, G.,** Separation of uranium, thorium and the rare earth elements by anion exchange, *Anal. Chem.*, 36, 850, 1964.
525. **Elderfield, H., Hawkesworth, C. J., and Greaves, M. J.,** Rare earth elements geochemistry of oceanic ferro-manganese nodules and associated sediments, *Geochim. Cosmochim. Acta,* 45, 513, 1981.
526. **Faris, J. P.,** Spectrographic determination of scandium, yttrium and the rare earths in thorium, *Appl. Spectrosc.*, 12(6), 157, 1958.
527. **Edge, R. A.,** Combined anion exchange-spectrographic procedure for the determination of trace amounts of samarium, europium, gadolinium and dysprosium in thorium compounds obtained from monazite, *J. S. Afr. Chem. Inst.*, 21(2), 127, 1968.
528. **Edge, R. A.,** Combined anion exchange-solvent extraction procedure for separating trace amounts of europium, gadolinium, dysprosium, samarium and erbium from thorium tetrafluoride, *Anal. Chim. Acta,* 28, 278, 1963.

529. **Huff, E. A.,** Trace-impurity analysis of thorium-uranium and plutonium-thorium-uranium alloys by anion-exchange partition chromatography, *Anal. Chem.*, 37, 533, 1965.
530. **Joshi, B. D., Patel, B. M., and Page, A. G.,** Anion-exchange separation and spectrographic determination of rare earth metals in plutonium with lithium fluoride-silver chloride carrier, *Anal. Chim. Acta*, 57, 379, 1971.
531. **Bornong, B. J. and Moriarty, J. L.,** The spectrophotometric determination of yttrium and rare-earth elements in cast steels, *Anal. Chem.*, 34, 871, 1962.
532. **Bryan, R. G. and Waterburg, G. R.,** Potentiometric titration of cerium and manganese following separation from plutonium and cobalt, Report LA-3402, U.S. Atomic Energy Commission, 1966.
533. **Savage, D. J. and Drummond, J. L.,** Separation of neodymium for the determination of the burn-up of nuclear fuels, Report TRG 1496(D), U.K. Atomic Energy Authority, 1967.
534. **Krtil, J., Bulovic, V., Mencl, J., and Moravec, A.,** Separation of stable isotopes of neodymium from irradiated uranium for burn-up determination, *Radiochem. Radioanal. Lett.*, 9(5-6), 335, 1972.
535. **Ahluwalia, S. S. and Korkisch, J.,** Determination of cerium in ferrous alloys, *Anal. Chim. Acta*, 31, 552, 1964.
536. **Cordova-Orellana, R. and Lucena-Conde, F.,** Coulometric determination of cerium(III) in alkaline solutions with electrogenerated octacyanomolybdate(V), *Talanta*, 18, 505, 1971.
537. **Green, H.,** Determination of cerium in cast iron, steel and iron-silicon-magnesium alloys, *Metallurgia*, 76, 223, 1967.
538. **Willis, R. R.,** Method for the determination of cerium in plain carbon and low-alloy steels, Research report, British Steel Corporation, GS/EX/100/70/C, 1970.
539. **Kashuba, A. T. and Hines, C. R.,** X-ray fluorescence ion-exchange method for determination of lanthanum, cerium and praseodymium in carbon steels, *Anal. Chem.*, 43, 1758, 1971.
540. **Myasoedov, B. F., Kulyako, Yu. M., and Sklyarenko, I. S.,** Polarographic determination of europium in americium in an acetonitrile medium, *Radiochem. Radioanal. Lett.*, 10(2), 123, 1972.
541. **Heumann, K. G. and Trettenbach, J.,** Trace determination of lanthanum, cerium and neodymium in inorganic matrices by mass-spectrometric isotope-dilution, *Fresenius' Z. Anal. Chem.*, 310, 146, 1982.
542. **Onishi, H. and Banks, C. V.,** Spectrophotometric determination of scandium with arsenazo I, *Anal. Chim. Acta*, 29, 240, 1963.
543. **Faris, J. P. and Warton, J. W.,** Anion-exchange resin separation of the rare-earth elements, yttrium and scandium in nitric acid-methanol mixtures, *Anal. Chem.*, 34, 1077, 1962.
544. **Strelow, F. W. E.,** Quantitative separation of samarium from neodymium by anion-exchange chromatography in dilute nitric acid-methanol, *Anal. Chem.*, 52, 2420, 1980.
545. **Chen, Y.-M. and Liao, S.-C.,** Anion exchange separation of rare earth metals in methanol-nitric acid mixed solutions. I. Study of the distribution coefficients and column elution method, *J. Chin. Chem. Soc. Taipei*, 19(3), 93, 1972.
546. **Ross, R. and Römer, J.,** Separation of terbium, gadolinium and europium in milligram amounts. II. Separation by anion-exchange, *Z. Phys. Chem.*, 233(1-2), 25, 1966.
547. **Ross, R. and Römer, J.,** Separation of Sm, Pm, and Nd by anion exchange in mixed solvents, *Isotopenpraxis*, 3(5), 197, 1967.
548. **Edge, R. A.,** Removal of cerium, neodymium, praseodymium and lanthanum from erbium, dysprosium, gadolinium, europium, and samarium by anion exchange, *Anal. Chim. Acta*, 29, 321, 1963.
549. **Kakihana, H. and Kurokawa, K.,** Elution of gadolinium, terbium, dysprosium, holmium, erbium, thulium, ytterbium, lutetium and yttrium from anion-exchange resin columns with use of ethanol-formic acid-nitric acid mixtures, *Jpn. Analyst*, 23(11), 1315, 1974.
550. **Fritz, J. S. and Greene, R. G.,** Separation of rare-earth metals from other metal ions by anion-exchange, *Anal. Chem.*, 36, 1095, 1964.
551. **Melsom, S.,** Precision and accuracy of rare earth determination in rock samples using instrumental neutron activation analysis and a Ge(Li) detector, *J. Radioanal. Chem.*, 4, 355, 1970.
552. **Chase, J. W., Winchester, J. W., and Coryell, C. D.,** Lanthanum, europium, and dysprosium distributions in igneous rocks and minerals, *J. Geophys. Res.*, 68(2), 567, 1963.

552a. **Hamajima, Y., Koba, M., Endo, K., and Nakahara, H.,** Determination of lanthanoids in Japanese standard rocks by radiochemical neutron-activation method, *J. Radioanal. Nucl. Chem.*, 89(2), 315, 1985.

553. **Roelandts, I.,** Methodology for neutron-activation analysis of geological materials, *Analusis*, 6(1), 2, 1978.
554. **May, S. and Pinte, G.,** Neutron-activation determination of rare earth elements in uraniferous rocks, *J. Radioanal. Nucl. Chem.*, 81(2), 273, 1984.
555. **Nomura, K., Mikami, A., Kato, T., and Oka, Y.,** The determination of scandium and gold in meteorites, tektites and standard rocks by neutron-activation analysis with an internal-reference method, *Anal. Chim. Acta*, 51, 399, 1970.
556. **Brunfelt, A. O., Roelandts, I., and Steinnes, E.,** Determination of rubidium, caesium, barium and eight rare earth elements in ultramafic rocks by neutron activation analysis, *Analyst (London)*, 99, 277, 1974.

557. **Brunfelt, A. O., Roelandts, I., and Steinnes, E.,** Some new methods for the determination of rare earth-elements in geological materials using thermal and epithermal neutron activation, *J. Radioanal. Chem.*, 38, 451, 1977.
558. **Broekaert, J. A. C. and Hoermann, P. K.,** Separation of yttrium and rare earth-elements from geological materials, *Anal. Chim. Acta,* 124, 421, 1981.
559. **Ahrens, L. H., Edge, R. A., and Brooks, R. R.,** Investigations on the development of a scheme of silicate analysis based principally on spectrographic and ion-exchange techniques, *Anal. Chim. Acta,* 28, 551, 1963.
560. **Bando, S. and Imahashi, T.,** Determination of scandium, chromium, iron, cobalt, zinc and antimony in deposits and air-borne particulate matter by neutron-activation analysis, *Jpn. Analyst,* 20(1), 49, 1971.
561. **Kameda, K., Folsom, T. R., Rock, R., and Mohanrao, G. J.,** Radiochemical analysis of cerium-144 together with cesium-137 in seawater, *Report* IMR-TR-922-62-F, U.S. Atomic Energy Commission, 1962.
562. **Higano, R., Nagaya, Y., Shiozaki, M., and Seto, Y.,** On the artificial radioactivity in seawater, *J. Oceanogr. Soc. Jpn.*, 18, 34, 1963.
563. **Chau, Y.-K. and Wong, P.-Y.,** Determination of scandium in sea-water by atomic absorption spectrophotometry, *Talanta,* 15, 867, 1968.
564. **Taylor, R. W.,** Determination of promethium-147 in environment samples, Report DPSPU 64-30-20, E.I. du Pont de Nemours and Company, Aiken, S.C., September 1964.
565. **Hadzistelios, I. and Papadopoulou, C.,** Radiochemical separation and determination of europium by lithium-drifted germanium detector in biological tissues, *J. Radioanal. Chem.*, 36(2), 427, 1977.
566. **Rybak, B., Joffre, J., and Boivinet, L.,** Micro-determination of lanthanides in blood plasma, *C.R. Acad. Sci. Ser. D,* 270(1), 220, 1970.
567. **Yamada, Y.,** Activation analysis in agricultural studies in Japan, in Proc. 8th Conf. on Radioisotopes in Japan, 1967, 725.
568. **Chitambar, S. A. and Mathews, C. K.,** Determination of rare earth impurities in high-purity uranium by isotope-dilution analysis, *Fresenius' Z. Anal. Chem.*, 274, 9, 1975.
569. **Blank, G. R., Twitty, B. L., and Dingeldein, W. H.,** The chemical separation of rare earths from uranium — a radioactive tracer evaluation, Report NLCO-940, National Lead Company of Ohio, 1965.
570. **Birks, F. T., Weldrick, G. J., and Thomas, A. M.,** Emission-spectrographic determination of impurities in uranium-233 dioxide after pre-concentration of the rare earth elements by ion exchange, *Analyst (London),* 89, 36, 1964.
571. **Ejaz, M.,** Chemical separation of cerium(III) fission nuclides from neutron-irradiated uranium using N-oxides as solvents, *Mikrochim. Acta,* 2(3-4), 265, 1976.
572. **Albu-Yaron, A., Mueller, D. W., and Suttle, A. D.,** Chemical separation of cerium fission products from microgram amounts of uranium, *Anal. Chem.*, 41, 1351, 1969.
573. **Rizvi, G. H., Chaudhuri, N. K., Rastogi, R. K., and Patil, S. K.,** Anion-exchange separation and determination of yttrium in uranium oxide-yttrium oxide mixture, *J. Radioanal. Chem.*, 78(2), 241, 1983.
574. **Foti, S. C. and Wish, L.,** Rapid carrier-free determination of individual rare-earth metals by ion exchange at room temperature, *J. Chromatogr.*, 29(1), 203, 1967.
575. **Laverlochere, J. and May, S.,** Determination by neutron activation of some impurities in beryllium, Report DPC/PCA/CCA/62-240/SM/OK, Centre d'Etudes de Grenoble, France, Commissariat a l'Energie Atomique, November 16, 1962; *Bull Soc. Chim Fr.*, No. 3, 457, 1963.
576. **Křivánek, M., Kukula, F., and Vins, V.,** Neutron-activation determination of impurities in gallium, *J. Radioanal. Chem.*, 1(3), 219, 1968.
577. **Kiesl, W., Sorantin, H., and Pfeifer, V.,** Determination by activation analysis of impurities in aluminum. III. Determination of iron, cobalt, zinc and scandium, *Microchim. Acta,* No. 5 and 6, 996, 1963.
578. **Lo, J. G., Ke, C. N., Tanaka, S., and Yeh, S. J.,** Neutron-activation analysis of trace metals in aluminum, *J. Chin. Chem. Soc. (Taipei),* 24(1), 21, 1977.
579. **Allen, J. F. and Pinajian, J. J.,** A Sr-87m generator for medical applications, *Int. J. Appl. Radiat. Isot.*, 16, 319, 1965.
580. **Roberts, J. E. and Ryterband, M. J.,** Ion-exchange separation and spectrophotometric determination of cerium in cast iron, *Anal. Chem.*, 37, 1585, 1965.
581. **Fedorov, A. A., Ozerskaya, F. A., and Linkova, F. V.,** Determination of micro- and macro-amounts of rare-earth elements, *Sb. Tr. Tsentr. Nauchno. Issled. Inst. Chern. Metall.*, No. 31, 197, 1963.
582. **Fokina, L. S., Fedorova, N. D., Kurbatova, V. I., Nikulina, I. N., and Zhilkina, S. P.,** Determination of cerium in refractory alloy with use of chromatographic separation, *Tr. Vses. Nauchno Issled. Inst. Stand. Obraztsov Spectr. Etalonov,* 5, 118, 1969.
583. **Steinnes, E.,** Determination of rare-earth additives in steels by neutron-activation analysis using Ge(Li) γ-spectrometry, *Radiochem. Radioanal. Lett.*, 28(2), 175, 1977.
584. **Berman, S. S., Duval, G. R., and Russell, D. S.,** Spectrophotometric determination of scandium in copper with xylenol-orange, *Anal. Chem.*, 35, 1392, 1963.
585. **Wilkins, D. H. and Smith, G. E.,** Ethanol-water-hydrochloric acid eluents in anion-exchange separations, *Talanta,* 8, 138, 1961.

586. **Hamaguchi, H., Watanabe, T., Onuma, N., Tomura, K., and Kuroda, R.,** Neutron-activation analysis for scandium, *Anal. Chim. Acta,* 33, 13, 1965.
587. **Banks, C. V., Thompson, J. A., and O'Laughlin, J. W.,** Separation and determination of small amounts of rare earths in uranium, *Anal. Chem.,* 30, 1792, 1958.
588. **Hamaguchi, H., Ohuchi, A., Shimizu, T., Onuma, N., and Kuroda, R.,** Separation of scandium from yttrium, rare-earth metals, thorium, zirconium, uranium and other elements by anion-exchange chromatography in ammonium-sulfate media, *Anal. Chem.,* 36, 2304, 1964.
589. **Strelow, F. W. E.,** Separation of scandium from yttrium, lanthanum and the rare earth-metals by anion-exchange chromatography of the sulfate complexes, *J. S. Afr. Chem. Inst.,* 17(2), 114, 1964.
590. **Oesugi, K.,** Spectrophotometric determination of scandium with Eriochrome brilliant violet B, *Anal. Chim. Acta,* 49, 597, 1970.
591. **Alimarin, I. P., Yakovlev, Yu. V., Miklishanskii, A. Z., Dogadkin, N. N., and Stepanets, O. V.,** Determination of dysprosium, europium, samarium and gadolinium in high-purity yttrium-oxide by neutron-activation analysis, *J. Radioanal. Chem.,* 1(2), 139, 1968.
592. **Minczewski, J. and Dybczyński, R.,** Determination of rare earth materials by neutron-activation and ion-exchange chromatography. Analysis of spectrally pure erbium oxide, *Anal. Chem. (Warsaw),* 10(6), 1113, 1965.
593. **Wódkiewicz, L.,** Determination of rare earth-metals by neutron-activation and ion-exchange chromatography, *Nukleonika,* 12(1-2), 93, 1967.
594. **Schoebrechts, F., Merciny, E., and Duyckaerts, G.,** The extraction of complexes of tervalent lanthanoids with different aminopolyacetic acids on a pre-conditioned ion-exchange resin, *J. Chromatogr.,* 79, 293, 1973.
595. **Wódkiewicz, L. and Dybczyński, R.,** Anion-exchange behavior of the rare earth metal complexes with trans-1,2-diaminocyclohexane-NNN'N'-tetra-acetic acid, *J. Chromatogr.,* 32(2), 394, 1968.
596. **Abdel-Rassoul, A. A. and Mohammad, S.,** Preparation of carrier-free scandium-48 from pile irradiated vanadium targets, *Z. Anorg. Allg. Chem. (Leipzig),* 330(1-2), 96, 1964.
597. **Yajima, S., Shikata, E., and Yamaguchi, C.,** Group separation of fission products with an anion-exchange resin in the oxalate form, *Jpn. Analyst,* 7(11), 721, 1958.
598. **Klerkx, L.,** Separation of strontium-90 and yttrium-90 by means of an anion-exchange resin in the oxalate form, *Bull. Soc. R. Sci. Liege (Belgium),* 33(3-4), 199, 1964.
599. **Maura, G. and Rinaldi, G.,** Detection of transition metals by formation of peroxy compounds and sorption on a chelating resin, *Anal. Chim. Acta,* 53, 466, 1971.
600. **Zhang, X.,** Spectrographic determination of trace rare-earth metals in nuclear-grade calcium after extraction-chromatographic separation on a resin containing PMBP, *Fenxi Huaxue,* 12(2), 130, 1984.
601. **Gaikwad, A. G. and Khopkar, S. M.,** Cation exchange separation of yttrium in mixed solvents, *Indian J. Technol.,* 23(7), 269, 1985.

Index

INDEX

A

B

C

D

E

F

G

H

I

K

L